CURRENTS

Contemporary Directions in the Visual Arts

second edition

Howard Smagula

PRENTICE HALL, Englewood Cliffs, New Jersey 07632

Library of Congress Cataloging-in-Publication Data

Smagula, Howard J.
  Currents, contemporary directions in the visual arts.

  Bibliography: p.
  Includes index.
    1. Art, Modern — 20th century — Themes, motives.
2. Art — Philosophy. I. Title.
N6490.S565 1989   709′.04   88-32542
ISBN 0-13-195595-0

Editorial/production supervision: *Carole Brown*
Interior design: *David Holbrook*
Cover design: *Lanning Stern*
Manufacturing buyer: *Raymond Keating*
Cover art: *Poverty Is No Disgrace*, 1982, by David Salle.
Used by kind permission of David Salle and Mary
Boone Gallery.

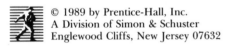 © 1989 by Prentice-Hall, Inc.
A Division of Simon & Schuster
Englewood Cliffs, New Jersey 07632

Printed in the United States of America

10 9 8 7 6 5 4 3 2

ISBN 0-13-195595-0

Prentice-Hall International (UK) Limited, *London*
Prentice-Hall of Australia Pty. Limited, *Sydney*
Prentice-Hall Canada Inc., *Toronto*
Prentice-Hall Hispanoamericana, S.A., *Mexico*
Prentice-Hall of India Private Limited, *New Delhi*
Prentice-Hall of Japan, Inc., *Tokyo*
Simon & Schuster Asia Pte. Ltd., *Singapore*
Editora Prentice-Hall do Brasil, Ltda., *Rio de Janeiro*

for my parents and my son Stefan

Contents

ACKNOWLEDGMENTS   vii

INTRODUCTION
The Individual Artist and the Present Moment   1

1   Art in Public Places   5

2   **CONTEMPORARY PAINTING**
Personal Mythology, Illusion, and Synthesis   36

3   **PRINTMAKING**
Art in the Age of Mechanical Reproduction   102

4   **FORM, ENVIRONMENT, PROCESS**
The Context for Contemporary Sculpture   138

5   **PHOTOGRAPHY**
Landscape to Narrative   179

6 PERFORMANCE/VIDEO
The Parameters of Time and Space   *214*

7 SPACES
Sculptural Events, Earthworks, Environments   *253*

NOTES   *315*

TIMELINE
Developments in Art from 1940 to the Present   *317*

INDEX   *323*

Acknowledgments

With the publication of this second edition of *Currents* I remain indebted to friends whose help and support with the first edition meant so much to me. Long-time friends and colleagues Steve Arkin and Howard Bloch were generous with their time and encouragement during initial readings. Bernard Chaet also offered helpful advice at the early stages of manuscript development; his continued concern and interest in the work of his former students is legendary in the contemporary art world.

As with the first edition, the greatest help with the preparation of the manuscript of this second edition came from Kathryn Hughes. Rarely have I encountered anyone with her keen editorial ability, native intelligence, and communication skills. She is what any author needs and wants—a reader of great perception and consummate good sense.

Bud Therien, my executive editor at Prentice Hall also needs to be thanked for his ongoing support and belief in this project. Production editor Carole Brown took the rough manuscript and photographs and with exceptional grace and cheer transformed the raw materials into the finished book you now hold in your hands.

Final thanks need to be extended to all of the artists who appear in this edition. Through their generous help with information and illustration gathering they contributed much to its successful completion.

H.J.S.

Introduction

The Individual Artist and the Present Moment

Ours is an age of personal synthesis. We face the most multidimensional, complex, and contradictory era the world has ever known; individuals must now determine for themselves their own set of values and beliefs. No one system, mode of thought, or methodology seems to provide all of the answers to the intricacies of present-day life.

Nowhere is such disarming diversity more manifest than in the world of contemporary art; all-encompassing movements like Abstract Expressionism appear to be a thing of the past. And, while every few years some new aesthetic is touted by galleries and critics (witness "Decorative and Pattern" styles and "Neo-Expressionism"), for the most part the art world is made up of individuals pursuing individual directions and themes. No longer does one particular aesthetic philosophy dominate to the exclusion of all others. On the contrary, there exists such a dazzling array of possibilities that the casual observer (and sometimes the seasoned veteran) often feels confused and lost. Some view this spectrum as an indication of the art world's lack of direction and belief. Others, however, see the range of possibilities as a source of cultural richness. Is this diversity, then, a symptom of cultural malaise or new-found strength, aesthetic confusion or meaningful *redirection*?

Responding to the diverse and personal nature of contemporary art, some recent criticism has prematurely declared the end of Modernism. I say "prematurely" because what many consider to be the essence of Modernist thought—a unified aesthetic

and purity of form—reflects only two aspects of a much broader sensibility. Also basic to this philosophy is the dream of individual freedom—freedom to pursue one's own direction and develop one's own personal vision. Contemporary artists today represent any number of meaningful directions, mixing historic styles with new concepts, traditional values with radical forms.

Fifty years ago, when the New York Museum of Modern Art first opened its doors, an adversary relationship existed between Modern art and established cultural institutions. Since that time, however, progressive art has emerged from the rarified atmosphere of the museum into the mainstream of life. What was once the concern of an elite minority has now entered the consciousness of the world in a way undreamed of by its early supporters. This integration and acceptance can be seen in the shape of the buildings in which we live and work, the chairs on which we sit, the design of magazines that entertain and inform us, and the museum exhibits that often outdraw popular sporting events. Interestingly, this cultural exchange is now a two-way street. Popular, or mass culture, now plays a significant thematic role in the work of many important young artists. The range of work we see about us is actually a result of the success of Modernist thought, not its failure.

The challenge of this and the next decade will not be the quest for newness or aesthetic "advance" but the establishment of a body of work that can communicate with an enlarged public. This new perception of the audience is an idea that significantly differentiates avant-garde figures of the past from leading artists of our own time.

Thanks to the pervasive influence of the media and educational exposure, large numbers of people have acquired an abiding interest in and knowledge of the visual arts. Consequently the climate of acceptance has dramatically changed. Today's audience is more willing to participate with the artist in exploring new modes of thought, feeling, and perception.

In response to these evolving conditions within the art world and society in general, *Currents* focuses on individual contributions rather than group movements. An in-depth case-study approach is used throughout the book to provide information about personal sources and processes of artmaking. Outstanding artists from many aesthetic backgrounds have been profiled, offering the reader significant examples of contemporary approaches. This method addresses several important issues in today's art: It clarifies and puts into perspective a field characterized by diversity; and it offers insight into how artists create as well as what they produce.

With so many accomplished artists active today the problems of selection were agonizing. By no means do the individuals profiled in *Currents* represent the only significant people working now. Because of methodology and space limitations, many important artists could not be included in the main text. The choice of artists in this book embodies a personal interpretation of recent developments in the art world. I sought as broad a spectrum of artists as possible —from senior figures who work in traditional genres to younger people who explore new media. *Currents* does not attempt to present the latest finds or fashions in the marketplace; discovering new artists is the province of monthly journals rather than books. It is hoped that through a better understanding of selected individuals a greater understanding of all contemporary artists will prevail.

Although group movements and rigid media affiliations no longer define the art world as they once did, artists do share mutual concerns often determined by the materials they use. Therefore, chapters have been organized around media such as painting, printmaking, photography, and performance/video. It is important to note, however, that many artists today work in a variety of forms and so cannot be defined by any one discipline or tool. Painters also make sculpture, printmakers paint, and video artists might draw. Many artists find their métier in combining various materials in new and exciting ways; Lucas Samaras, for example, makes constructions that involve painting, drawing, photography, and sculpture. Synthesis is a primary concern of contemporary art and is expressed in many ways: relating and connecting one's life with artistic processes, combining unique materials in relevant ways, and juxtaposing traditional concerns with present-day perceptions.

The first chapter, "Art in Public Places," establishes the background and context of art intended to address the public in buildings, streets, and parks rather than inside museums. For many people this is their first encounter with a work of contemporary art. Chapters 2, 3, and 4 profile artists working within established media such as painting/drawing, printmaking, and sculpture; although the individuals in these sections work with traditional materials, their responses are often anything but traditional. Chapter 5 presents the work of five photographic artists and examines the aesthetic basis for contemporary photography. The performance/video modes outlined in Chapter 6 are among the newest of Modern artforms. Time and theatrical experience play important roles in these controversial media. In many ways they embody and reflect various aspects of contemporary culture such as popular music, television, and modern theatre. Chapter 7, "Spaces: Sculptural Events, Earthworks, Environments," explores various ways art now interacts with everyday life. Some of the artists in this section work directly in the landscape, producing Herculean pieces that transform our perceptions of the natural world; others construct interior spaces that express inner psychological concerns (using the room as a metaphor of the mind). Some artists use urban spaces to produce temporal artworks that take people by surprise.

Although *Currents* deals essentially with work of the last twenty years, in order to give the reader a broader historic perspective, a timeline charting developments in art from 1945 to the present has been placed in the appendix. Included in this section is a reading list which provides guidance for further study.

As can be seen from this brief summary, there is little in the visual world that contemporary artists do not find worth exploring. *Currents* presents many divergent ideas, methods of working, and processes in order to convey the rich resources of the contemporary art scene. Interpretations and critical opinions are offered as examples of how one might begin to approach and evaluate this work in a thoughtful and considerate way. In essence, this is what the personal nature of most contemporary art is about: learning to see, think, evaluate, and form individual perceptions. It is my hope that *Currents* will further develop this sensibility.

1

Art in Public Places

In the spring of 1967, Henry Geldzahler, former curator of contemporary art at New York's Metropolitan Museum, happened to visit the city of Grand Rapids, Michigan, on a lecture tour. While being shown the city he casually mentioned to his guide that the new town plaza would make a splendid site for a large outdoor sculpture.

At that time Geldzahler was acting as the first director of the National Endowment for the Arts (NEA). The plaza he visited was at the heart of an ambitious urban-renewal program. NEA's new public-art program converged with Grand Rapids' redevelopment effort to catapult this city into the national spotlight as one of the outdoor sculpture centers of America.

Two years later, *La Grande Vitesse* (illustration)—a 43-foot-high, lobster-red, metal sculpture by Alexander Calder—was unveiled in this city's center plaza. Geldzahler's original remark eventually led to a $45,000 matching grant from the National Endowment's Art in Public Places Fund, thus starting Grand Rapids' romance with progressive sculpture in community spaces. This city's story serves to dramatize the convergence of a growing public interest in the arts with governmental responsiveness to that interest. Since the midsixties, public art has enjoyed unprecedented growth, support, and acceptance throughout the country. Recently, local businesses and national corporations have joined with the federal government in championing the cause of this new, accessible art. Both government and business have recognized the need to make high-quality, public works of art available to many people in various geographic settings.

The history of the federal government's role in public art is itself interesting. Soon after the National Endowment for the Arts was created by an Act of Congress in 1965, the first chairperson, Roger Stevens, formed a series of panels to address the needs of the various arts. The Visual Arts Panel met in March 1966 at the Museum of Modern Art and heard Rene d'Harnoncourt, the museum's director at the time, propose that a National Award of Excellence be created to honor outstanding living artists. But official government medals were rejected as too "elitist." D'Harnoncourt's suggestion was to make the award in the form of a major public commission. The cost would be shared by federal, state, and local agencies. This program attempted to avoid imposing a governmental "Big Brother knows best" approach which would give the community little say in the selection process and tend to alienate the citizenry.

According to the guidelines of the NEA's Art in Public Places Program, the initial request for assistance in funding and selection of large-scale public art must come directly from the community. In 1966 the National Council of the Arts—a committee set up to advise the chairperson of the Endowment—recommended that $140,000 in matching funds be allocated to enable three communities to commission major sculptural works for specific sites. Grand Rapids' *La Grande Vitesse* was the first work sponsored by this program; it set the stage for future public art sponsorship on the part of the federal government.

Calder's sculpture is an outstanding example of public art, the NEA reasoned, because this work of a renowned artist engaged the community's imagination in various ways. It answered to the special concerns of many different groups: Urban renewalists were happy with the increased public awareness and support it lent their rebuilding efforts; social activists and cultural leaders envisioned the plaza and sculpture as a rallying point and staging ground for open-air theatre productions, fairs, and band concerts; townspeople liked the national recognition and sense of pride their city gained from the extensive publicity; resident artists in the community saw it symbolically as a sign of deliverance from the aesthetically unenlightened views of "the locals." Although not everyone was convinced of its artistic merit, it did have a curious and lasting effect on the public consciousness. Proof rests in the remarkable fact that to this day, *La Grande Vitesse* has remained untouched by the "creative-art" activity of aerosol-spray-can–wielding vandals usually prone to desecrate prominent public structures.

No one doubted the importance of this sculpture in obtaining national recognition for the city and focusing local energy towards the reconstruction of its decaying

Alexander Calder, *La Grande Vitesse* (photo by Jim Gerritsen; courtesy of the Estate of Alexander Calder).

downtown district. President Ford, a native of Michigan, witnessed the effect of this sculpture and became a staunch supporter of the Endowment's Art in Public Places Program. In 1975 he was quoted as having said, "At the time I didn't know what a Calder was." (He wasn't alone in this respect. Grand Rapideans are still known to ask one another in all honesty, "Which artist did the Calder?") President Ford went on to assure the members of Congress that "Calder in the center of an urban redevelopment area has really helped to regenerate a city."

After the initial unveiling and success of *La Grande Vitesse*, Calder was invited back to design a painting for the roof of the Civic Building (illustration) and was later commissioned to do a small tactile sculpture that would make art experiences available to the blind. Not only did Calder make further contributions but a variety of other artists were also commissioned to create prominent public works of art.

Perhaps the single group most responsible for this sustained art activity is the Women's Committee of the Grand Rapids Art Museum. Several years after the Calder debut, this group applied for and was awarded an NEA Museum Program grant of $8900, which they parlayed, through generous private contributions, to a grand total of $50,000. With these funds, a nationally recognized exhibit called *Sculpture off the Pedestal*, featuring the works of thirteen nationally prominent sculptors, was staged and ran for three months. Employing an effective grass-roots, personal touch to the exhibit, each committee member was assigned to one of the participating sculptors and acted as a liaison between the artist, museum, local industry, and city. All but one of the artists visited Grand Rapids during the show, and they usually stayed at the homes of committee members and happily interacted with townspeople in ways that would be unlikely within the usual context of most museum exhibits. The city's businesses were remarkably generous: Seven of the artworks were fabricated on site with the financial help of local industry, and the remaining five were transported to Grand Rapids free of charge by local trucking firms. A handsome catalogue with an introduction by Barbara Rose, a well-known writer and scholar of contemporary art, was published, providing the finishing touch to an exhibit notable as much for its geographic location and low cost as for the quality of the work. Once again, this remarkable city as been a flagship leading the way towards assimilation of tough, challenging, pub-

Alexander Calder, rooftop painting, Civic Center, Grand Rapids, Michigan (photo by Jim Gerritsen; courtesy of the Estate of Alexander Calder).

lic artworks into mainstream American life.

During the spring of 1974, as a postscript to the *Sculpture off the Pedestal* exhibit, Robert Morris, a significant figure in the "materials and process" school of art, constructed a public earthwork (where the landscape itself is rearranged) on a hill overlooking a ball field near the outskirts of downtown Grand Rapids (illustration). Also funded by the NEA, this sculptural piece could not be further off the pedestal and was an appropriate addition to a challenging group of outdoor works. Morris' project took the form of a gigantic asphalt "X" that not only conceptually branded the landscape and visually ordered the hillside but also provided the added benefit of halting soil erosion. In one swift "X marks the spot" gesture, the artist related traditional ideals of art — to awaken our perception and vision — with current concerns for the well-being of our natural environment. This "usable" artwork functions in many ways: Children use the asphalt surfaces as bike paths to the playing field below; when base-

ball games are played it functions as a spectator perch; the whole piece functions as a drainage system preventing floods; and perhaps most important, at least to the city officials who gave their approval for this unusual use of public land, it provides antisoil-erosion advantages in a cost-effective way. Robert Morris observed at the time, "I think it's time artists were more involved with planning and, for the most part, rebuilding of cities." Grand Rapids proves that this kind of awareness is no longer limited to the wishful thinking of isolated artists.

Grand Rapids' commitment to public art did not come about without a large measure of public bewilderment and stormy debate. Even though today the Calder image is publicized on official city stationery, street signs, and even sanitation trucks, feelings are mixed as to the value of *La Grande Vitesse* as a work of art. Regardless of their evaluation, however, few citizens fail to take pride in what Calder's sculpture has done for the city. Much of the public debate has

Robert Morris, *Grand Rapids Project* (copyright Robert Morris/VAGA, New York, 1989).

centered on the age-old argument of "They spent my hard-earned tax money on that monstrosity?" But it is just this kind of lively controversy that provides the opportunity for a community to openly reexamine its goals and priorities. Thus public art works avoid the detached realm of classical aesthetics and enter the real world of social issues and debate.

Planning and funding public art are not always easy tasks; many different concerns must be balanced. Sometimes communication breaks down between the artist and the sponsoring agency. These misunderstandings can confuse and alienate everyone, thus making acceptable solutions to complex problems even more difficult. On September 18, 1974, five years after the unveiling of the first NEA-funded artwork, sculptor Mark Di Suvero was awarded a General Services Administration (GSA) "one percent for art" contract to construct a large outdoor sculpture for the Vandenberg Center, a new federal building being constructed in downtown Grand Rapids. Thus

began a modern-day version of *Rashomon* —a story which offers several versions of the truth.

At the time of the GSA contract signing, Di Suvero was no stranger to Grand Rapids, having first visited the city to enter a piece in the *Sculpture off the Pedestal* show. The monumental sculpture which he installed then was cryptically titled *Are the Years What? (for Marianne Moore)* (illustration). This dynamic, angular arrangement of brightly colored steel "I" beams, jutting out in many directions, evidently impressed some local people; after the jurying for the Vandenberg Center competition, Di Suvero's entry was earmarked as the judges' first choice. Before discussing what then ensued, it is helpful to understand the legal restrictions placed on the sculpture selections of the GSA. Under a relatively recent law, a fixed percentage of the costs budgeted for federal buildings is set aside for acquiring artworks that could be placed in, on, or outside of these structures. This program has been generally referred to in the

Mark Di Suvero, *Are the Years What? (for Marianne Moore)*.

past as the "one percent program"; more recently, however, "deflation" has eaten away at this figure until it stands now (according to the latest GSA fixing) at exactly .375 percent. Despite the dwindling amount of funds available, a grant total of $40,000 was earmarked for the commissioning of a Di Suvero sculpture for the Vandenberg Plaza.

The National Endowment for the Arts appointed a selection committee of three distinguished local art leaders — Fred Myers, Grand Rapids Museum director, Ron Watson, an art professor, and Mary Ann Keller, a prominent local art collector. This group consulted with the architect of the Center and submitted the names of three acceptable sculptors to the GSA's Fine Arts Design Panel for approval. Ellsworth Kelly, David von Schlegell, and Di Suvero were nominated. No doubt having shown in the recent outdoor sculpture show helped Di Suvero's cause; a strong lobbying effort was mounted by leading citizens and, not surprisingly, he was their choice. His maquette, or model, for the proposed piece was then submitted and approved. After Di Suvero signed the contract he was paid almost half of the commission fee, about $20,000. After that, according to the GSA, he was not heard from for over a year. When he did surface, in the summer of 1976, with the finished piece, titled *Motu Viget* (illustrations), the agency was dismayed to discover it bore little resemblance to his approved model. The GSA was not only disappointed with the aesthetic change, but also by what they took to be a potentially hazardous element now incorporated into the sculpture's design. Di Suvero had hung a huge truck tire from the overhanging beams enabling, and encouraging, people to swing freely over the hard surface of the plaza. The costs of laying a cushioned surface beneath this swinging gondola were high. Also, this addition would visually disturb the plaza. In the opinion of the GSA at this time, such drastic changes and their ramifications constituted a breach of contract, so they decided not to pay Di Suvero for the work.

Other factors clouded the dispute which then followed. The GSA was disturbed to learn that the sculptor was paid an additional $10,000 from private sources to add to his government fee of $40,000. In the eyes of the administrators this was an unauthorized and questionable practice on the part of the artist. Publicly, the GSA insisted that "bad business procedure" was the issue at stake. Privately, some officials conceded that the artistic value of the piece was compromised by the "playground equipment" appearance of the new design. The bottom line for the government was: If artists accept large sums of money for commissions, and employ accountants and lawyers, they must then accept the responsibilities of business transactions.

When the GSA announced its intent to cancel the contract, Di Suvero's many supporters in Grand Rapids organized to defend him. Many people reasoned that an artist is not an ordinary government contractor like McDonald-Douglas or Lockheed, supplying components machined to specific tolerances, but a creative human being responding to changing feelings, ideas, and situations.

An even bigger issue, in their eyes, was the government's attempt to impose inflexible procedures on an artist creating a unique work of art. Besides, townspeople argued, the artist did remain in contact with the regional branch of the GSA in Chicago, which neglected to inform headquarters of changes, delays, and problems. As to the impropriety of the additional $10,000, large-scale Di Suveros were selling in the $100,000 range so there was no question of overpayment.

In order to grasp the complexity of the issues at stake, it is worthwhile to mention the artist's reasons for making the changes. During the interim between the GSA contract and the execution of the Grand Rapids piece, Di Suvero had several large public shows in Paris and New York which caused him to question the effectiveness of the public art he had previously produced. Di Suvero realized that participatory experience should be an important aspect of his work; consequently, the "gondola" was added to the plan.

Mark Di Suvero, *Motu Viget.*

In order to preserve Di Suvero's revised work, local townspeople circulated petitions and started letter-writing campaigns, and newspapers in the region championed the cause of artistic integrity. But support was not unanimous by any means. One local citizen wrote to *The Grand Rapids Press* suggesting that in place of the controversial sculpture each townsperson bring a rock to the plaza and place it on the pile; a plaque would be placed next to the mound reading, "From one small rock a mountain grew. The strength and wisdom of Grand Rapids." Something was happening in Grand Rapids!

The Di Suvero story does have a happy ending; pleased with the concern of townspeople, the GSA gave in and permitted the contract to stand. Grand Rapids added another important piece of sculpture to its rapidly growing collection, and Di Suvero left Michigan a bit more wary of contractual agreements. Grand Rapids, Michigan, may not be typical of most mid-American cities of the same size; nevertheless, this case stands as a prophetic example of the acceptance large-scale public art will continue to enjoy in the future. Thanks to governmental support such as the GSA's "one-percent" program and the NEA's Art in Public Places program, more and more people throughout America are experiencing the works of artists who are interested in reaching the community.

Large-scale monumental sculpture, most of it a far cry from the public-park bronze horse and general statuary of the past, has recently become a permanent cultural fixture of the urban landscape. In fact, this contemporary renaissance of public art is the first major revival of civic art in nearly fifty years. Not since the pre-World War I Neoclassic school of art inundated the landscape with public sculpture has there been such an outpouring of artistic activity. Except for a brief, socially inspired experiment in mural painting conducted by the WPA Federal Art Program in the thirties, the concept of public art, for the most part, has remained dormant.

The current popularity of large-scale public sculpture expressed by the Grand Rapids' story actually had its genesis in the midfifties. When David Smith's remarkable collection of large, welded sculpture appeared on the art scene, people began to recognize the potential excitement of outdoor monumental art. Smith's habit of placing completed works in the large, open fields surrounding his Bolten Landing studio (illustration) in New York State convinced certain museum directors and public officials that this type of art could function effectively outside of museum walls. In fact, these large, gleaming, stainless-steel structures — sculptural equivalents to the dynamic, splashy Abstract Expressionist canvases then in vogue — could only be imagined in outdoor settings. By the mid-sixties large sculpture was around — many artists were now fabricating enormous pieces out of many new industrial materials — but widespread acceptance of this art and an effective delivery system were yet to come. People needed time to think about and ponder the scale and possible meaning of these puzzling works. A more pragmatic problem that remained to be solved was a financial one. Where would the money come from to pay for these public artworks?

By the late sixties, art and the economy were so prosperous that answers to these and many other questions seemed to be forthcoming. Government during this era was perceived as the guardian of the public interest. Everyone knew (although vaguely perhaps) that art was "good." And who was better prepared to present it to the public on a national scale than the federal government? It was out of these beliefs that the NEA formed.

The official mission of this relatively new federal agency is worth mentioning. The explicit goals of the Endowment are to foster individual creativity and excellence, to aid in the commissioning of public artworks throughout a wide geographical area, to help with training of talented artists, and to assure that all Americans have access to high-quality art experiences. This policy is an enormous undertaking for an agency that started out in 1965 with the small — by government standards at least — federal appropriation of $2.5 million. It is significant, however, that the arts were *officially* sanctioned as significant contributors to our nation's well-being. Until recent across-the-board budgetary cuts, Endowment appropriations had steadily grown until a funding plateau of $167 million was reached in fiscal year 1981; out of this total the visual arts' share came to approximately $7.1 million.

By now well into its second decade, the Endowment has achieved an enviable record in the eyes of many observers. According to a recent Ford Foundation report on governmental arts policy, the NEA has achieved more impact per dollar than any other governmental agency. Many citizens, art professionals, and Congresspeople concur with these findings. The arts, in fact, are becoming more and more of a sacred cow, as conservatives and liberals alike have become generous in their praise and cautious with their criticism. Questions are, however, raised by some critics regarding the use of public money: Have too many "elitist" institutions been funded? Has the broad outreach policy resulted in a lessening of "quality" in the art programs that are supported? Is too much money wasted, as some critics contend, on an unwieldy administrative structure? And does the Endowment really know what it has accomplished with all those millions of dollars?

Pleasing the broad constituency that exists in America is an impossible task. According to a 1979 Endowment self-study, about 20 percent of the applicants in the visual-arts category merit an award. Funds are available, however, for only 3 percent of the applicants, leaving 17 percent with legitimate complaints.

Unlike the sciences, which have been generously funded, particularly when the magic words "cancer research" and "national defense" are invoked, art and the humanities have always been the poor stepchildren as far as public monies are concerned. At the signing of the act that created the NEA and NEH on September 29, 1965, President Johnson observed that in the past, scientists always seemed to get the pent-

David Smith, *Wagon II* (photo by Ugo Mulas; David Smith Papers, on deposit at the Archives of American Art, Smithsonian Institution).

houses, while the artists and humanists lived in the basement.

Developments such as the scare of radiation poisoning at Three Mile Island and the overuse of dangerous chemicals in our immediate environment have made us realize that scientific and technological advances do not necessarily contribute to the improvement of our lives. Perhaps society is just beginning to realize the life-affirming benefits of the arts. Similarly, megabuck social programs have not begun to address the real causes of poverty and social malaise. The great war on poverty was lost before it began. Many urban ghetto dwellers cynically perceived the lunar landing as a thinly veiled military mission and sideshow distraction from unsolved problems here on earth. The NEA has sought to counter the failures of science, technology, and governmental social programs by providing art experiences that could inspire our lives, provide the basis for contemplative reflection, and enable us to share in a sense of community and belonging. Towards the end of making these experiences available to a varied cross section of American citizens, the NEA has made a concerted effort to direct its resources towards projects that are geographically diverse. During the 1976 fiscal year, NEA awarded only 25 percent of the grant money requested by New Yorkers, as compared with 75 percent sought by Alaskans. California received 22 percent and Idaho 75 percent. This policy gets enthusiastic support from Congresspeople who are concerned with state issues and low marks from critics who feel that the Endowment's blanket coverage defeats one of the primary goals of the agency: to foster artistic excellence in America. According to this second camp, the Endowment has strayed from its commitment to artistic excellence by responding far too easily to political pressures. Defenders of the Endowment's policies are quick to point out, however, that a commitment to high-quality art is only part of the founding credo. Because the people of the United States, by tax support, fund this operation, the agency has an obligation to make the benefits of the arts available to as many individuals as possible, regardless of their geographical location. Admittedly, the dual aims of providing high quality and reaching a mass audience present great problems; the architects of the Endowment's structure have, in all fairness, tried to balance these issues.

Contemporary art, once the exclusive domain of privileged families like the Whitneys, Vanderbilts, and Rockefellers, has in recent times achieved a much broader impact on our society than the staunchest proponents of modern art fifty years ago could have imagined. Not only do bold, unconventional works of contemporary sculpture stand in public squares throughout America as beacons pointing towards new ways of thinking about ourselves, but modernist philosophy and ideas have quietly infiltrated the companies we work for and even the environment of our homes. This invasion, the legacy of the early avant-garde, can be seen in the layout of the magazines that inform and entertain us, the design of new consumer products, and the seductive, glossy, four-color ads that stimulate desire and keep the economy moving.

Until recent years, art was viewed by hardnosed businesspeople as superficial, a little crazy, and *unprofitable*. These views have changed; many corporations now look upon contemporary art as good business.

The new attraction of business towards the arts is timely. It occurs at a time when a ceiling has been placed on government patronage and when many citizens have begun to question the fiscal priorities of the federal government. There is growing support for the idea that we should encourage the arts to rely more on philanthropic and corporate support. Realistically, however, we cannot expect business to take up all the slack. Because corporations are usually located in large, urban centers, geographically remote areas will find it most difficult to maintain previous levels of professional art activity in the coming years. What is more realistic is to expect that the art-going public will have to personally finance many of the experiences that directly benefit themselves. According to a recent Harris poll, 89 percent of Americans sixteen years of age or older (this amounts to about 130 million people) feel that the arts play a positive role in their personal and community lives. Even more surprising is that 64 percent (about 93 million citizens) are willing to pay an additional $5 annually on their income tax to support these activities. As the Harris survey raised the figure for the hypothetical contribution, support dropped—but not as much as might be expected. Forty-seven percent said they would dig still deeper into their pocketbooks and voluntarily tax themselves $25 a year for the arts, and a substantial 36 percent of the sampling would even pay $50 yearly towards the support of a national arts fund.

Although we can expect limited business spending for the arts in the next decade, there is one company that stands out as a model of what corporations might do. Fluctuating social conditions in recent years have made the business community acutely aware of the positive role the arts can play in the life of the community. In fact the maintenance of a free and open society might depend in large part on the sort of public dialogue the arts seem to bring about. Artists and businesspeople are learning to bury past prejudices and openly learn from each other.

The Philip Morris corporation, for example, has demonstrated the kind of mutual understanding that can exist between the business community and the arts. Philip Morris' entry into the sponsorship of controversial art exhibits dates all the way back to 1965 when the company underwrote a small, high-quality traveling show of contemporary Op and Pop art (Op refers to works that incorporate various optical illusions; Pop alludes to the popular and banal imagery found in these paintings). At modest cost, sixteen museums throughout America were able to present this prepackaged show to a cross section of geographically diverse audiences. Longtime observers of the art scene were suspicious, at first, of this new business-art alliance; but during its two-year run, it received praise from audiences as diverse as *The New York Times* and the Vallejo, California, *Chronicle*.

Some wire-service financial journalists were openly puzzled that a corporate giant would make itself so vulnerable to consumer and stockholder opinion. "What could be gained by these risky shows?" they openly wondered in syndicated print. It was the *kind* of art shown that really baffled the newspaper writers.

George Weissman is a central figure behind Philip Morris' successful sponsorship of new art. He is a new breed of businessperson—keenly aware of the corporation's social responsibility to the public. Along with these concerns Weissman was also interested in the rewards art could bestow on the corporation through improved plant architecture and the use of contemporary art in work areas to create and sustain an atmosphere of quality consciousness, innovation, and pride.

As the company's cultural spokesperson, Weissman answered those early critics who questioned its role in supporting controversial art exhibits:

We are innovators. We started out with Johnny, the little page boy who called for Philip Morris. Then we were the first with the Flip-Top box. Then we were the first with the plastic cigarette package. I don't know how we can tell what these picture exhibitions will do to help our product. We sponsor some big television shows, but our sales are consistently good even in the weeks when these shows aren't on. So it would be difficult to say.[1]

Of course, the sponsorship of prime-time television shows by tobacco companies is now a thing of the past. Federal legislation in the midsixties severely limited this industry's ability to freely advertise its goods. No doubt the Surgeon General's official warning about the potential hazards of cigarette smoking contributed to Philip Morris' decision to "innovate" through art activity. Not only would cultural support create an image of community responsibility but it might conceivably aid them in promoting product recognition during an era of increased trade restrictions. These self-interests do not necessarily cast doubt on the sincerity of the company's social commitment. The business community has a greater awareness today of how it fits into the "ecological" life of a community. In order to enjoy the financial profits of residency business might have to contribute to the community's on-going social life and well-being.

Many artists, some of them avowed radicals in the sixties, now feel that corporate money has become necessary for maintaining a high level of artistic activity in times of economic constraint. In fact, some art professionals believe that corporations should be exploited for the public good.

There are monetary limits, of course, beyond which any fiscally responsible company will not venture; but, even more importantly, there are ideological and philosophical limits which carefully circumscribe just what a corporation will support. For instance, companies are less likely to fund individual artists and politically active neighborhood arts groups—as the National Endowment does. Public recognition and the association with quality are high on the list of business priorities when it comes to cultural contributions. Usually these public-relations goals are achieved by supporting old master painting shows and "blockbuster" exhibits like the King Tut show at which thousands of people reverently filed through top-security museum corridors, gazing simultaneously at millions of dollars worth of gold and art. If one looks at Philip Morris' history of contributions to the arts, this company, at least, appears to be genuinely interested in artistic innovation and challenge. Over a decade before contemporary avant-garde activity was legitimized by stepped-up museum activity and media attention, this company sponsored several important exhibits that delved into issues and aspects of modernist art.

Between 1969 and 1970 the Philip Morris company sponsored a provocative series of international exhibitions. Especially noteworthy was the show *Live in Your Head—When Attitudes Become Form*, a fresh sampling of new-format work, including conceptual art, earthworks, and process art, that opened at the Kunsthalle in Bern, Switzerland, under the sponsorship of Philip Morris Europe. Another show, *New Al-chemy: Elements-Systems-Forces*—an exhibit presented by Benson & Hedges of Canada (a Philip Morris subsidiary)—showcased the work of four artists exploring investigative processes that involved scientific methodology and the relationship of ecology to aesthetics. The works in this show all reflected belief that works of art are not merely beautiful objects but might be encoded systems of conceptual and philosophical thought that could, in an alchemical way, transmute natural phenomena into works of art.

Perhaps the most thought-provoking and imaginative show of this contemporary art series was the exhibit simply titled *Air*. This relatively small show by eleven artists first opened in Australia under the local business sponsorship of Philip Morris (Australia) Ltd., and then embarked on an international tour. As its title implied, all of the events and artworks in this show dealt with the invisible presence of the element that imperceptibly surrounds us from birth and sustains our biological life process. The artists' use of air—basically unseeable and untouchable—expressed their belief that art should be more than a purely visual experience. It should also enlarge our conceptual and perceptual understanding of the natural wonders in the world.

Bruce Nauman's compelling environment created a powerful effect of animism, a notion that all inanimate objects in the world possess a hidden natural life or soul. In this specially constructed room, the walls rhythmically swelled in and out as if they were alive and breathing. Hans Haacke, an artist who works with information systems and natural phenomena, built a special wind chamber which mechanically simulated the sensations of certain weather effects. These two artists, along with the others, wanted to bridge the gap between art and life, knowledge and feelings, and to ultimately create special situations in which inner experiences could unfold in the viewer's mind.

It still remained a mystery to many people at the time why the Philip Morris corporation, or any company for that matter, would knowingly sponsor these controversial and challenging shows. The answer lies in George Weissman's initial perception of the contribution the arts could make to business. Ulrich Franzen, a designer who had worked on Weissman's home, was called upon to reorganize the fourth-floor office spaces of Philip Morris' New York headquarters. Weissman recalled:

We had very few private offices and everybody was sitting out on the open floor like a bank, and it was not working for a creative group. Rick (Franzen) started talking about creating atmospheres and environments in which people could be creative, number one. Number two, move away from the old tobacco traditions. . . . Show your people you're not afraid of change. After all, we were the sixth company in the industry. So we went ahead and had Rick re-do our offices. . . . It was amazing how when we moved into the offices, the people dressed up to their environment. It was crazy, because you have no idea how dowdy people used to dress around here. There was privacy; yet there was open communication. There was a creative atmosphere. What we're saying internally is look, we're a creative marketing company . . . we're not a manufacturing company. . . . We are not afraid of new ideas.[2]

Despite cynical observers who view all big-business activity as an inherently oppressive, self-serving, and immoral force, the fact of the matter is that arts organizations everywhere are depending more and more on corporate support. During this era of diminishing government contributions and a drying-up of private philanthropic funding due to changing tax laws and inflation, business is a prime resource. Admittedly, no one gives something for nothing, least of all corporations that must answer to thousands of shareholders. But what society gains from this public-relations–minded effort might well be worth the modest return in goodwill it secures for the corporation. After all, there are far more direct forms of advertising and public-relations methods which only benefit the Madison Avenue firms that implement them. Without progressive companies like Philip Morris many important cultural events and art exhibits would never have been presented.

One recurrent fear is that corporate sponsorship of art will inevitably lead to curatorial control and manipulation. Although this could happen, there is also evidence that corporate sponsors are sensitive to this problem and will exercise appropriate restraint. Peter Frank, associate editor of *National Arts Guide*, has written about the high degree of independence he enjoyed while organizing an exhibit of contemporary art for the Exxon Corporation that opened in the fall of 1981 at the Guggenheim Museum in New York City. During his entire tenure as curator, corporate directors refused to even meet with him about the choices he was making, maintaining that the selections were entirely up to him; even the disbursement of fees and expenses were paid through the museum (Frank claims they would not even issue him a service station credit card during this period).

Few companies are as selfless as to refrain from mentioning their names in the credit; but, on the other hand, companies may not be the devious manipulators some fear them to be. Actually, many museum directors have found that corporate sponsors interfere less in curatorial decision making than do their own trustees. A corporation's interest in developing an image of itself as a civic-minded community citizen seems a reasonably small string attached to usually quite generous and much-needed support. Rapidly changing economic, social, and political conditions have combined to modify old, perhaps out-of-date notions of government and business as inherently hostile towards the arts.

New alliances are forming in this country and they are transforming old polarities that have existed between intellectual and capitalist, artist and corporate executive. Seymour Martin Lipset, a well-known political sociologist, reports that according to his recent survey, almost 90 percent of American university professors are at least sympathetic to America's form of capitalism. This is certainly quite an ideological reversal from the midsixties. Have they been mugged, as one neoconservative critic wryly suggests, by reality? This does not mean, however, that artists, liberals, and the intellectual community have become, all of a sudden, unqualified supporters of nineteenth-century capitalism. But, in the face of economic stagnation and a reevaluation of governmental intervention, many people are not as willing to condemn business interests today as intrinsically evil.

No doubt a similar transformation is taking place within the corporate community. Top-level management has to some extent relinquished its archaic picture of the artist as a slightly crazed, financially irresponsible social misfit. Things are not like they used to be. Some artists are even buying three-piece suits today (a necessary business "costume," they reason) and are appearing at board meetings to present their ideas and debate the cost-effectiveness of various art projects. Businesspeople are learning from artists also. Some corporate types have discovered the delights of renovated, warehouse loft living and "artistic life styles." With their relatively substantial incomes they can usually outbid the artist for studio and living spaces. Despite a little resentment on both sides, these groups are discovering what each has to offer the other. We are more aware today of the artistic needs of business as well as the business aspects of art. This new symbiosis of art and commerce could, during the next decade, secure a firm ideological and financial base for improvements both can make towards the quality of our life.

At the same time it is by no means far-fetched to believe that this art activity will subtly affect the corporation itself, modify its social priorities, make it more responsive to individual feelings and needs, and perhaps make it less determined to pursue short-term monetary gains at the expense of long-term social goals. No doubt the worlds of art and business will become even more interdependent in the future, creating new opportunities and limitations.

It is a rare city in America today that cannot boast of at least one contemporary, large-scale, usually abstract public sculpture that is situated in a conspicuous location. One of the functions of these sculptures is to relieve the austere tedium of third-generation, international-style architectural celebrations of concrete, steel, and glass. These new sculptures are also perceived as signs of "good taste" and cultural literacy on the part of townspeople and local cultural leaders.

Few people today who consider themselves in any way sophisticated and socially aware would run the risk of being branded philistines by criticizing the installation of some innocuous steel monster, of dubious artistic merit, in a town square or civic center. In many cases the angry souls who complain may be right (although sometimes for the wrong reasons); often these sculptures are not distinguished works of art. Art may be more fragile than we think. The need to make allowances for the complex environmental social context (into which the piece must fit) creates difficult, sometimes insurmountable, problems for the artist and patron.

Modern art, until fairly recently, has existed exclusively within the realm of an upper-middle-class support system of museums, dealers, and patrons who possess the education, interest, and financial means to support their fancy. Avant-garde art has rarely had to relate to the diverse audiences and physical environments in which it now finds itself immersed. The appreciation of this kind of art has usually required contemplative if not focused attention, precisely the opposite of what it receives within the context of bustling public squares, malls, and crowded city streets.

Public art, because of its overtly public context and diverse audience, sometimes becomes more of a social artifact than work of art; it is subject to modification from the start in an attempt to please everyone or, perhaps more accurately, to offend no one. Artists and art professionals, steeped in the traditions of Modernism, are keenly aware of this paradox: How can an aggressively contemporary work of art, dependent in large part on its ability to express a very personal statement and challenge convention, be selected by a committee of diverse members, installed in a public square, exposed constantly to capricious debate, and not only survive but prosper? The artist and critic Brian O'Dougherty expressed the incongruity inherent in this situation when he stated that, "Finding a work of advanced art outdoors is like running into a Vassar girl working the street."

Despite these limitations, some artists have responded to the special challenges of public art in meaningful ways. Claes Oldenburg, Beverly Pepper, and George Segal have managed to remain true to their aesthetic interests and create works of public art that relate particularly well to the environments for which they were designed.

Claes Oldenburg

Claes Oldenburg's ironic and controversial Pop monuments seem, in a delightfully perverse way, to be especially appropriate for the public sector. They have entered urban environments from Philadelphia to Des Moines amid public appreciation, outcry, and, I suspect, much private amusement. His monumental sculpture returns to a culturally insecure society its popular icons, wryly cloned, enlarged by industrial means to heroic proportions, and mutated into benign monsters that sit quizzically in public squares. After years of constructing life-size papier-mâché sculpture of household objects and soft vinyl sculptures of commonplace goods, Oldenburg had the opportunity in 1969 to fabricate a large, soft sculpture called *Giant Icebag* for an exhibition at the Los Angeles County Museum of Art. This marked the beginning of his outdoor sculpture phase. He quickly followed with *Lipstick*, an unmistakably phallic, 24-foot monument placed in Yale University's Hewitt Quadrangle during May of 1969 (perfect seasonal timing for the unveiling of this contemporary Priapean maypole). A 45-foot-high clothespin has recently been installed in Philadelphia; made of cor-ten steel—a special alloy well suited to outdoor application—it corrodes only to the extent of forming a handsome, self-protective layer of warm, brown rust.

Oldenburg, an early proponent of Pop art, got his start in the public-sculpture business when he was invited by Maurice Tuchman, senior curator of Modern art at the Los Angeles County Museum, to be one of eighteen artists to collaborate on an exhibit with industrial and technical companies located in southern California. Tuchman, who had recently arrived to the area at this time, was impressed with the futuristic nature of Los Angeles. The main industries there were either research oriented or connected to aerospace production, or involved with the vast entertainment business. Tuchman conceived of artists working in these corporate facilities as if they were their studios and making use of new technical and informational resources.

As a conceptual basis for his scheme, Tuchman cited three important Modernist movements that lent historic support to his undertaking: the Italian Futurists, a group that believed the industrial machine, with its awesome power and enormous speed, was the key to twentieth-century meaning and beauty; the Russian Constructivists, who sought to unite the practical needs of life with the spiritual concerns of art and in so doing bring about a new social order; and the Bauhaus, an art school in Dessau, Germany, that worked to bring about a meaningful alliance between mass-production methods and aesthetically advanced design.

Tuchman conceived of his art and technology show as a powerful vehicle to extend the interests of these early Modernist groups into the late twentieth century.

In November of 1968, he presented his ideas to the museum's board of trustees, who raised serious doubts about the aims of the project and questioned the museum's ability to raise the necessary capital. Tuchman was persistent, however, claiming he personally would raise most of the funds, and repeatedly emphasized this main point: America needs institutions that are responsive to the interests of a broad range of society; in this particular project not only would the cause of art and artists benefit, but also the corporation, through its involvement with highly creative artists, could stand to gain new, important social perceptions.

Tuchman's arguments to the business community revolved around three main points. First, business contributions to the arts, compared to its medical and educational funding, were woefully inadequate and should be expanded. This would pay a dividend to the companies in the form of improved cultural life of the community which, in turn, would help to attract and hold valuable corporate personnel. Second, as part of the agreement between corporate sponsors and artists, the company would receive one major work of art from the collaboration. Tuchman was quick to point out that someday this might be worth a considerable amount of money, perhaps much more than the initial investment. The third and most important point, this enterprising curator argued, was that through corporate exposure to imaginative individuals, new pathways of creative thought would be opened.

After several years of planning, negotiations, and fund raising, the *Art and Technology* show—which eventually involved more than twenty artists working with thirty-seven companies—opened in the spring of 1971 to enthusiastic but mixed reviews. Some critics assailed what they thought to be a dubious mating of art and technology and referred to it grimly as a "shotgun wedding." But crowds came in record numbers to witness the high-tech visual wonders of argon and helium neon lasers, fog machines, 23 feet of curved wall covered with 640 specially shaped mirrors, rear-projection screen "kinetic painting," and Claes Oldenburg's gargantuan 30-foot-high *Giant Icebag* (illustration), which undulated and squirmed by means of a hidden hydraulic system mounted within the vinyl-covered structure.

The procedure for matching an artist's interests with corporate resources and facilities was a critical aspect of the program. Each artist selected for the collaboration was initially presented with documentary material on several corporations that might be of special interest. Having read this information the artists were encouraged to personally tour the industrial plants before deciding which company would best suit their needs.

Considering Claes Oldenburg's past interest in popular American imagery, the choice for his sponsor was an easy one—Walt Disney Productions. The communication between the high priest of popular imagery and the leading corporate purveyor of "Prime-Time Pop" was amiable at first, but eventually the Disney people decided against fabricating the giant icebag and hastily withdrew from the partnership (written into the contract was the proviso that either the artist or sponsor could back out of the agreement at any time). Gemini G. E. L., a large Los Angeles publisher of artists' prints, was persuaded to take over the project and construct the full-scale version of the giant icebag, saving the artist and museum from headaches of equal proportion. Despite the necessity of switching sponsorship in midstream, the Paul Bunyon-sized icepack was completed well in advance of the museum's schedule; in fact, it was one of eight projects from the *Art and Technology* exhibit that previewed a year earlier to enthusiastic crowds at *Expo 70* in Osaka, Japan.

Oldenburg's next grand-scale sculpture, *Lipstick* (illustration), was dedicated to our society's fixation on mass-produced trivia and designated for a site about as far removed from Disneyland as you can get—Yale University. Commissioned a year earlier by a group of Yale faculty, students, and alumni, *Lipstick* was rolled onto Hewitt Quadrangle on May 15, 1969, without permission or permits from the university. It stood in close proximity to the elegant Beinecke Rare Book Library, which housed in a glass case an important monument of its own, a precious copy of the original Gutenberg Bible. Yale's idea of culture—expressed in the pseudo-Greek architecture, esoteric learning, and prestigious cultural artifacts like the Bible—contrasted enormously with this illegally installed phallus on a half-track. Of course, this unseemly juxtaposition was precisely Oldenburg's point.

The 1969 to 1970 academic year was a period of political turmoil in New Haven and the nation; Bobby Seale, a political activist of the era, was on trial there, and the possibility of city- and university-wide rioting was a constant threat. The fear of cam-

Claes Oldenburg, *Giant Icebag* (copyright Claes Oldenburg/ VAGA, New York, 1989).

Claes Oldenburg, *Lipstick Ascending on Caterpillar Tracks* (Yale University Art Gallery: Gift of the Colossal Keepsake Corporation).

pus civil war prompted Kingman Brewster, president of Yale at the time, to suppress personal and institutional doubts about the artistic merits and appropriateness of *Lipstick* and graciously accept this controversial gift on behalf of the university.

Oldenburg admitted that the combination of caterpillar tractor bottom, raising visions of military tanks, and pop-erotic imagery of the lipstick case was meant to stimulate public debate. In a curious way the piece illuminates the passions of the time and brings to mind the widely circulated media images of flowers in gun barrels and popular slogans such as "Make love not war." *Lipstick* expressed, in Freudian terms, some of the artist's hostility towards institutions in general and Yale in particular.

A statement by Oldenburg in the December 1968 *Yale Alumni Magazine* touched off the events leading to this iconoclastic sculpture and event. The artist claimed in an interview that he spent four years at Yale without being able to remember very much about the place. Soon after the magazine article appeared, Herbert Marcuse, a leftist philosopher very much in vogue at the time, made the mistake of claiming that if any of Oldenburg's fantasy monuments were ever erected it would signal the end of civilization as we knew it. For many students and faculty this challenge was too good to pass. Maybe Marcuse was right after all. The "Colossal Keepsake Corporation" was promptly formed to commission such a sculpture, and soon enough money was raised to pay for the fabrication. Wheels were turning that might propel our civilization towards its gloriously apocalyptical end. But there were many disappointed sponsors on May 15, when the sculpture was unveiled; despite a prophetic warning from a respected cultural philosopher, Yale weathered the storm and the republic has gone on to withstand far more serious challenges. Today, Oldenburg, with a refreshed memory of school days, has stated half apologetically that, "In those days you couldn't say anything nice about universities, so whatever I said was twisted around."

Within a year political slogans appeared to deface this public monument, and soon after it was quietly removed by the same corporation that was responsible for its noisy and greatly heralded appearance.

Today, *Lipstick* has been rescued after years of storage at a nearby art foundry and now stands proudly resurrected in the courtyard of one of Yale's residential colleges. The passage of time has quietly obliterated the bitter political foment that heralded its birth in the late sixties. Most viewers today tend to read it as a witty and bold statement of modern sculpture; this encourages a broad range of discussions about its intent, meaning, and worth.

Oldenburg's ironic and witty monuments inevitably contribute to this kind of on-going discourse; whatever you think of their artistic merits, the audacious nature of their imagery and scale forces us to consider and question societal ethics and cultural values. Thematically, they function in quite a different way from most contemporary sculpture and probe issues of contemporary life that are ignored by many artists. Unlike many civic monuments Oldenburg's pieces are deliberately designed to humorously provoke a complacent population into new modes of thinking about a wide range of cultural, environmental, and social issues.

The razor-sharp political edge of early works like *Lipstick* is no longer prominent in Oldenburg's recent work, but his monuments still invite and perhaps court a form of bemused, benevolent controversy. In fact, one of his most effective public monuments to date helped celebrate, in an ironic way, America's Bicentennial. The commercial overtones of this patriotic event were predictable, yet mind-boggling — flag-bedecked kitchen towels, bald-eagle stationery, and even red, white, and blue "stars and stripes forever" underwear (which prompted a few social critics to caustically refer to the year-long party as the "buy-centennial"). Oldenburg helped bring this already active feeding frenzy of commercial kitsch to a climax by unveiling on July 4, 1976, a 45-foot-high cor-ten steel clothespin (illustration) outside of a newly constructed office tower in downtown Philadelphia.

The developers of this building project,

in compliance with the city's "one percent for fine arts" ordinance (in 1951, Philadelphia was the first municipality to enact such a law), were interested in a sculptural work that would attract the attention and interest of crowds that passed through their busy intersection every day. Because of the immediate recognizability of his imagery, Oldenburg was the first choice of art collector Jack L. Wogin, the developer of this urban center. Oldenburg told Wogin that beside the "New-Realist" aspect of this household item, he was also thinking of the formal relationship the clothespin bore to the European sculptor Brancusi's elegantly understated *The Kiss*, which incidentally happens to be owned by the Philadelphia Museum of Art. *Clothespin* restates, in late-modern terms of scale and context, some of the formal elements of *The Kiss*, making at the same time, a wry comment in terms of thematic content. The reaction at the time of unveiling was predictable; many people felt the sculpture mocked them. Today, however, most citizens of Philadelphia view the Pop icon with tolerant amusement, if not outright affection.

Claes Oldenburg, *Clothespin* (copyright Claes Oldenburg/VAGA, New York, 1989).

Beverly Pepper

Beverly Pepper's large-scale, precariously balanced, brightly painted steel structures evoke feelings and sensations of a different order from those evoked by Oldenburg's comments on American culture. One might argue, however, that the two artists are motivated by similar perceptions of cultural values. Even though the former may appear to be "abstract" and detached, the emotional intentions of her work are very concrete. Pepper believes that in today's busy world, people are so distracted by the insatiable demands of career, the need for status, and the acquisition of consumer goods that they have simply lost touch with their feelings. Pepper's intention is to elicit feelings by the use of specific shapes, colors, and materials. What the artist means by the term *feeling* is the total, unique experience the spectator takes away from the piece. Consequently, visual, tactile, and emotional experiences provide the thematic basis for her work. Whereas cultural iconography forms the basis for Claes Oldenburg's work, emotional–perceptual experience is the aim of Pepper's efforts.

According to this artist, public sculpture should be more than large-scale decoration meant to show good taste. It needs to be something that is physically and emotionally involving. Pepper believes in works of art that can be walked on or crawled through as well as looked at — in other words, structures that can be experienced from many angles and attitudes.

As we have mentioned before, public sculpture usually functions in busy urban environments which make heavy demands on our psyches and sight; sculptural pieces must not only compete spatially with an active site but must also offer unique experiences and, in Pepper's own words, "feelings." Pepper's sculpture, juxtaposed as it usually is to an architecture that is designed to serve practical needs, attempts to offer us special experiences far removed from the everyday world. But there is a useful element to her work that goes far beyond what we usually think of as "useful." In a speech delivered at Dartmouth College in 1977, Pepper clearly outlined her views on the potential meaning of public art:

Monumental sculpture is directed to man and his perception, not to function. There is no need to adapt. No need to use the work. The only inescapable need is to relate and create an interaction between man and aesthetic experience. Which here becomes a social act. Perhaps this explains a great deal of my belief in the positive value of public large scale works, where the spectator is forced into a confrontation with the materials — hopefully facing them without preconceived ideas. In that instant, the work is potentially an extended human experience, as well as my isolated human expression.[3]

Many of Pepper's works express, through overhanging forms that jut out for great distances, a precarious and delicate sense of balance. *Perazim* (illustration), an 8-foot by 18-foot-long painted steel structure, is a good example. Psychologically and visually, it appears to be on the very edge of change. This effect appeals to the artist very much; in a recent interview she maintains that this balance is what living is about.

Beverly Pepper's personal origins go back to Brooklyn, New York, where she was born and where she grew up. After completing studies at the Art Students League and Pratt Institute, she, as many American art students did during the midforties, left for the European art capital, Paris, to study with renowned French artists Fernand Léger and André Lhote. Pepper's artistic work at this time took the form of abstract, vaguely organic watercolors. After several years spent back in America, in 1949 Pepper left to settle more or less permanently in Europe. The availability of affordable assistants, trained in sculptural procedures, was a major factor in her decision to live and work in northern Italy. Pepper attributes the first works she did there, Social-Realist paintings, to the unbelievable poverty she witnessed in postwar Europe. They were awful paintings, she recalls, but they did make her aware of her feelings towards people and got her thinking about how art could play a more socially active role.

Her introduction to sculpture came about in a curious way. While an addition was being built onto her house in Italy, a group of trees were cut down to make room for the improvements. Lying on the ground in random distribution, the felled trees seemed to suggest and invite sculptural possibilities. At this time Pepper had very little experience in sculpture; so rather than laboriously carve the trees with mallet and wood chisels, she shaped them with electric drills and a large hand saw. They ended up looking like traditionally carved wood sculpture, even though contemporary tools had actually modified this ancient craft.

Although she has lived and worked for many years in Italy, where artistic traditions and craft have been kept alive for centuries, Pepper's experience in America with new tools and materials has played a role in the development of her work. In 1966, while staying in Watermill, Long Island, Pepper grew tired of the vacation routine, visited the shop of a local blacksmith, and asked if she could work with him. The smith thought she was crazy, but, intrigued with her energy and enthusiasm, finally allowed her to work there.

Using bits of scrap metal lying around the shop, Pepper forged and joined them into intriguing configurations. Something was not quite right with their finish, however. She tried painting them and changing the surface tones; nothing seemed to work. One day a customer came in with a piece he had just chromed. The shiny chrome surface was a revelation to Pepper. When she returned to Italy she completed a whole series of neon and chrome-plated works as a result of her chance apprenticeship with the Watermill blacksmith.

Fortuitous circumstances in life play important roles in the development of many contemporary artists. Beverly Pepper feels that these unexpected events (like a meeting with a certain individual who suggests a particular process or material) lend excitement, adventure, and new perceptions

Beverly Pepper, *Perazim* (photo courtesy Andre Emmerich Gallery, New York).

to her professional and personal life. The story of her participation in the Spoleto Sculpture Festival further illustrates this point. Giovanni Carendente, the organizer of the festival, asked Pepper if she knew how to weld and, if so, would she come for a few months to live and work in a factory town near Spoleto. At this point she did not know how to weld, but she was excited by the opportunity to do a large-scale piece and accepted the offer. Early the next morning, Pepper showed up at an iron-monger's shop and offered to pay him to accept her as his apprentice. After two months of sweeping the floor and being involved in every aspect of metal welding and shaping, she was ready for Spoleto. The organizers thought the Communist town of Piombino would appreciate her blend of art, idealism, and hard work, particularly since she was a woman. Pepper had a wonderful time there constructing her first monumental sculpture.

The dilemma of just what constitutes effective large-scale public art is of great concern to Pepper. Some observers feel that we do not need more art set in public places, but rather a new form of "public art" that relates to people more on their own terms. This is a complicated and difficult question

that calls for accurate definitions of key terms such as *people* and *public art*. Personally, Pepper feels that you cannot make an effective work of art *for* the public; if this is attempted, compromise will weaken the work beyond recognition. People's reactions should be taken into consideration but not catered to. New experiences and unusual forms may not be enjoyed or appreciated at first, but this teasing of perception is precisely the point behind Beverly Pepper's public sculpture. Besides, she feels, you cannot control or predict how a diverse group of individuals will respond, so it is pointless and frustrating to try.

Not too long ago Dartmouth College found itself suddenly immersed in a spirited controversy over just these issues. This neatly landscaped New Hampshire campus, snow-bound five months of the year, rose as a body to question and protest the installation of Beverly Pepper's sculpture *Thel* (illustration) on one of their favorite lawns. From the earliest stages of planning, *Thel* was designed with the conservation of lawn area in mind. Nevertheless, students complained about how their "beach" (a student nickname for the Wheeler lawn site) was being taken from them. Pepper responded by saying that she wasn't robbing

Beverly Pepper, *Thel* (photo courtesy Andre Emmerich Gallery, New York).

them of their beach but giving them deck chairs for it. True to her intentions, she incorporated a complex interior crossribbing structure into the piece, thus creating inviting spaces in which the students could sit. Seen on a moonlit night during the dead of winter, *Thel*'s white enameled steel surface magically blends with the quiet, austere New England landscape to create an otherworldly, romantic image of unparalleled beauty. Although partially camouflaged in winter, when the spring thaw approaches, it manifests a different "self" from its winter image and eventually stands dazzling white against verdant green grass. During the late spring months it becomes a popular location for outdoor seminars and a lounging site for students preparing for final exams.

Citing *Thel*'s eventual acceptance and contribution to life at Dartmouth, Pepper feels strongly that public artworks, to achieve their highest ambitions, must remain true to the artist's inspirations and sources. Good works of art must express the unexpected — not safe and easily acceptable conclusions; only then can art enable us to freshly perceive our environment and make us aware of processes of change at work in nature and in our lives. In recognition of this goal Pepper has written:

I feel my work, when successful, does more than emphasize an emotion, more than extend the landscape, more than compound an exterior environment. It should take the spectator from viewing to entering — and with entering, to feeling.[4]

George Segal

George Segal, for many years an outstanding sculptor of indoor environmental groupings and individual pieces, has recently entered the arena of corporate- and government-sponsored, monumental public art. Highly personal in nature, Segal's work represents a strong departure from most large-scale outdoor art. This work projects an introspective, lonely, and almost fragile mood; plaster and bronze figures are placed surrealistically in settings that make use of real doors, sofas, and scaffolding. The elegantly crude finish of the bodies plays off nicely against the heightened reality of "real" objects in the world. At first glance these metaphorically fragile and ephemeral works seem unlikely candidates for the grand scale and social dynamics required of contemporary public artworks. But in a particularly relevant and ironic way this quality of psychological openness, human scale, and strangeness makes them highly appropriate. Because

George Segal, *The Restaurant* (courtesy Sidney Janis Gallery, New York).

Segal's normal-sized figures are frozen, isolated, and carefully composed with the materials of the world, they relate in a very direct way to people of varied backgrounds.

In his first permanent outdoor piece, called *The Restaurant* (illustration), which was commissioned by the GSA for a new Social Security building in Buffalo, New York, Segal created a strikingly authentic scene that sets up visual and conceptual dynamics with the real restaurants and coffee shops a short distance away. Made of real bricks, bronze, cement, steel, aluminum, tempered glass, and a fluorescent light, *The Restaurant* is arresting in its presence. Through the plate glass window a single figure may be seen sitting at a solitary table. *Outside* — which is the other side of the wall — two other figures are captured in bronze, one in the process of walking by *The Restaurant*, the other leaning against the wall of the structure while absorbed in contemplative thought. There is no sense of security in the ambience of this piece. It looks at once reassuringly lifelike and unsettlingly real.

In recent years commissions for public sculpture have occupied more and more of Segal's time and thinking. After twenty years of sculpting, for the most part individual works for private sale, he finds constructing public works a rejuvenating experience. In Segal's mind it becomes imperative, in these new outdoor pieces, to consider other people's feelings and opinions. This new activity has led to increased personal growth for the artist and the opportunity to interact, through his art, with many more people and groups.

Largely because of his use of the figure and his interest in environmental settings, Segal's art thrives particularly well in public settings. In fact, it is probably more effective here than in the isolated and rarified atmosphere of museums and galleries. First of all, when viewed in public places, his haunting white figures take on new identities. We relate to them in personal ways that would not be revealed to us if they were seen in traditional art-world settings. Second, they communicate with people who, for one reason or another, might never visit a museum or commercial gallery.

The Commuters (illustration) is a large-scale

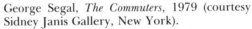

George Segal, *The Commuters*, 1979 (courtesy Sidney Janis Gallery, New York).

Richard Serra, *Tilted Arc* (photo by Glenn Steigelman Inc.; copyright Richard Serra/VAGA, New York, 1989).

environmental sculpture completed by Segal in 1979 and installed in a rather unique setting — the bustling Port Authority center in New York City. Every day, tens of thousands of commuters and long-distance bus passengers pass this life-sized bronze and mixed-media installation. *The Commuters* reflexively comments on the situation in which many people find themselves while at the Port Authority Building: waiting to board a bus. This sculpture — located in a central waiting room — perplexes and amuses countless patrons every day as they confront painted bronze figures frozen in attitudes similar to their own. This interplay of life and art is the real subject matter of *The Commuters*. For years, public transportation systems have fascinated Segal. He believes they provide opportunities and situations in which people consistently turn mentally inward and become openly introspective. For this reason Segal finds places like the Port Authority Building particularly suited to the comtemplation of sculpture.

Segal finds nothing surprising in the recent popularity of public art. He believes "people are hungry for some common belief. Everyone wants to feel attached to something larger than himself." Segal's environmental sculptures make it possible for us to gain a different perspective on our collective roles in twentieth-century life. Waiting for the bus in the same room with *The Commuters* allows us to gain insights into

some of the behavioral modes we are forced to assume in the modern world.

THE "TILTED ARC" CONTROVERSY

Despite the acceptance of public sculpture in general, there have been recent cases in which popular sentiment has turned against prominent works of art installed in public places. The most notorious and agonizing instance was the political storm that centered on Richard Serra's sculpture *Tilted Arc* (illustration).

Installed in front of the Jacob K. Javits Federal Building in downtown Manhattan in 1981, this massive wall of leaning steel cuts incisively across the broad plaza greatly modifying the view and physical movement of anyone entering or leaving this busy government complex. *Tilted Arc* was immediately met by hostility from office workers and federal employees who use this plaza during the course of their work day. Following a series of complaining letters to newspapers, sentiment against this sculpture quickly led to the circulation of a petition demanding its immediate removal; soon a total of 1300 federal-complex employees had signed the document. This reaction was surprising to everyone who had participated in the sculpture's selection process. After all, the committees were well aware of Serra's international reputation in the world of art. But the audience of this public sculpture seemed to have little reverence for the opinion of "experts" and museum directors. What they did know was that an anonymous group of individuals connected with the government had selected this sculpture and placed it directly in their path as they came to work and went home. *Tilted Arc* seemed to focus adverse opinion against it in direct proportion to its huge mass and imposing scale.

To better understand this controversy it is necessary to examine the specific selection process used in awarding the commission to *Tilted Arc*. In 1979, as part of the General Services Administration's Art in Architecture program, Richard Serra was chosen by a panel of art professionals—Robert Pincus-Witten, Suzanne Delahanty, and Ira Licht—to submit a sculpture proposal for the Javits Federal Building on Foley Square in New York City. After review of Serra's plan, Roland G. Freeman, the head of GSA, gave final approval for *Tilted Arc*.

In retrospect, some difficult questions should have been asked in order to evaluate the suitability of the sculpture to the site. Certainly Serra's standing as an artist was never in question—at least where art professionals were concerned. Few of his fellow artists have contributed as much as he to the language and sensibilities of contemporary sculpture. Serra's proposal, it seems, was primarily chosen on *aesthetic* grounds; but when considering public sculpture, one cannot easily ignore the *social* environment in which the artwork must live. Unlike a museum where people are there presumably because they choose to be, work like *Tilted Arc* functions in the most public space imaginable: an open square of precious city space surrounded by the high-rise buildings of lower Manhattan. Serra's massive sculpture, by dominating the plaza, seems to have greatly affected people's territorial sensibilities in this crowded city.

Before we go into issues that were brought up at the public hearing, it is both interesting and revealing to hear Serra's explanation of the ideas behind *Tilted Arc*:

The Federal Building didn't interest me at first. It's a "pedestal site" in front of a public building. There's a fountain on the plaza, normally you would expect a sculpture next to the fountain, so the ensemble would embellish the building. I've found a way to dislocate or alter the decorative function of the plaza and actively bring people into the sculpture's context. I plan to build a piece that's 120 feet long in a semi-circular plaza. It will cross the entire space, blocking the view from the street to the courthouse and vice versa. It will be twelve feet high and will tilt one foot toward the Federal Building and the courthouse. It will be a very slow arc that will encompass the people who walk on the plaza in its volume. . . . After the piece is created, the space will be understood primarily as a function of the sculpture.[5]

The domination of space that Serra sought was no doubt perceived by the office workers as an expression of dictatorial control and power. *Tilted Arc* became, perhaps undeservedly, a tangible symbol for various social ills and psychological forces the public felt they could do without: the unremitting power of the federal government which commissioned this sculpture and controlled their daily work lives; the dehumanizing vastness and anonymity of New York City itself; and the ability of a suspect "Modern" artist to change or affect, with no prior explanation or dialogue, where they walked and what they saw.

The *Tilted Arc* controversy came to a head three years later in the early part of 1985. Urged by the constant lobbying of Judge Edward D. Re, chief justice of the United States Court of International Trade, William Diamond, the New York administrator of the GSA, called for three days of hearings in March of that year to decide the fate of *Tilted Arc*. Serra was irate that the government would even *consider* removing his sculpture. He steadfastly maintained that he had fulfilled his part of the contract and, furthermore, that *Tilted Arc* was specifically designed for this site and could not simply be moved elsewhere.

The hearing turned into a lively debate between foes and defenders; all witnesses dutifully took turns at the microphone and, depending on their position, either decried the sculpture as "nihilistic, rusting junk" or cited it as "a beautiful, well-conceived work, involved in the plaza and its surroundings." As is often the case in polarized disagreements such as this, little real communication was fostered between the general audience and the professional art community. The crucial battle for public understanding and sympathy was lost well before the hearings began.

By the end of the hearings—extended to allow everyone a chance to speak—the GSA decided to remove the sculpture; the decision was made over Serra's strong objection that such an action was not only illegal but tantamount to destroying the work. Despite this ruling the case of *Tilted Arc* is far from over. The relocation guidelines drawn by the GSA require that a suitable alternative site be found, subject to the approval of an NEA-appointed panel that can work with the artist. Since Serra has publically vowed not to cooperate with any attempt to move his sculpture, *Tilted Arc* is likely to remain in place for quite some time—perhaps forever.

The failure of *Tilted Arc* does not reside in its aesthetic value but in the lack of any attempt to explain the position of the artist and his work to the public prior to its installation. Although this sculpture has been placed in a public space, little sensitivity toward the legitimate needs of the public was shown by its sponsors. Information sheets could have been handed out, and meetings between the artist and building workers could have been planned. Had the public been prepared in any of numerous ways this unfortunate confrontation might have been avoided.

Much testimony delivered by art professionals during the hearings centered on the autonomy of artists and their "freedom" to pursue their aesthetic interests. While these issues are tied inextricably to much twentieth-century art, *Tilted Arc* raises the question of public rights as it concerns public art, particularly art funded by the government and not private sources. There is no doubt that this environmental sculpture makes a powerful and effective statement in terms of Modernist aesthetics. But what remains in question is the right of this sculpture to aggressively assert itself upon this public plaza (as Serra said "the [plaza] space will be understood primarily as a function of the sculpture"). No one ques-

tions the aesthetic freedom of the artist today, but what about the rights of the public? Ostensibly, the sculpture is there for them.

Surely an ethical response to the problems of quality public art demands a mediation between the interests of the artist and the needs of the public. What has become clear from this unfortunate experience is that creating public art is very different from making private sculpture that is sold to collectors and museums. The audience needs to be brought into the art-making process from the beginning and made to feel that they have a stake in its outcome. Admittedly, this is not always easy to do without compromising the artist's long-standing concerns and values.

Some good, it is hoped, will come out of controversies like this one which pit artists and art supporters against the "public." At the very least both sides may have gained some understanding about each other's interests and needs. With any luck, *Tilted Arc* should remain in place permanently—to remove it would set a dangerous precedent for future programs of public art. But, the continued presence of this controversial sculpture might serve as a constant reminder to those people who are entrusted with the administration of public art programs that the social consequences of their actions are as important as their aesthetic decisions.

SEATTLE'S PUBLIC ART PROGRAM

While Grand Rapids, Michigan, might be credited with starting the Art in Public Places movement, today the leading center for public art appears to be Seattle, Washington. Because of this city's one-percent-for-art ordinance enacted in 1973, and the statewide one-half-percent law of 1974, the Seattle area is a model of what can be achieved with adequate funding and cooperation between artists, planners, architects, and local and state officials. Seattle is alive with art. Installations by art-world luminaries such as Robert Rauschenberg and Frank Stella grace the local airport, built in

1973; more recently between 1976 and 1983, an ambitious series of site-specific sculptures and environmental artworks have been built on the grounds of Seattle's National Oceanic and Atmospheric Administration. Works of public art can be seen in a wide variety of sites in this well-kept Pacific Northwest city—in fact, it would be hard to spend even a short time in Seattle without running into a few notable outdoor art works.

Seattle has, in large part, achieved what other city planning commissions only dream about: the integration of art within the fabric of everyday life. For instance, Nathan Jackson's cast-iron hatchcovers (illustration) are located throughout the downtown area. These utilitarian items provide a reminder of Seattle's special traditions to residents as well as visitors, by drawing upon the tribal motifs of Native Americans indigenous to this coastal region.

One remarkable feature of this work was its cost-effectiveness. At little added expense these iron hatchcovers have meaningful, expressive designs instead of anonymous mechanical patterns that vary little from city to city. The Indian motifs embossed on each hatchcover are only one aspect of the grand design of this artwork. Just as important is the way in which this composition shows up repeatedly all over Seattle, transforming an otherwise invisible element into an intriguing reminder of this city's unique traditions and history.

Jack Mackie's bronze and concrete *Dancer's Series: Steps* (illustration) is another imaginative, low-cost work of art that has the potential to surprise and delight Seattle's inhabitants. In 1982 eight different dance-step sequences were inlaid on sidewalks throughout the city. Clearly numbered steps teach the willing participant a variety of ballroom dance steps. Teenagers unfamiliar with these classic dances can turn off their "boom-boxes" and perhaps give these old-fashioned dances a whirl. For elderly people, *Dancer's Series* might evoke nostalgic memories of their youth; and if the bus they are waiting for proves long in coming, they may even cautiously use the numbered bronze steps to refresh their technique in performing these once-familiar dances. There is an openness and warmth to much of Seattle's public art that seems directed toward making the urban envi-

Nathan Jackson, hatchcover, 1976 (photo courtesy Seattle Arts Commission. Commissioned with Seattle City Light 1% for Art Funds).

Jack Mackie, *Dancer's Series: Steps*, 1982 (photo courtesy Seattle Arts Commission. Commissioned with Seattle Engineering Department Funds).

ronment more humane and liveable. What this type of art might lack in monumentality and "seriousness" is far outweighed by the contribution it makes to the city's quality of life.

One of the main reasons for the success of Seattle's art program has been the early stage at which artists enter the design process. Rarely are they brought in as an afterthought. Also, the public is kept well informed about projects through every stage of their development. Thus, few works are presented without some kind of audience preparation. Consequently, a large and well-informed constituency for public art has blossomed in Seattle which further enhances the receptivity for challenging new works.

Not all of this region's public art projects are as small and intimate as the hatchcovers and bronze dance steps. One of the most ambitious, long-term outdoor sculpture projects to be seen anywhere can be found in the Seattle area. Over a seven-year pe-

riod, artists of national reputation have been commissioned by the regional headquarters of the National Oceanic and Atmospheric Administration (NOAA) to build large-scale environmental works on the grounds of their spectacular oceanside site.

Douglas Hollis' *Sound Garden* (illustration) is a sound sculpture that takes advantage of the wind that blows constantly across this waterfront location throughout the year. Mounted on a series of steel towers, wind-activated organ pipes create changing sound patterns reminiscent of the haunting calls whales make to one another underwater. For quite some time Hollis has been interested in sculpture that makes use of wind and other natural forces. Even before you reach the sculpture site and see the various artworks there, you *hear* Hollis' piece. The strange wails of the wind-activated organ provide an ever-varying audio background for the other environmental sculptures.

All of the artists chosen to construct pieces at NOAA were sensitive to the natural

Douglas Hollis, *Sound Garden* (courtesy of Max Protetch Gallery).

Scott Burton, *Viewpoint*, 1983 (courtesy of Max Protetch Gallery).

characteristics and beauty of the ocean-front location; and all were interested in making use of some indigenous feature in order to heighten the visitor's experience of this special place. Scott Burton's environmental sculpture *Viewpoint, 1983* (illustration) was designed with the idea of giving people visual and physical access to the water. Burton, known for his minimalist sculpture in the form of utilitarian furniture, had granite benches made for this work and landscaped the area with native boulders, small trees, and shrubs. *Viewpoint* offers the visitor more than a visual experience—participants can sit and talk with one another on the granite benches, listen to the sound of the water, and look out over the ocean as clouds race by. All of these elements happily combine in Burton's "user-friendly" design.

From the experiences of the last two decades, it is obvious that public art presents both extraordinary challenges and unique opportunities. Artists working in this sector must walk a fine line in terms of balancing their own personal aesthetic needs with those of the public. For this reason effective, socially responsive public art may be more difficult to produce than we had imagined. But when various elements fall into place —the right artist, the right site, the right work—art in public places has unique, transformative power that can change aesthetic environments into supportive social situations.

2 Contemporary Painting

Personal Mythology, Illusion, and Synthesis

When Jackson Pollock, a gifted young painter from Cody, Wyoming, moved to New York City in the early thirties he helped reverse a trend in American art towards a narrow provincial realism, a dominant mode of the time. American Realist painting of idealized rural life often pandered to popular levels of taste and did little to establish an original cultural identity for American art. The pervasive interest in regionalism can be attributed, in part, to the revival of isolationist feelings after the defeat of President Wilson and his policy of internationalism. Foreign, "abstract" art became associated with a host of social and political ills that were antithetical to "wholesome American values." Moral decadence, Communism, even the threat of homosexuality were associated with the European avant-garde at this time.

The cause of American Realism was fervently championed by the Art Students League in Midtown Manhattan. The artist John Sloan served as director, and Thomas Hart Benton as one of its most popular teachers. Benton had keenly honed an aesthetic to fit his ideology: Painting should be comprehensible to large numbers of people, not the privileged few; in this way it could express the democratic principles upon which America was founded. Ideally, this art would glorify national values by depicting a variety of geographic settings from Maine to California. Above all, he felt that native painting should stem the tide of radical, abstract, "functionless" art and reject "aesthetic colonialism."

As a young westerner in New York, Pollock studied at the League with Benton; his early paintings of seething, baroque landscapes reveal this influence. He did not, however, entirely share Benton's fear and distrust of Modern, nonimagistic art. Soon he was avidly studying the work of Picasso and other contemporary pioneers. Pollock was able to fuse an indigenous feeling for the western landscape with sophisticated concepts and techniques of the European avant-garde. In a curious and highly creative way, Pollock—a homegrown loner and maverick—was able to synthesize a passion for the grandeur and openness of the "American Experience" with seemingly arcane aspects of experimental European art.

Together with a growing band of young European and native-born American artists (usually of emigrant parents, though), Pollock helped lead the way towards the establishment of a unique art movement— Abstract Expressionism. Ironically, many of the major contributors to this "native" form were emigrés like John Graham, a Russian; Willem de Kooning, a Dutchman; Arshile Gorky, an Armenian; and Hans Hoffman, a German. These artists were unwilling to restrict themselves to illustrating scenes from American urban and rural life. They longed for the heroic style of historic masters and the intellectual fervor of pioneers like Cezanne, Matisse, and Picasso. The important role American painting enjoyed in the late forties and fifties derived mainly from these European artists, who started arriving in the thirties.

Abstract Expressionism succeeded in combining the best aspects of the Old and New Worlds to create a truly distinctive American style. New York City provided a special stimulus to the avant-garde artists; here they could create and take part in a new tradition—the tradition of the new. In their work they consciously turned the aesthetic limitations of America into positive assets: If American art lacked refined sophistication, they would paint with an almost brutal directness; if most of the work looked contrived and "overfinished," they would create spontaneity; and most importantly, if native art was isolated and idiosyncratic, they would develop its individuality as a positive strength. It was this last characteristic that set the stage for the emergence of a far-reaching "individualism" that took root in the late sixties and flourished in the seventies.

Abstract Expressionism developed into a major school of art that enjoyed worldwide recognition and reigned supreme from the fifties until the early sixties. Pop art was one of the few major movements to successfully challenge the aesthetic preeminence of Abstract Expressionism. In many ways Pop art was opposed to the underlying features and beliefs of the early "New York School." Pop generally used clear and simple imagery, whereas Abstract Expressionism employed very little. Subtly, Pop art seemed to express some of Thomas Hart Benton's notions about "populism"—superficially perhaps—by creating an artform that employed pictures that the average person, with little knowledge of the history of Modern art, could relate to.

Minimal art was the next major art movement to emerge on the scene after the dramatic success of Pop art. It was a highly refined reductivist school of sleek forms, isolated objects, and serial structures that made extensive use of high-tech tools, materials, and methods. Minimalism signaled the end of an era whereby specific styles of artmaking dictated current fashions. This hierarchical system is now a thing of the past, much like the fashion world of the fifties and early sixties, when a handful of Paris designers decided the hem length for

millions of women. Since Minimalism there have been occasional flurries of new movements in the art world—Conceptual art, Process art, Photorealism, New-Image, and Neo-Expressionism—but nothing like the mass movements of Action painting and Pop.

Contemporary painting today is characterized by individualistic pursuit of a broad variety of thematic interests and aesthetic persuasions. Following an era that emphasized adherence to movements and nonnarrative themes, many painters are returning to simple handcrafted methods and are telling stories. Individual passions, private rituals, cultural myths, and ideological dreams are fabricated out of paint, canvas, collaged images, calligraphic marks, and many nontraditional materials and processes. Contemporary painting is as complex and multidimensional as the contemporary world. To a great extent individualism has become the "movement" followed by most modern artists. Making one's way in the art world consists of staking out and defining specific aesthetic territory.

If you are looking for stylistic order and overall coherence, the world of painting is not a particularly easy place through which to wander today. But, if we forget about movements and which styles are in fashion, we are free to contemplate each individual's contribution and can more easily appreciate the unique insights artists have to offer us.

Andy Warhol

Many aspects of Andy Warhol's untimely death in February of 1987 seemed like the sensationalistic media events that were so much a part of his personal and professional life for the last twenty-five years. Warhol was the first household name in art since Picasso, and his death was front-page material in many newspapers throughout America and Europe. After quietly entering the hospital for a routine gall-bladder operation, Warhol was discovered dead the next morning of an apparent heart attack.

Numerous stories appeared in the press conjecturing what might have happened on that fateful day. During his life Andy courted the media, and the media was, in turn, infatuated with this mysteriously vacant individual, who first became known by painting soup cans, dollar bills, and row upon row of Coke bottles. Warhol's media life did not end with the coverage of his death. Articles continued to appear for weeks, reporting the autopsy results and the progress of a medical investigation into the cause of his death. Warhol died bathed in the same media spotlight he had courted throughout his life. Warhol was in love with the power of the communication industry and its ability to transform and elevate anything; through his media-inspired work he proved that even the most banal subjects could achieve superstar status. Warhol's genius as an artist lies in his early recognition of how popular culture in the twentieth century was greatly affecting the way we think, feel, and see.

Compared to other events in his curious life, Warhol's demise was paradoxically quiet and peaceful. Warhol's first brush with death occurred on Monday, June 3, 1968, when he was shot and critically wounded, at point-blank range in his silver-foiled studio, by Valeria Solanis, a deranged, would-be scriptwriter and actress. The scene that followed could have been used as an image in one of his "disaster" paintings or as a plot in one of his underground films that mixed sham violence with the patently absurd. But there was nothing fictive about the pain and shock he experienced as he was rushed to a New York hospital, where a team of surgeons worked five and a half hours in a desperate effort to save his life.

The headline of the next day's *New York Daily News*—"ACTRESS SHOOTS ANDY WARHOL"—was an unsettling reminder of one of his earliest paintings, *Plane Crash*. Completed in 1963, it is simply a carefully copied, 8-foot-high blowup of another *Daily News* front page. The painting shows a burned and crumpled airplane wing with the legend, "129 DIE IN JET!" (In a bizarre coincidence this work, dated June 3, 1963, predated the Warhol shooting head-line five years to the day.) The much-talked about gap between art and life was almost too narrow for Warhol: On the night of the shooting the doctors only gave him a fifty-fifty chance for survival.

When Fred Hughes, a long-time associate of Warhol's, hurriedly called the ambulance service after the shooting he heard a laconic voice on the other end say that if he wanted the emergency siren to sound on the way over it would cost $15 extra. The senseless elements of his life seemed to fit the contradictions of his art, and vice versa. Although we may view them as fictional, Warhol's artworks are powerful, mute reminders of the violence and absurdity that surround our lives.

Another friend of Warhol's at the studio that day remembers pleading with Solanis not to also shoot him. As she was leaving the building she pointed her loaded revolver directly at him and stared as if in a daze. Just as she seemed about to pull the trigger, the elevator door near her opened and she stepped in. Three hours and ten minutes after the shooting Valeria Solanis walked up to a rookie police officer on his Times Square beat, handed the gun to him, announced, "I am a flower child," and surrendered. The news media that Warhol had so carefully courted and manipulated for years had a field day with the shooting: Tabloids played up the sensationalistic aspect of the gruesome event and supplied millions of Tuesday morning subway riders with all the lurid details; a local television station came on the air with commentary about how this might be a big Pop-art joke and if Warhol recovered he would probably laugh with us all the way to the bank.

Many people, threatened by Warhol's powerful social aesthetic, even felt he brought this on through his strange attitudes and behavior. Beneath the humor was the insecure feeling people had that Warhol's art, with its images of ordinary products, was secretly making fun of them.

This enigmatic artist's personal background is revealing and might help account for the meaning of his endless rows of painted soup cans and popular film stars.

Publicly Warhol offered little help: "I'd prefer to remain a mystery," he said, "I never give my background, and anyway, I make it all up different every time I'm asked." In the past Warhol purposely clouded his personal history and consequently tried to deny the reality of his past. Lila Davis, a friend who went to art school with him at Carnegie Tech, recalls him as a shy, withdrawn boy from a relatively poor immigrant family. Warhol's father—a coal miner and construction worker, who died when Warhol was nine—came to America in 1912, like many eastern Europeans before him, to earn a fortune and send for his family. It took nine years of his laboring at a variety of menial jobs before his wife could join him.

Warhol was born in Pittsburgh, Pennsylvania, sometime between 1928 and 1930; possibly out of vanity or the desire to befuddle the press, he was purposely vague about the date. Taken by itself this elusiveness is insignificant. But seen within the context of his background and personality—working-class origin and extreme shyness—it provides a key that helps us unlock the meaning of his art. To an awkward, uneasy child of foreign parents, the mesmeric beauty of endless rows of canned food became emblematic icons of compelling beauty, a beauty which signified wealth, abundance, and psychological security. His movie star portraits symbolized a mass yearning for social power and status far beyond the reach of millions. And his *American Death* series focused on the kind of death most familiar to those stuck on the bottom rung of the social ladder: electric chairs, suicides, and car crashes.

"Don't look beneath the surface," Warhol cautioned about his paintings, "there's nothing there." Perhaps he was also referring to the media images on which many of his artworks are based and from which they are derived. He called attention to these fictitious four-color dreams designed to placate a population, divert our attention from social realities, and lull us into a baleful sleep. Degradation, terror, and death lurk here, too. One look at Warhol's most powerful and compelling image *Marilyn*

(colorplate) confirms everything: The out-of-register, smeared, red lips; the garish cadmium-yellow wiglike hair; lavender skin, reminiscent of a poorly tuned color TV; outlandish Las Vegas green eyeliner; and the quintessential element of the whole piece, the rabbitlike teeth that viciously sneer and totally dominate this contemporary death mask. Warhol celebrates beauty in death, death in beauty. *Marilyn* simply stares at us, half real, half fake.

In 1949, after graduating from the painting program at Carnegie Tech, Andrew Warhola—he soon changed his name to Warholl and eventually shortened it to Warhol—arrived in New York City and within a few years established himself as one of the top illustrators on Madison Avenue. One schoolmate remembers Warhol as one of the most uncommunicative people on campus but with a naive gift, close to genius, for achieving the right look on his art assignments without the slightest verbal ability to explain why.

Wearing a pair of dirty sneakers and his best chino pants, Warhol regularly made the rounds of magazine editors and ad agency art directors. In between visits potential clients received beautiful and cleverly made cards reminding them of his presence and artistic abilities. One thing Warhol was very single-minded about was his ambition to make a mark for himself in the art world. Several years after graduating from art school he was commercially successful, illustrating shoes for I. Miller and designing stationery for Bergdorf Goodman, television weather charts for CBS, album covers for Columbia, and Christmas cards for Tiffany.

But the financial aspect of success was only part of what Warhol sought: He yearned for the public recognition that only the fine arts could bring. At this time artists like de Kooning, Pollock, and Rothko were earning unprecedented media attention. He frequently visited the contemporary galleries, bought a few small drawings, and began to feel confident of his own ability to make paintings as good as anything he saw. After all, he told himself, he had studied painting at Carnegie Tech. Graphic design was just a means to earn some money and get started in the fine-arts world. Once a financial base had been established through commercial art, his transition from illustrator of shoes to painter of Pop icons was remarkably quick.

The influence of Robert Rauschenberg and Jasper Johns on Warhol's budding career cannot be overestimated. They were indisputable leaders who showed a way beyond the limits of Abstract Expressionism and opened painting up to greater literary, symbolic, and imagistic content. Johns and Rauschenberg also worked in the commercial art world — as window designers — but unlike Warhol they limited their work hours and devoted considerable time to painting and drawing. What particularly intrigued Warhol about their work was the way they combined various aspects of graphic design — simple, easy-to-comprehend forms — with vigorous but controlled brushwork. Both artists managed to synthesize expressionistic painting techniques with new philosophical ideas. After seeing shows of their recent work, Warhol recalls saying to himself, "Gee, I can do that." Soon he did.

One of the earliest Warhol paintings, *Del Monte Peach Halves*, 1961 (illustration), is a direct predecessor of his later highly controversial and notorious Campbells soup can paintings. In *Del Monte Peach Halves* the brushwork of the can and label area ranges from imagistic clarity to obscurity: The first half of the product name is legible but the second half is indistinct; the peaches are painted in a "realistic" mode but the paint drips down from the upper label making us question the identity of the fruit. Are we to believe in the abstract painterly elements of the painting or the iconographic image the canned food represents? In this early work, the artist wants us to notice and accept the paradoxical elements of both. Early Warhols seem to be more sure of their imagery — popular, easily identifiable, "nonartistic" — than they are of their means. *Before and After* (illustration), produced in 1962, reveals that Warhol soon gave up the idea of expressionistic painting in favor of anonymous looking enlargements of commercial cartoon imagery. Both Warhol and Roy Lichtenstein, another Pop artist working in this imagery, delighted in and made use of simple drawings that might be used for matchbook and yellow page advertisements. *Before and After* shows the promised results of cosmetic surgery, commonly referred to as a "nose job," whereby a prominent aquiline nose is transformed into a presumably more desirable form. The paint carefully keeps within the bounds of the figure and does not show the slightest inclination towards "expressive" paint handling. Warhol's sharp, culturally ambivalent, banal wit — the potent trademark of his mature work — is firmly established in this early painting.

Some of the most amusing paintings of Warhol's formative years are the highly successful *Do It Yourself* series. Looking exactly like greatly enlarged versions of hobby and dime store paint-by-number sets, these paintings mimic the kind of art projects amateurs create through rote number systems and proudly hang in the den or television room. In *Do It Yourself* (illustration) — a familiar landscape scene — the filled-in colors stand out in marked contrast to the unfinished white areas of the canvas that contain the numbered color code supposedly used to complete this "artwork." To achieve an authentic look of mass-produced "art," printed numbers were transferred directly to the canvas by means of a heat-embossment process. Even at this early stage Warhol slyly comments on the role mechanical reproductions have played in social perceptions of art.

The year 1962 was productive and crucial for Warhol. Pop art was beginning to make a big splash on the New York art scene but rather than helping him secure a gallery and sales, the emergence of this movement seemed to have just the opposite effect. Leo Castelli looked at his "cartoon" paintings and showed him Roy Lichtenstein's comic strip series; both Warhol and Castelli agreed that Lichtenstein's paintings were better. Warhol was depressed and despondent. Henry Geldzahler, a friend of Warhol's who was then curator at the Metropolitan Museum, advised him to develop new images but he could not suggest any.

RIGHT: Andy Warhol, *Del Monte Peach Halves* (© The Estate and Foundation of Andy Warhol, 1989/ARS New York).

BELOW: Andy Warhol, *Before and After* (Geoffrey Clements Photography: © The Estate and Foundation of Andy Warhol, 1989/ARS New York).

Murial Latow, the director of a failing gallery, told Warhol one evening over a drink that she had an idea for him but it would cost money to hear it. Warhol got out his checkbook and promptly paid her $50. In an amusing essay about Warhol called "Raggedy Andy" Calvin Tomkins relates the following dialogue:

"All right," Murial said. "Now, tell me, Andy, what do you love more than anything else?"

"I don't know," Andy said. "What?"

"Money," Murial said. "Why don't you paint money?"[1]

Warhol agreed it was a fantastic idea. Cash in hand, Latow threw in a free bonus idea; why didn't he paint something so familiar nobody saw it anymore, like a Campbells soup can? Warhol's life was changed that evening. The next day he bought a huge supply of art materials and painted the first pictures of money and soup cans.

At this time everything was laboriously copied by hand; later he discovered silk screening and speeded up the process.

Despite Warhol's best promotional efforts, no gallery slot in New York was available to show his new work. Irving Blum, director of the Ferus Gallery in Los Angeles, liked the paintings and agreed to show thirty-two Campbells soup can paintings that July. The exhibit opened to quite a bit of hostile criticism and ridicule. A neighboring gallery even went so far as to place a pyramid of Campbells soup cans in the window with a sign that said, "The real thing for twenty-nine cents." Warhol's cans were selling for $100 each. Despite the enormous price difference, the Warhol cans were a better buy and would be worth thousands of dollars each today. Although criticized for their visual "sameness," Warhol offered us diversity in the form of flavor choice: Split Pea, Chicken, Vegetable, Cheddar

Cheese, Scotch Broth, and Chili Beef. Six of the small paintings sold but were later repurchased by the gallery owner to form a complete set.

By the end of 1962 Warhol was exclusively using the silk-screen process to produce his exhaustive series of mechanically reproduced icons of movie stars, endless rows of Coke bottles, soup cans, and dollar bills. There was something compelling and hypnotic about the endless repetition of goods, people, and even historic paintings like the *Mona Lisa*. *Thirty are Better than One* reads the title of a painting incorporating multiple images of this historic and widely known masterpiece. The title might reflect the quantitative concerns of a society that appears to be more interested in the mass accumulation of many products and cheaply made goods than in value and quality. If thirty are not necessarily better than one, they are certainly different. The reiterative insistence of thirty enigmatic smiles accounts for a quantum leap in meaning. We are not seeing the *Mona Lisa* thirty times over, we are experiencing an event that comments on essential aspects of Modernist life.

Rauschenberg and Warhol both developed interests in silk-screening at about the same time. This reproductive process enabled Rauschenberg to greatly enlarge the scale of some small photocollages on which he had been working. Warhol used this method to achieve an impersonal, "cool," and printed look at great savings of time. This widely used commercial printing process is no more than a sophisticated stenciling technique. Over a rectangular wooden frame, a fine wire mesh is stretched; the stencil, in Warhol's case a photographically prepared one, then adheres to the screen. Certain areas of the stencil are clear or open. Paint or ink is forced through these sections with a rubber squeegee. Other places on the screen are blocked, which prevents paint from passing through. Although this printing process is quite simple, many things can go wrong to distort the image. If too much paint is pushed through or the squeegee is pulled unevenly across the screen, the image will appear too dark or uneven. Sometimes a fast-drying ink will clog the mesh and obstruct the flow. Warhol made creative use of all these "problems," striving for great variety in the potentially identical images. A wonderful quality of handmade imperfection exists in all of his photographic paintings.

Nowhere is this deliberately accidental approach used to better advantage than in his *Marilyn Monroe Diptych* (illustration) of 1962. Because of the photographic process, the same image can be varied in scale and used in different artworks. Despite Warhol's avowed commitment at this time to "cool," emotionless art—a famous quotation has him saying he would like to be like a machine because machines can't feel—this painting stands as one of the most thematically charged paintings of the twentieth century.

Reflecting on Marilyn Monroe, whose unhappy life and eventual suicide are known to millions, this painting is a moving meditation on the contradictory themes of beauty and ugliness; fame and obscurity; public adulation and a lonely, pathetic death. The left section of the diptych reveals a cosmetically pristine Marilyn, projecting the image created by the media and worshiped by the public. The colors are slightly garish but appropriately beautiful. Luxurious chrome-yellow hair crowns a slightly off-key pink-fleshed face. No two silk-screen images are exactly alike, however; subtle manipulation of paint and squeegee pressure create unique differences among the twenty-five images. The right panel dramatically shifts into a somber range of blacks and greys. In this section "Marilyn" runs the gamut from obscure darkness to the faintest "ghost" image. The changes occur gradually and sequentially like a section of motion picture film that shows a fade-in and fade-out. If the left panel presents the public image of Marilyn Monroe, movie star, the right section touchingly portrays "Norma Jean." In this monochromed panel, Warhol suggests the constant insecurity that haunted Miss Monroe's relatively short life. Norma Jean, in her vulnerability and frailty, is juxtaposed with Marilyn's stellar beauty, remoteness, and abstract desirability. De-

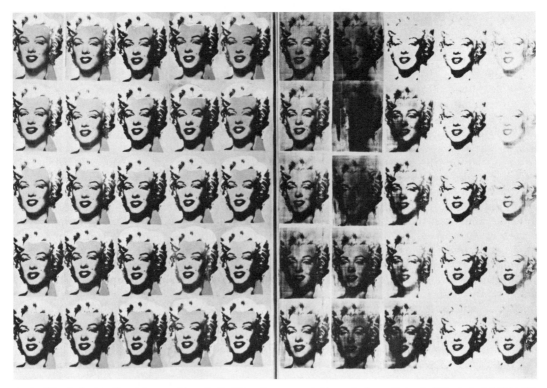

Andy Warhol, *Marilyn Monroe Diptych* (© The Estate and Foundation of Andy Warhol, 1989/ARS New York).

spite the persistence of fifty images (or perhaps because of them) Warhol creates such a powerful hallucinatory effect that we begin to wonder if she ever really existed or was just the collectively conjured dream of a sex-and-love–starved generation of media junkies.

The image repetition and reiteration that occurs in this series is an important thematic, philosophical, and visual device in Warhol's work. The paintings metaphorically mimic motion pictures, an artform that started out as a popular entertainment vehicle. Both Warhol's paintings and the cinema make use of subtly changing sequential images that are inherently reproducible through mechanical means. There are differences, of course; the images in a film are shown to us in a strict linear mode, one at a time, but Warhol's paintings enable us to see the whole structure at once. Also, film and silk-screen paintings contradict the his-

torical precedent of artworks' being "handmade," unique works and exist in a curiously modern state of multiple "originals."

Disaster and tragedy of a broadly social nature figure greatly in Warhol's work at this time. His *American Death* series (suggested perhaps by Geldzahler's early recommendation to do a painting based on the "129 DIE" headline) stands apart from the harmless deadpan humor and benign jesting of paint-by-number, money, and soupcan paintings. The relatively inoffensive images of movie stars, society figures, and product portraits gave way to a hellish vision of mass-media death and carnage. From 1962 to 1963 was a period of great productivity for Warhol; weekly he would arrive at the Stable Gallery with a roll of canvases under his arm. As fast as the amusing portraits of Troy Donahue and Elvis could be stretched, they were sold. But the disaster series was an entirely different matter. No one was laughing now. Death on a grand and gruesome scale was Warhol's new painterly obsession; photos of tragic auto wrecks, bloody race riots, the electric chair, food-poisoned tuna cans (a far cry from the wholesome, nourishing soup images he presented earlier), and suicide jumpers caught in midair confronted us greatly enlarged, often in cosmetically pretty colors.

Many of the original photographs were gleaned from weekly supermarket tabloids that pandered to the public's perverse taste for violence and death. Images of bodies draped over wrecked and twisted car doors, pinned under crushed automobiles, and impaled on telephone-pole spikes beside flaming wrecks were often accompanied in these newspapers with headlines such as "The Accident that Made the Cops Cry." Except for the electric-chair series this group of paintings never sold well in America. But for some reason (perhaps Europeans thought this depiction of inane violence was quintessentially American), they were popular abroad, particularly in Germany.

Bellevue (illustration) is a painting that makes use of another newspaper photo, an apparent suicide victim who jumped off a building. A uniformed patrolman stands ominously to the side while a team of police and hospital attendants administer to the unknown woman on the pavement. Her skirted figure appears on the bottom right of the frame. This painting is particularly expressive of the cinematic elements that run through Warhol's mature work. The image, reading left to right, top to bottom, gradually moves from dark to light like a cinematic fade-in. As the rows proceed downward they tilt and begin to run off the canvas frame.

In the same series, *Suicide* is one of Warhol's most horrific paintings. It shows the black outline of a person in the midst of a suicide leap. The jumper is caught in midair by the newsphotographer literally seconds before death. This particular work, to use a filmic analogy, "freeze frames" the macabre event and sets it alongside a mysterious black section, empty except for a horizontal, white streak. This whole series

Andy Warhol, *Bellevue* (© The Estate and Foundation of Andy Warhol, 1989/ARS New York).

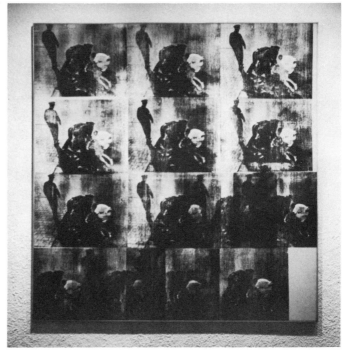

seems to comment on the fact that everything in America can be successfully marketed and sold—even death and dying.

One of the most fiendish looking devices built in recent times is the electric chair. Warhol makes devastating use of this particularly American instrument of death in *Lavender Disaster* (illustration). This painting depicts fifteen electric chairs that sequentially begin to darken and visually deteriorate as they reach the bottom row. Each near-empty room in the repeated image contains one "hot seat." The inhuman device itself takes on the archaic appearance of a medieval torture implement. There is something deadly about the calm, even-handed repetition and insistence of these fifteen pictures, but the choice of color for the painting is far crueler than the image. The insane juxtaposition of pretty cosmetic color with a fiendish instrument of institutional death achieves a visual and emotional power unequaled in contemporary art. Like Melville's use of reversed symbolism, where the color white—which normally expresses innocence and good—becomes the supreme manifestation of evil, Warhol's candy-colored vision of death creates a powerful short circuiting of logic and feeling. With all of the disaster series, we breathe a subliminal sigh of relief in the presence of these frightful scenes which are viewed from the relative safety of a gallery, museum, or book page.

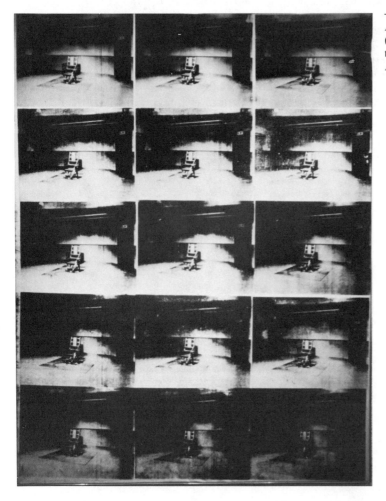

Andy Warhol, *Lavender Disaster* (© The Estate and Foundation of Andy Warhol, 1989/ ARS New York).

His paintings alone would assure Warhol a coveted place in the annals of contemporary art history. But his amazing and prolific output in other artforms raises him to levels of artistic achievement few artists have attained. Warhol's social observations, cultural themes, and artistic formats have predicted many of the directions art has explored into the eighties: multimedia events, the relationship of visual art to "new-wave" music, performance art, videotape, sound pieces, film, and even writing. By working in all of these varied forms, Warhol single-handedly redefined the role of the artist in late twentieth-century terms. He functioned as a medium letting the events of the day act on him and through him. His art is not so much an aesthetic product as it is a kind of personal process that encompasses in the broadest way his background, life, and feelings. Every artform he used—painting, film, performance, underground music, autobiographical writing—was a specialized tool for exploring and defining personal concepts and social perceptions. His "persona" is indelibly stamped on everything he did. Warhol was always the real work of art; paint, canvas, film, videotape, and words are the material means to make it manifest.

He was also one of the founders of the mixed-media school of art. Quite early in his career he worked with vanguard rock groups like the "Velvet Underground," producing a sound very much like today's "new-wave" music that mixes mundane events with sadistically primitive sounds. Many of these early music productions involved projected slides, live performance, lighting effects, and outrageously loud sound. In 1972 he produced a performance piece in London called *Pork*, which featured a variety of "freakish" individuals improvising a wide range of fantasy events. His filmwork reinvents the entire history of cinema, going from silent, static footage to technically proficient, dual-screen epics. Warhol even published a book in 1975, titled *The Philosophy of Andy Warhol (From A to B and Back Again)*, which is a personal, witty statement that anticipates the artist's emerging role as a writer. For years Warhol

carried around a small sound recorder, capturing elements of his life on magnetic tape; the artist viewed these recordings as documentary works of art.

Despite the enormous financial and critical success of his soup cans, movie-star portraits, and "cinematically" structured paintings, Warhol's artistic concerns in the midsixties turned more and more towards actual filmmaking. In 1971, eight years after making his first film, the Andy Warhol filmography numbered well over 250 titles. They varied in importance and length from the well-known, but seldom seen, eight-hour long *Empire* to ten-minute "one-reelers" that were never released to the public. Thus, Warhol established himself, at an early date, as an important figure in the field of personal filmmaking. His special examination of the roots of cinema predicted the current interest in early sources of art. In the process he redefined this relatively young popular artform in terms of a modernist aesthetic. Jonas Mekas, one of the most astute commentators on contemporary film, observed:

Andy Warhol is taking cinema back to its origins, to the days of Lumière, for a rejuvenation and a cleansing. In his work, he has abandoned all the "cinematic" form and subject adornments that cinema had gathered around itself until now. He has focused his lens on the plainest images possible in the plainest manner possible. With his artist's intuition as his only guide, he records, almost obsessively, man's daily activities, the things he sees around him.

We watch a Warhol movie with no hurry. The first thing he does is he stops us from running. His camera rarely moves. It stays fixed on the subject like there was nothing more beautiful and no thing more important than that subject. It stays there longer than we are used to. Long enough for us to begin to free ourselves from all that we thought about haircutting or eating or the Empire State Building; or, for that matter, about cinema. We begin to realize that we have never, really, seen haircutting, or eating. We have cut our hair, we have eaten but we have never really seen those actions. The whole reality around us becomes differently interesting, and we feel like we have to begin filming everything anew. A new way of looking at things and the screen is given through the personal vision of Andy Warhol; a new angle, a new insight—a shift necessitated, no doubt, by the inner changes that are taking place in man.[2]

Haircut, an early 33-minute, black-and-white silent film, changes our perception of time and reality only if we slow down our media-conditioned scale of pace and action. When this film was premiered, it outraged audiences and delighted the avant-garde establishment because it simply shows Warhol's friend, Billy Linich, receiving a haircut. There was no plot, no dialogue, no acting, no chase scenes, no psychological conflicts, and no symbolic "message." In fact this film can essentially be characterized by what it is not, rather than what it is. *Haircut* replaces all of the illusionistic, theatrical, entertaining, and diversionary activities of commercial film with a painfully slowed-down narrative of an ordinary haircut. The commonplace act of haircutting has been transformed by Warhol's vision into a minor miracle of heightened banality. Our internal clocks — set to breakneck speed in modern life — are slowed down and almost stopped by this curious film. *Nothing* is going on, and this, within the context of millions of hours of commercial footage when *everything* seems to happen, is by all definitions an event. There is an air of expectancy to these films that strongly suggests the possibility that anything can and perhaps will transpire. Viewing this strangely simple and haunting film projects us into a state of increased awareness of the world and the many ritualistic events that make up our lives but remain largely "unseen."

Empire, according to Warhol, is a film that portrays an eight-hour working day in the history of one of New York's tallest skyscrapers, the Empire State Building. Warhol envisioned this famous building as a "superstar" and documented its everyday life in one continuous and extremely long eight-hour "take." Few people, of course, have actually viewed this work in its entirety, but just knowing that this is an eight-hour film conceptually affects our perception.

Warhol never conceived of this film as something you watch from beginning to end. It was a filmic painting that slowly evolved in time. He was, after all, an artist with a painter's point of view, who became interested in filmmaking through his sequential structuring of silk-screen images.

This and other films like it come to life, and perhaps function best, when they are projected on the wall at a large party or opening. They work beautifully under these circumstances as cinematic paintings; no one is forced to captively and quietly sit in a darkened theatre for hours on end. In this case the clinking of cocktail glasses and the steady buzz of conversations provide an interesting sound track to this literal "moving-picture."

Empire and *Haircut* are strongly reminiscent of turn-of-the-century films that photograph commonplace activities with a stationary camera. "Interesting" action was not necessary then; the technological miracle of the new filmic process was enough. After these early static films Warhol went on to recapitulate the short but dynamic history of this popular artform. Soon people began to silently act out stories; then the camera panned to interact with the performers; spoken dialogue appeared in crude "talkies"; color footage with more sophisticated sound recording gradually replaced black and white footage; finally full-length dramatic scenarios were introduced and Warhol developed his own version of the Hollywood "star system."

In the early seventies, following a decade-long involvement with moviemaking that took him stylistically from early De Mille to late American International, Warhol returned to his principal arena of activity — painting. His new works anticipated the aesthetic concerns of a generation of artists who were to emerge years later. For instance the previously "cool," remote, photomechanically printed look of the sixties was superseded by a handmade, emotionally expressive, painterly style. By the eighties these general characteristics were strongly established within the art world. Warhol seemed to possess an unfailing intuition that told him when to act, what to do, and when it was time to move on.

Characteristic of his later paintings is *Portrait of Leo* (illustration), which shows Warhol's New York art dealer, Leo Castelli,

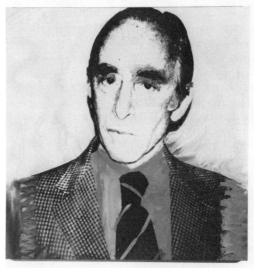

Andy Warhol, *Portrait of Leo* (photo by Eric Pollitzer; © The Estate and Foundation of Andy Warhol, 1989/ARS New York).

in two, side-by-side, 40-inch-square panels. As in his early paintings silk-screen photos are used. Instead of the central image and unifying solid-color background, however, this series features wildly gesticulating swirls and slashes of polychrome paint, recalling the early days of Abstract Expressionism. Warhol used identical images of Castelli but with widely varying brushwork in each panel. Before he silk-screened the photo-image on the canvas he individually underpainted each panel to achieve calculated differences. Thus he synthesized and connected two distinctly different approaches: the mechanically reproduced, "objective" image and hand-done, personally expressive brushwork. Unique and marvelously contradictory visual and conceptual overtones are set in motion by this juxtaposition. Despite the incorporation of abstract painterly elements, Warhol did not completely abandon sociological themes. The choice of pose in this painting is revealing, and perhaps a bit malicious on his part; Castelli's expression appears to be that of a shrewd and calculating businessperson. When asked about this effect, the dean of New York art dealers wryly smiled and ac-

knowledged that the portrait was, in his words, "savage, quite savage."

Are the rich different from you and me? Warhol's celebrity portrait series inevitably asks this question by exclusively portraying individuals who possess great wealth, power, and status. But Warhol was too busy lunching with Jackie O., attending elegant dinner parties, and weekending in the Hamptons to have much time to really worry about this question. In a way these were his people now. Gone were the mass-consumption goods, eerie electric chairs, and hideous car accidents. Andrew Warhola, the awkward, pale son of a Czech immigrant, traveled at record speed from the bottom of the social ladder to the pinnacle of New York's social and cultural elite.

But Warhol continued to do what he had always done with his art: document and observe the social milieu in which he was immersed and through which he was passing. His vision carefully avoids blatant social comment and judgmental opinion that would inhibit our own personal questioning and sense of wonder. The uneasy tension in Warhol's life was transferred to his high-society portraits. His cool, detached attitude and elusive personality partially masked a class-conscious anxiety that extended deep into his life and art. Warhol's paintings portray and expose the essentially cruel myth of "The American Dream": cruel because relatively few individuals will transcend the economic and social limitations imposed on them by circumstance of birth. His paintings reveal one of the best-kept secrets about America: A class structure, comparable to Europe's, that exists in our democratic, egalitarian society. Warhol was an escapee, like other upwardly mobile achievers, but untold millions remain behind. The way we view these painterly personifications of wealth and power might depend on our own social, cultural, and political consciousness. Rather than presenting propaganda messages, Warhol produced visually complex, psychologically layered portraits that were designed to confound, amuse, bewilder, and annoy. He was a lone dreamer, reflecting uneasily on sociological and cultural issues that few artists contend with as subtly and provocatively as he did.

In a rare and revealing account of the personal feelings and fears that conditioned his work, "Andy" wrote in his autobiography:

When I think of my high school days, all I can remember, really, are the long walks to school, through the Czech ghetto with the babushkas and overalls on the clotheslines, in McKeesport, Pennsylvania. I wasn't amazingly popular, but I had some nice friends. I wasn't very close to anyone, although I guess I wanted to be, because when I would see the kids telling one another their problems, I felt left out. No one confided in me—I wasn't the type they wanted to confide in, I guess.[3]

Warhol's early awareness of personal isolation and loneliness coupled with his sensitivity to his poverty was reflected later in his life by his collection of society "patrons" and the many portraits he would do of them.

Quite a few critics seemed to object to this apparent courting of wealthy clients. After his Whitney Museum exhibit, *Portraits of the '70s* in 1980, review response seemed particularly hostile. Most objections, however, seemed to be unrelated to purely aesthetic considerations of composition and color. These were salon portraits, critics maintained, paid for with cash and designed to reinforce the cultural ego of the wealthy patron. Perhaps these critics were really railing against the blatant admission by Warhol of what has been a historical fact for centuries: High art is essentially designed and destined for upper-class consumption.

But Warhol's courting of powerful patrons and his apparent indifference to economic and social imbalances may have masked feelings to the contrary, just as his denial of emotion was betrayed by the remarkably moving *Marilyn Monroe Diptych*. These society icons were entirely in keeping with his past concerns. For almost two decades, Warhol was a superb chronicler of late-modern American capitalism. His "blank slate" brand of genius functioned in the same way as the cameras and tape recorders he constantly carried with him, objectively and ubiquitously recording the ugly and beautiful, moral and immoral aspects of our technologically sophisticated society.

Perhaps the real meaning to many puzzling and mysterious aspects of Warhol's artistic legacy lies somewhere in the future. We may be too immersed in the present to clearly perceive it now. Barbara Rose, a critic of contemporary art, believes that ". . . someday his commissioned portraits will appear as grotesque as Goya's paintings of the Spanish Court. Like Goya, Warhol was a reporter, not a judge, for it was not obvious to Goya's contemporaries that they were deformed either."

Richard Estes

In 1972 Sidney Janis, a prominent New York art dealer, staged an exhibit that caused great repercussions in the art world. This show, called *Sharp Focus Realism*, became an important rallying point for many artists who for years had been quietly plying their trade as "realistic" painters. For the past two decades, they had been struggling against the total dominance Abstract and Pop art seemed to enjoy in the art market.

Sharp Focus Realism featured traditional, nonabstract, illusionistic painting with an emphasis on overall image, detail, and definition. Most of the paintings looked like gigantic, full-color photographic enlargements, which is not surprising, considering that a majority of the artists worked directly from color prints or slides. All of the work in this show rejected abstract approaches to painting and celebrated a return to historic concerns of illusion and image. After decades of paint patches, stains, swirls, and splatters, the effect of all these blown-up technicolor picture postcards was remarkable. There were no brush marks visible, only wall-to-wall, highly detailed, sharp-focused painted images. All of the paintings were devoid of obvious symbolism, social parody, and blatant references to commerical products. They were connected to the past New Realism only by their mutual interest in the reality of everyday life. But the sharp-focus painters viewed the world from a different perspective; they were in love with one of the most universally shared visual experiences of contemporary life: the photographic snapshot.

What they seemed to be aware of in their work was the realization that the photograph—commonly reproduced in the millions—has become a cultural paradigm for visual reality and profoundly affects the way we think, feel, and see. Contemporary art, they reasoned, could no longer ignore the special vision this optical-mechanical-chemical process made possible. No other way of seeing has gained the widespread acceptance and appreciation photography has garnered. One of the primary aims of the Photorealists is to use the objectivity of the camera to achieve a visual impartiality that will enable the artist and public to "review" and reconsider the visual world that constantly surrounds us.

Manet, Monet, Degas, and Courbet were nineteenth-century French artists who were also concerned with the visual poetry of everyday life. Photography was developed during their lifetimes and it was no coincidence that it had a significant effect on their art. For instance, some of the compositional devices of Degas, such as cut-off heads and tilted compositions—reminding us of modern snapshots—can be attributed to his early interest and pioneering work with photography. To nineteenth-century artists the camera possessed a magical ability to capture every detail and nuance of a scene. The instantaneous shutter could also arrest the fleeting moment, capturing people, places, and objects in a visual web of frozen time. There was no doubt about photography's importance to the visual arts, but until the Janis exhibit, contemporary painting did not reflect that importance in any tangible way.

Most of the contributors to the Janis show could be considered Photorealists in one way or another because they all dealt with an extremely acute—or sharp-focus—image that resembled a photograph. Quite a few artists in the exhibit relied on photography in one way or another to help them render their accurate images in paint. Some projected slides on the primed, white canvas and drew in the forms; others gridded the color print and "blew" it up section by section. A few artists considered photographs to be detailed notes that enabled them to form complete paintings from composite prints. Richard Estes is just such a painter and one of the most consistently effective and satisfying artists working within the genre of superreal, photographically inspired art.

There are two distinguishing characteristics of Estes' work that raise it above the mechanical limitations of this process. First of all, rather than copying the surface look of a color photograph—down to laboriously reproducing the grain of the film emulsion—Estes uses a complete series of photos to, as he says, aid his memory. Estes would be content to paint the scene from life if such a method were practical; all of his cityscapes are places of great activity during working hours. He is not interested in accurately translating photographic information into paint. Consequently, he avoids the flatness and one-dimensionality associated with artworks that faithfully copy color prints. Because of this, Estes' paintings have powerful visual lives of their own which spring from the amazing "truthfulness" and detail of the photographs while at the same time transcending their aesthetic limitations.

The second distinguishing quality of an Estes painting is its remarkable ability to mediate and synthesize two apparently opposite qualities in modern painting: realism and abstraction. Because he has not made a commitment to slavishly reproducing a detailed photograph, Estes' painterly simplifications and generalizations portray paint as "abstract" forms and shapes as well as create an overall illusion of "realism." On close inspection abstract qualities of these imagistic paintings are highly pronounced. By his consistent use of reflection—from mirrors, storefront windows, chrome car bumpers, and stainless-steel facades—Estes calls our attention to the many distorted abstract shapes and images we confront every day. He seduces us into accepting the dual reality of direct and reflected images.

Richard Estes comes from a background of traditional academic figure painting and drawing. One of the wonderful things about his schooling at the Chicago Art Institute, Estes recalls, was being able to walk from the classroom into the museum and view firsthand the figurative paintings of masters like Degas or El Greco. Estes is quick to point out his lack of interest in Abstract Expressionism and Pop: Much of it seems silly to a sensibility schooled in blue-chip European painting such as Seurat's *La Grande Jatte*. Estes' feelings were widely shared by fellow students in the early fifties; he can remember only a handful of students doing abstract work. There were only one or two instructors at the school during this era who openly encouraged their students to paint and draw in nonfigurative, experimental modes. Consequently, most of the work at the Art Institute centered around figurative painting in a traditional style. Estes characterized his work at this time as good but not exceptional; it was not the sort of painting you would want to hang on your wall; it was essentially student work.

The artist believes that most universities and art schools leave the student unprepared to deal with the reality of survival in the art world. The apprenticeship of learning both the craft and business side of painting is a long and complicated process for which schools rarely prepare art students. In Estes' case, twelve years elapsed between graduation in 1951 and his first one-person show at the Allan Stone Gallery in New York. Four years of schooling is not enough time to adequately train a professional painter. The developing artist needs a realistic plan for survival to grow and mature.

For Richard Estes employment in the graphic design industry was part of his master plan for survival: After a brief six-month visit to Manhattan, immediately after graduation, he returned to his native Chicago and soon found work as an illustrator for a magazine publisher. Like Warhol's experience in commercial art, this work taught him the practical realities of the art world. Also, his business experience provided a reasonably well-paying means of continuing training and exposed him to many new art techniques. He spent time on the job doing quick layout sketches with chalk and ink markers. In a short time his drawing and rendering abilities improved, and he learned how to work in a quick and effective way. Sometimes his office needed a drawing of a specific figure or hands holding a certain object. Rather than have someone hold the pose for a finished drawing, a Polaroid camera was available for photographic "note taking." Estes remembers that no one would dream of working from photographs in art school; it would have been like cheating. But no such moral dilemma existed in the business world where time, money, and deadlines were constant concerns. Through his commercial work Estes discovered that using photographs for visual references was a more accurate and effective method than the traditional art school "life-drawing" approach.

Although some of his work involved illustration, his main task was doing mechanical artwork such as paste-ups and color overlays. Technical jobs proved less demanding than original designing and left Estes with enough energy to do his own creative work in the evenings.

By 1965 — using $5000 in savings — he quit his commercial job to work full time on his art. After twelve months of uninterrupted work Estes managed to paint for another year by doing free-lance illustration on the side. This was an important time of development for him; eventually work completed during this period led to his first one-person show at the Stone Gallery. Paintings done during his sabbatical years are much more accomplished than the earlier ones executed after tiring eight-hour work days.

Automat (illustration), completed in 1967, was painted during his "holiday" from commercial work and is one of the first works to firmly establish Estes' mature style. Although it is executed in a decidedly realistic manner, it avoids looking like a painted photograph. Two years earlier, Estes did quite a different semi-impressionistic painting of two figures seated at a table in the same restaurant. But *Automat* is different; it was done with the aid of

photographs and has the clarity, coherence, and detail that typifies his future work. Painted from a birds-eye view, this painting — unlike early works that feature the figure — plays down the human element and focuses instead on compositional aspects of food, plates, coffee cups, napkins, and eating utensils. The figures are portrayed as subtly shaded masses of dark browns and cool blacks. His later paintings avoid figurative issues altogether. Most of his cityscapes reveal empty streets, sidewalks, and stores. The inhabitants of Estes' landscapes are alluded to but remain unseen. He explained this phenomenon in an interview with John Arthur:

If there is a figure (in the painting) . . . it becomes romanticized — a period piece like an Edward Hopper. It changes one's reaction to the painting and destroys the feeling of it to put a figure in because when you add figures then people start relating to the figures and it's an emotional relationship. The painting becomes too literal whereas without the figure it's more purely a visual experience. . . . I don't want any

Richard Estes, *Automat* (Anonymous, Private Collection).

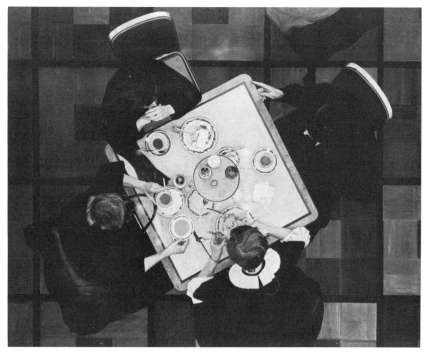

kind of emotion to intrude. When I look at the painting — the color and the forms — everything is so beautiful that I'm not really concerned about who the people are or what they're doing.[4]

Estes plays down narrative elements and overtly humanistic overtones in his work; he is interested in the visual spectacle of enormous detail and baroque renderings of multisurfaced reflections. All of the high-density trappings of a technically advanced civilization are presented to us — neon signs, large plate-glass surfaces, parked cars, multistoried buildings — but the inhabitants of this fast-moving and crowded world are mysteriously hidden. An unsettling, pervasive silence emanates from these paintings. Because of this puzzling contradiction, his urban landscapes, in spite of their meticulous attention to realistic detail, become dreamy urban landscapes of *unreality*. Estes is technically challenged by the task of organizing copious detail, and aesthetically transfixed by the opulent beauty of reflection piled upon reflection. His mind becomes completely engaged with the organizational act of painting; everything in this enormous puzzle must fit together and visually "work": buildings, empty streets punctuated with litter, and congested store windows.

Before Estes begins a new painting, he spends a few days walking around the city taking photographs more or less at random. After the rolls of film are processed and proofed, if anything interesting shows up in the contact sheets he returns to this site to make a more detailed record of aspects that interest him. For a painting called *Downtown* (illustration), for instance, the composite photographs alone number sixty or seventy. Estes usually shoots five or six rolls of 120 film — twelve exposures to the roll — plus three 4 by 5s. These large-format negatives pack an enormous amount of sharp, well-defined visual information that aids the artist's process of pictorial editing and interpretation. Up to eighty photographs may be taken for one particular painting. Of course, not all of them will be directly used, but having this amount gives him options and spare shots in case of an emergency.

Most of Estes' photographic note taking is done on Sunday; the traffic and crowds prove to be too heavy during the weekdays. During the winter, when a rare sunny day occurs on a weekend, Estes abandons pre-

Richard Estes, *Downtown.*

vious plans and sets out with his camera, tripod, and a generous supply of film. Sometimes even weather conditions become a matter of artistic concern in contemporary art.

Once the site and visual scope of the cityscape are determined, the finished "size" of the painting depends on various factors—including how Estes feels at the moment. If his previous two paintings have been small, a larger canvas might be in the offing. No particular value is placed on large canvases. The artist enjoys working on big and small formats. Even his larger paintings are not terribly grand by today's standards; they are very much easel paintings comfortably measuring 5 or 6 feet at the most. In the case of *Downtown* Estes happened to feel like working on a big canvas, so he did.

After he studies the chosen selection of composite photographs, he sketches a loose and very rough drawing on the canvas with a thin wash of water-based acrylic paint. Just the most general compositional aspects of the painting are outlined at this stage. Once Estes feels happy with the overall approach, then he proceeds to work on smaller details. The painting develops in a holistic way from the general to the specific, rather than from one completed section to another. In this way effective control is maintained over color balance, tonality, and composition. After this acrylic underpainting accurately develops most of the detailed visual organization, Estes switches to oil paint because of its inherently richer tonal and color range. From this point on, his task consists of careful adjustment of paint values and even finer realization of detail. If the painting were photographed in black and white at the completion of the acrylic underpainting stage, it would look very much like a finished painting. Estes believes:

The oil gives a greater depth and more control of gradations of color, blending for example. It's a much richer look than acrylic. Acrylic is easier to work with; it's good for underpainting and dries quickly, so it's easy to make big corrections. With oil you have to know pretty much what should be there because of the differences in the characteristics and drying times of the colors.[5]

Sequential photographs (*Baby Doll Lounge*, illustrations) reveal the progressive development all of his paintings undergo. A first stage shows only a generalized, sketchy and rough outline of street perspective and building placement. Subsequent stages develop with increasing sureness and accurate color, detail, and overall composition. Unlike many Photorealists, Estes does not grid the painting to transfer photoimages to canvas or use a slide projector and color transparencies. His working method employs the advantages of photographic note taking with traditional and historic processes of oil painting. Perhaps because of this remarkably effective synthesis of old and new methodology, Estes' work achieves a look and feel that few artists working in this genre attain. Placed next to painted versions of Kodachrome slides and color prints, his paintings have a quality of light, space, and "believability" that totally transcend the accuracy of image and form found in other hard-core Photorealists. Estes' works are not imitations of photographs. Significantly, there is no fuzziness in his paintings associated with enlarged reproductions. He observed:

Even with a 4 × 5 negative, a photograph would be a bit fuzzy blown up to this size. The paintings are crisp and sharp. I think with painting it's a problem of selection and imitation, but it's never a problem of creation. It's wrong to think that anyone ever creates. At best one selects new imagery. I can select what to do or not to do from what's in the photograph. I can add or subtract from it. Every time I do something it's a choice, but it's not a choice involving something creative or reproductive. It's a selection from the various aspects of reality. So what I'm trying to paint is not something different, but something more like the place I've photographed. Somehow the paint and the intensity of color emphasize the light and do things to build up form that a photograph does not do. In that way, the painting is superior to the photograph. I think that for figures it would be better not to use photographs. There's far more information if you have the person sitting there. You really don't know what a person looks like from a photograph. The reason I take a lot of photographs is to make up for the fact that one photograph really doesn't give me all the infor-

Richard Estes, *Baby Doll Lounge* (photos courtesy of John Arthur).

mation I need. Also, the camera is like one eye and it really deals only with values. And painting is trickery, because you can make people respond by guiding their eyes around the picture. The photograph doesn't do that because a camera doesn't have ideas. It can only reproduce, so you have to use a lot of trickery.[6]

One of Estes' most successful early paintings taking this technical approach — *Escalator*, 1970 (colorplate) — clearly states many of his mature thematic concerns including urban emptiness and a great love of mirrored reflections.

A wide-angle lens perspective dominates this silent scene of a deserted downtown office building. The centrally placed escalator entrance visually invites us to step onto this moving staircase. A labyrinthian, complex, and baroque space encompasses the whole scene; another escalator appears in the upper right hand of the painting going in the opposite direction. No figure is pres-

ent but the painting projects the message that these means of human transport operate efficiently nevertheless. The painting's chromatic range is close-hued, almost monochromatic, and chillingly effective in the mood it sets: anonymous, lonely, ennui-ridden, but beautiful. Large, flat, stainless-steel surfaces prismatically reflect the surroundings. The pattern of the tread grill is echoed by the iron railing directly above. Although it is not a tour de force orchestration of detail, as Estes' recent "street" paintings are, this piece effectively reconciles "realist" and "abstract" elements in a compelling way. The strong mood and theme make this one of his loveliest paintings. Both Edward Hopper and Richard Estes perceive downtown America as an essentially solitary, physically imposing, and wistful place.

A year later *Helene's Florist* (illustration) represents the other end of the composi-

Richard Estes, *Helene's Florist* (courtesy of The Toledo Art Museum, Toledo, Ohio).

tional spectrum as far as sheer visual detail is concerned. An enormous amount of visual and verbal information bombards our senses in the form of advertising signs, architectural patterns and grids, organic flower forms, reflections, and mysteriously darkened interior spaces. This piece takes its title from a prominently featured store sign that appears at the top of the painting. Many other signs appear and dominate the composition: "barber shop," more specifically "Billy's Barber Shop," phone numbers "EN2-7909," and bits and pieces like ". . . D SHOP," "P. 267," and "OP." The details of human activity and occupation contained in these written announcements are significant; even a list of daily lunch specials in the coffee shop window is readable on the original painting. The lighting on this urban outpost is dramatic, complex, and somber. Horizontal rays of sunlight, which suggest an early morning hour, filter through and pierce the canyons of this technological jungle. Some areas of the street to the left remain in shadow, their neon signs glowing through the darkness. Inside the barber shop and florist, seen through open doors and plate-glass windows, mysteriously no one seems to be present, minding the stores. Huddled together outside in a geometric configuration are the lush and lovely rows of flowers offering lyrical counterpoint to the metallic architecture and littered sidewalks.

Another element, besides the missing presence of people, that figures greatly in Estes' mature works is his profound and abiding love of plate-glass window reflections. Without the camera's effortless ability to record and document these subtly shaded, infinitely complex plays of light on reflective surfaces, his thematic vision would be severely limited. One could not imagine Estes setting up his easel on the corner of Lexington Avenue and 53rd Street even on a relatively quiet Sunday. If he could, there is the overwhelming factor of the hundreds of hours it takes to develop a painting, not to mention wind, dust, dirt, sleet, rain, and snow. Painting these kinds of scenes from life is out of the question; his camera is an important but silent partner in these en-

deavors. The technological process of photography is so compelling in its ability to provide accessible, accurate visual information for Estes that we might even consider it as part of the subject matter of the painting. At the very least it is a tangible part of his process of painting.

Paris Street Scene, 1972 (illustration), is a prime example of how a whole world of reflections is made "visible" by the use of a camera. Almost half of the surface area in this canvas is taken up by a large storefront window that faithfully mirrors the Parisian street scene to the left. There is no clear dividing line between these two elements —image and reflection; only the most subtle variation of color contrast and tone separates them. Estes prepares us in this painting to doubt what we see and question what we know. The power of his paintings emanates as much from this chimerical uncertainty as it does from the presence of lucid form, light, and space.

In *Ansonia* (illustration), painted five years later in 1977, Estes continues to cloud the issue of "reality" by painting the reflected images in tones that are more intense and vivid than the nonreflected, directly viewed world. The mirrored sky is an even deeper blue, the reflected buildings are as clear and sharp as the "real" ones. Only the tilted and fractured framework of the image and the intrusion of elements inside the store offer substantial clues about the veracity of this scene. By reinterpreting photographic information rather than copying photographs wholesale, Estes can editorialize and reverse certain visual elements. Usually, reflected images appear faint and less distinct but in some of his paintings he contradicts this generality.

In a curious and creative way, Estes, and some of his fellow Photorealists, have halted and reversed the flow of time and art history through their paintings. In a way they "aesthetically" move forward by going back. Of course they can no more literally go back than we can summon to our present time the physical presence of Leonardo, Rembrandt, and Cezanne. Contemporary artists cannot return to a preliterate, pretechnological, and preurbanized time. We

Richard Estes, *Paris Street Scene* (Collection of Sydney and Frances Lewis).

Richard Estes, *Ansonia* (photo by Geoffrey Clements; Collection of the Whitney Museum of American Art, New York, Gift of Frances and Sydney Lewis).

see things differently because our knowledge and beliefs are substantially different. An artist's "vision" is not entirely dependent on optical factors alone but relies on personal feelings, societal attitudes, and cultural experiences. Even though, in a sense, Estes would like to return to the past, he must instead bring the past to the present and view it through a twentieth-century artistic sensibility. So he reinvents and recasts this mythical past of realism in essentially modern terms.

In a sense, a process of reversed phylogeny is at work in Estes' art. Any notion of "progress" is summarily dismissed by his painting process and aesthetic philosophy. His artworks begin with modern attributes — loose brushwork, spontaneity, a casual look — and seem to proceed to the past via a trans-time expressway to the carefully considered, pristine, three-dimensional, and rational works of Vermeer and Canaletto. Although it would be a mistake to view Estes entirely in terms of past "illusionist" art, there is a marvelous element of historic reinvention present in all of his mature work. Seen on these terms, his paintings become a form of eclectic, modern illusionism, borrowing inspiration from the past, while retaining a firm hold on present-day sources. Above all, his work is uniquely modified and transformed by one of the most recent and profound inventions to ever influence how and what we see: photography. Despite stylistic appearances that relate to Realist art of the past, his work definitely draws its inspiration and methodology from twentieth-century sensibilities and technology. His paintings discover, and celebrate, the hidden beauty and poetry of urban streets, empty office buildings, rows of telephone booths, and flowers. Even his titles reflect this essentially contemporary vision and reconfirm his standing as a modern artist: *Baby Doll Lounge*.

Shusaku Arakawa

For some time now Shusaku Arakawa has been making paintings that skillfully blend words, images, and concepts, breaking down the usual distinctions between them and leaving us slightly bewildered and vastly amused. He is a modern-day master of semantic innuendos and ironic verbal and visual non sequiturs (looking at his work and considering its literal meaning leaves us open to cerebral short-circuiting). The paintings abound in clever, whimsical, and sophisticated jokes that make us question what we think we know. Despite their seemingly intimate involvement with mathematics, modern physics, and the relatively new science of linguistics, they function quite outside of any logical system of Western thought. This is precisely one of Arakawa's major artistic and philosophical points.

His inspirational heroes appear to range beyond the usual artistic influences of other painters. Allusions to the work of Lewis Carroll, Ludwig Wittgenstein, Gertrude Stein, and Hegel run through his exercises of sense and nonsense. All of these literary and philosophical writers played with the innate structure of language and the semantics of meaning.

In Arakawa's paintings, the inscrutable wisdom of the East meets the level-headed logic of the West and combines to create new pathways of thinking and experience. Many of his ideas seem to ideologically parallel traditional Japanese "koans" — Zen teaching parables that cannot be understood by means of Western-style logic. Arakawa's thematic sources are an unusual blend of European and Oriental thought and methodology. Done up in Buddhist style are verbal dialogues that stem from Western philosophical ideas, particularly Hegel's. This famous philosopher argued "that thought in its very nature is dialectical, and that, as understanding, it must fall into contradiction."

A wonderful example of this poetic union is a painting titled *Untitled 1969* (illustration), which incorporates stenciled letters on a 4-by-6-foot empty canvas. Clearly inscribed on the surface of the vacant canvas is the solitary statement: "I have decided to leave this canvas completely blank." Of course the act of painting this sentence on the canvas changes the meaning of the language, and vice versa. Another painting incorporates the legend, "Say one, think two." Both of these conceptual pieces play with meaning and logic in the same way our minds play with and meditate on life and living; they are lessons in self-conscious reasoning and function as cerebral exercises that force us to question thought itself.

On November 13, 1969, Arakawa's obsession with this kind of linguistic discourse provoked an incident at the Dwan Gallery in New York that had some amusing legal ramifications. *Untitled Painting, 1969* (illustration), valued then at about $2000, was stolen from Arakawa's show there. It bore the tantalizing statement: "If possible steal any one of these drawings including this sentence." Obviously well-prepared art thieves left behind a statement to the effect that they were only carrying out the instructions contained in the painting. In an interview with the art critic Lawrence Alloway soon after the theft, Arakawa stated, "My first reaction was very surprised. Then I felt angry at the situation. Then minutes later I was strangely excited. I talked to the secretary at the gallery explaining the painting. It was as if I was explaining it to myself. Then I felt very good about it."

A few days later Arakawa received a telegram from the art thieves reading, "Drawing safe. Work completed." On December 20, the Wadsworth Atheneum in Hartford, Connecticut, took possession of a package containing the stolen painting. Although the thieves thought of their prank as a collaboration with the artist, Arakawa felt otherwise. He pointed out that they had totally misread the work and what it meant; the legend on the canvas said, ". . . steal any one of these drawings [several areas in the painting contain linear drawings] including this sentence." It is physically impossible to lift the drawing from the surface, and of course it is literally impossible to steal a sentence. Nowhere does Arakawa say steal this painting. Much of Arakawa's work points towards the ambiguity of language and the imprecision of meaning communicated through it.

The origins of this particular painting are fascinating and reveal Arakawa's fortuitous methods of operation. He wrote the verbal statement months before he made the painting; although he liked the sentence, he could not find a suitable image to accompany it. Then he saw Bresson's classic French film *Pickpocket* and became fascinated with the idea of stealing.

Eventually Arakawa got to meet his conceptual "collaborators." He remembers feeling somewhat shy and awkward in their presence, but they were delighted to meet him and did not seem a bit embarrassed by their illegal action. As a postscript to this odd assembly he wryly observed, "In a way meeting is a little bit like stealing." Since then Arakawa has made several versions of this notorious piece, but he is quick to point out that in the future he hopes people will steal with more discretion, or at least read the instructions carefully.

The artist continued his droll and amusing brand of poetic philosophy with *Mistake*. A written statement on the lower central portion of the canvas reads, "The letters on this work Mistake have an average height of 5'6" (5'5"–6'2") and an average weight of 145 lbs (110 lbs–190 lbs)." "M-I-S-T-A-K-E" begins to take on a feeling of solidity because of the architectural rendering of the letters. But this illusion is not faithfully carried out; some of the letters seem weighty and constructed of heavy stone whereas others appear weightless and ephemeral. At first glance many of Arakawa's paintings appear to be clear and logical statements but upon closer inspection they function entirely within arbitrary, nonsensical systems of signs and thought. *Mistake* continues to confound and tease us; words like *sun, chair, mountain, dust,* and *work* appear throughout the painting and several arrows direct us towards small cryptic marks on the canvas.

Shusaku Arakawa's personal experiences in life relate to the duality evident in his work. He was born in Nagaya, Japan, in 1936 but immigrated to New York City as a young man, where he has lived and worked since the early sixties. Science fascinated him during his childhood, and as a college student he studied medicine and mathematics in Tokyo. As he learned more and more about the methodology of sci-

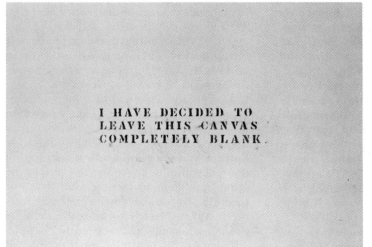

Shusaku Arakawa, *Untitled 1969* (courtesy of Ronald Feldman Fine Arts, New York).

Shusaku Arakawa, *Untitled Painting, 1969* (photo by Joseph Szaszfai; courtesy of Wadsworth Atheneum).

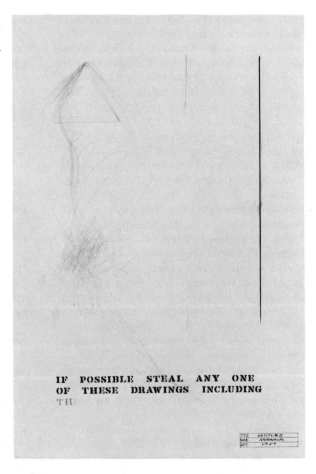

ence, he came to realize that its claim to objective truth and logic was based on self-referential systems and arbitrary signs. Soon Arakawa became distrustful of the exaggerated claims made for rationality and "order," maintaining instead that there was just as much wisdom in certain modes of nonlogical thought. The rich complexity, structure, and patterning of life—with its turnabouts, twists, quirks, and dead ends—seemed to have no place in the logic-suffused world of twentieth-century science. Also the intellectual delights and wisdom of humor, ironic wit, and absurdity seemed banished from this grimly serious branch of knowledge. But despite his apparent irreverence and distrust of logical thought, he is anything but antiscientific. In fact he very much considers his artwork to have "a particular degree of eccentric parallel" to quite a lot of recent experimental scientific thought. Arakawa—whose sensibility is poised between Eastern meditative thought and Western positivistic thinking—has formulated his own brand of humorous metaphysical questioning through his paintings. He realizes, as the early twentieth-century linguistic philosopher Ludwig Wittgenstein did before him, that we can have no substantive understanding of meaning without examining the language used to contain and express it. Arakawa's painterly speculations on seeing, semantics, and reason demonstrate a personal search for a primal dialogue with meaning and contemporary aesthetics.

Soon after he moved to Manhattan, Arakawa met and eventually married Madeline Gins, a professional writer and poet. This partnership has played a significant role in the development of his philosophy and his art. Gins supported and shared his speculative linguistic interests; eventually their professional relationship led to one of the most remarkable collaborations in recent times between a visual artist and a writer: publication of *The Mechanism of Meaning*, a work which the two co-authored.

The book begins with a two-page preface that outlines in an amusing way the authors' preoccupation with vision, thought, and language:

If we had not been so desperate at that time, we might have not chosen such an ambitious title for this work. Yet what else would we have called it? After all, the phenomena we were studying were not simply images, percepts or thoughts alone. Our subject is more nearly all given conditions brought together in one place. . . . The vagueness of the term was suitable. Meaning might be thought of as the desire to think something—anything—through; the will to make sense out of the ever-present fog of not-quite-knowing; the recognition of nonsense. As such it may be associated with any human faculty. . . . We hope future generations find our humor useful for the models of thought and other escape routes that they shall construct![7]

The Mechanism of Meaning is divided into sections, but these divisions are not like ordinary chapters in a book, which have a sense of logical and orderly progression. Each subdivision can be seen as an enclosed fugue on a certain "philosophical" theme. The section headings have titles like, *Meaning, Reassembling, Reversibility*, and *Feeling of Meaning*; the book ends didactically with a chapter called *Review and Self-Criticism*. Reproductions of Arakawa's paintings are interspersed with anecdotes, photographs, and textbooklike drawings. This is not the usual art book monograph with reproductions and text. Because of its overall structure and commitment to concept it functions more like a work of art itself.

Arakawa and Gins lay linguistic traps with the curious text and images. As we leaf through this book—reading, looking, and wondering—the snare is tripped. We are caught in a complex quagmire of visual puzzles, illogical statements, and shifting patterns of meaning and form. A tone of half-serious, self-effacing humor is predominant throughout the work. On page 91 an ominous, morguelike photograph of a prone, shrouded figure is accompanied by two statements. The top left corner reads, "Forget any form"; the bottom right reads, "Forget any non-form." In a caption below this picture Arakawa and Gins discourse: "We are told to forget any form. All right. Then it is non-form we must forget about when viewing this photograph. This brings confusion. Of course, neither is possible. At least not absolutely. And each suggestion (command?) makes the other less possible. What kind of nonsense is this?"

Benign confusion and nonsense are two important themes that consistently run through Arakawa's work. Arakawa seeks to break up and reform old language patterns that impede the way we think and see. With Arakawa, puzzlement opens the path to discovery and fresh insight.

The interfacing of cultural thought that occurs in Arakawa's painting is by no means limited to his work alone. Over the last fifty years there has been a massive exchange of goods, ideas, and culture between the East and the West. Tokyo looks more "Western" and industralized than New York. Oriental

philosophy and meditative forms of relaxation enjoy widespread popularity among American artists, college students, and segments of the public. Japanese cars and Oriental electronic goods can be found in abundance throughout Europe and the Americas. Through increased trade, understanding, and cultural exchange the world has profited immensely from the confluence of these two currents of thought and tradition. The Gins–Arakawa partnership represents the meeting and fusion of Oriental–Occidental modes of thought. Out of this marriage comes a new — less ethnocentric — understanding of various attitudes, viewpoints, and perceptions of the world.

One of the funniest artworks in the book is *Ambiguous Zones of a Lemon* (illustration).

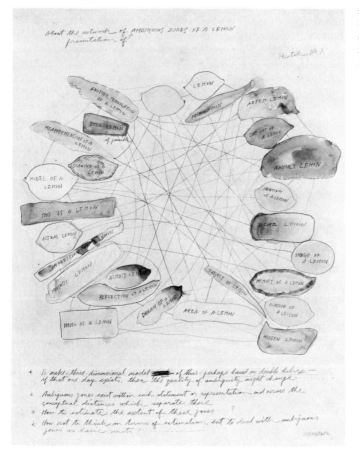

Shusaku Arakawa, *Ambiguous Zones of a Lemon* (photo by Walter Russell; courtesy of Ronald Feldman Fine Arts, New York).

It is certainly one of the most visually appealing works, but its real effectiveness lies in its effortless mixing of language, concept, and image. At the very top of a circular collection of multicolored shapes — all carefully titled — is a traditional rendering of a lemon. Weblike lines link the various shapes to each other in this circular collection. *Ambiguous Zones of a Lemon* reveals the way language becomes a verbal catalyst for images and concepts in Arakawa's work. Each shape, no matter how unlemonlike it is in form and color (some are painted purple, dark blue, or red), is referred to as some kind of a lemon: *animal's lemon, memory of a lemon, dream of a lemon, impression of a lemon, still lemon (if possible)*, and *past lemon*. The litany of lemons continues to include an elongated tan rectangle labeled — suggesting the shape does not look like a lemon — *this is a lemon*. As if things were not delightfully confused enough, Arakawa includes, at the upper left-hand corner of a yellow square, a tiny arrow which points to a pale green splash of color labeled *sound*.

Chapter 12, titled "Feeling of Meaning" features an unusual painting called *Panel No. 3* (illustration). In it appear four identical photographs of the artist collaged to the surface of the canvas. They are head and shoulder shots, harshly lighted, that bear a great resemblance to post office "wanted" posters. Reading left to right, the first of these images is unlabeled; the second says *mad* underneath; the third, *madder*; and the fourth, *maddest*. In a rectangle below the photographic grouping, defined

Shusaku Arakawa, *Panel No. 3* (photo by Walter Russell; courtesy of Ronald Feldman Fine Arts, New York).

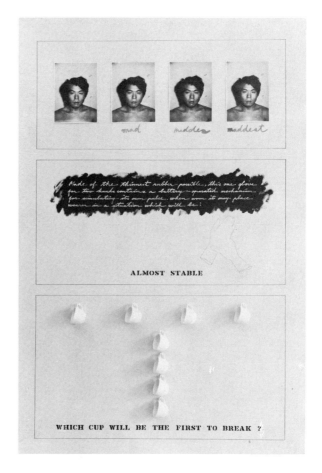

by a thin black line, is a loosely painted dark area with a sentence stenciled on it alongside a simple mechanical drawing of two joined gloves. It reads, *Made of the thinnest rubber possible, this one glove for two hands contains a battery-operated mechanism for simulating its own pulse. When worn it may place wearer in a situation which will be: ALMOST STABLE.* On the lower rectangle, eight pure white ceramic coffee cups are hung on hooks. Tersely written under the cups is the question, *Which cup will be the first to break?* A certain psychological tension based on the fragility of the mugs is created by this statement. To a great extent the painting is very much about verbal and visual juxtapositions and contrasts: photographs with descriptive captions that contradict the apparent truth of the images (either the words or the pictures are lying). "Abstract" paint splotches combine with pseudotechnical statements that try to relate the two but inevitably trail off into nonsense; and three-dimensional, dime-store cups hang above a pessimistic question about their fate. All of these elements relate to each other in Arakawa's arcane world precisely because they appear to have nothing in common. He forces us to compare and consider things we ordinarily would not. It is infuriating to try to interpret this painting, at least by logical means.

In a painting called *Construction of the Memory of Meaning, Panel One* (illustration),

16 CONSTRUCTION OF THE MEMORY OF MEANING

A STUDY OF MEMORY: ITS OPERATIONS, ITS SCOPE, ITS ROLE IN THE REALIZATION OF MEANING. TOWARD THE CONSTRUCTION OF A TOTAL SITUATION IN WHICH MEMORY CAN REMEMBER ITSELF (ITS OWN OPERATIONS)

NIGHT

Shusaku Arakawa, *Construction of the Meaning of Memory, Panel One* (photo by Walter Russell; courtesy of Ronald Feldman Fine Arts, New York).

Arakawa addressed the issues of memory and consciousness in an unusual way. Most of his works exhibit an unusual blend of serious-sounding rhetoric and absurd punch lines. This painting is no exception, but plays down some of his zanier elements and seems to be inspired by science-fiction writing and the encoded notebooks of Leonardo da Vinci. Three vertical mirrors dominate the composition of this particular canvas. Inscribed across the mirrors is a strange form of cryptic writing. The undecipherable text is written in silver paint and is actually a personal letter addressed to his mother. The way to decode the letter is to view it in a mirror (like the secret notebooks of Leonardo). It reads:

January 4, 1915

Dear Mother:

It snowed all through the night. It's so difficult to recognize New York these days. It's all silver. This morning I went to buy bread but it was already sold out. Anyway my work continues. I'm working day and night. Please don't worry about my health. When the project is finished I will send you a ticket to come and see it.

My friends keep telling me what a sad time we live in, how dark these ages are, I agree but I must have some hope so I cannot stop, I will not be disappointed unless I die. (Don't worry I don't feel that way again.)

Last month, I began the most complicated section of the project. I am determined to find out how (if not why) our brain functions. Hundreds of people have tried before but still we only know a small percentage of one percent.

Thank you for the pillowcases. They smell wonderful. I have to go to sleep now (up tomorrow at 7). I will write again soon. Will you write soon? Please take care.

Love,

Arakawa[8]

Arakawa, a modern-day immigrant, conjures in our minds the thoughts of a visitor from another place and time, living and working in a foreign land. This autobiographical material is connected to certain elements of Leonardo's life: His historic notebooks sought to discover the hidden workings and anatomical secrets of the human body. When a mirror is used to decipher the language, an infinitely deep illusionistic space is created that visually places us inside the painting. Many aspects of the work seem to allude to a "time-space" warp that confuses conventional perceptions of place and event. The reference to the "silver snow" of New York relates in a literal way to the silver writing on the surface of the mirror and even the hidden silver backing of the mirror itself. Images of ourselves confront us in a strange way and play with our memories. An unusually rich layering of visual devices and literary references distinguishes this particular piece.

Arakawa's recent work, without abandoning its verbal concerns, adopts a less sparse visual look and takes a more painterly approach to color. One of the most amusing and successful pieces in this series is *Courbet's Canvas* (colorplate). The size of the stencil used for the writing is larger than usual and the letters are loosely filled in with many colors. Arakawa makes great use of chromatic contrast in this piece; some of the colors stand out sharply against the white ground, others are pale and difficult to read. A thin, black mechanical line almost completely bisects the painting on a diagonal and joins a triangular shape wedged against the top corner. Appropriately, Arakawa philosophically muses on color itself: "We are told to forget about grey. All right. Then it is non-grey we must forget about when viewing this painting. This makes me angry. Of course neither is possible. At least not absolutely. And each suggestion (command?) makes the other less possible. What kind of nonsense is this? I'm so confused I'd like to forget the whole thing." On the lower right and upper left corners are two "grey scales" (the kind used in photography), running in opposite directions. One is labeled "forget any grey," the other "forget any non-grey." A hopeless and confounding task.

Despite Arakawa's open humor, his paintings are visually demanding and require careful study from the viewer. A complete understanding of Arakawa's philosophical sources is not really necessary, however, to appreciate the subtle nuances of visual design and contradictory humor found in his work. These are sophisticated, literate, and complex works of art that function on many levels. How far you take them depends on your personal interest. Arakawa is delighted that they can and do provoke strong feelings. On page 22 of *The Mechanism of Meaning*, a story appears about a painting in which the word "mistake" is scrawled over a series of important mathematical equations as if to cancel their scientific validity and truth. While workers were installing an Arakawa show at the Nationalgalerie in Berlin, in 1972, the director of the museum stopped in front of this particular work, carefully read it, and became furious. "How dare this artist write 'mistake' over the formulas of physics upon which our world is built." If assistants had not calmed him, the story goes, he might have damaged the painting in his rage. When Arakawa was informed of the event, he was pleased and said that he wanted to thank the director for "having been concerned enough to make even more energy of meaning." A small photograph of a monkey swinging on a rope appears beneath this tale in the book.

Recently Arakawa has produced a series of drawings, prints, and paintings based on a passage from *A Treatise of Human Nature* by the eighteenth-century Scots philosopher David Hume. In 1979 an enormous 20-foot-long painting called *A Forgettance (Exhaustion Exhumed)* (illustration), which incorporated a passage from Hume's book, was exhibited at the Ron Feldman Gallery.

The background of this billboard-sized canvas varies from dark grey to light pink; the same variance of tone occurs in the stenciled sections, making parts of the Hume passage difficult to read. A vivid red triangle is placed directly in the center of the canvas, obscuring some of the language. To the left of the triangle a plank riddled with bullet holes is fixed to the surface. Basically the Hume text deals with the faculties of concept and percept: "When we remember the past event, the idea of it flows in upon the mind in a forcible manner; whereas in the imagination the perception is faint and languid." Hume and Arakawa both believe in the primacy of the idea over the ephermerality of perceptual memory.

Arakawa deliberately confuses the art-viewing public with his painterly signs and paradoxical language. He does this from the position of one who is genuinely puzzled and amused by the elusive nature of language, meaning, and thought. He is a master manipulator, revealing in canvas after canvas that our perception of the world is largely dependent on thought and that thought is significantly modified by language. Arakawa's art is in a constant state of flux. His paintings are kinetic mechanisms of a sort, slowly turning and enmeshing us in a tangled web of visually beautiful forms and philosophical speculations that skillfully investigate language and the meaning of meaning.

Shusaku Arakawa, *A Forgettance (Exhaustion Exhumed)* (photo by Eeva-Inkeri; courtesy of Ronald Feldman Fine Arts, New York).

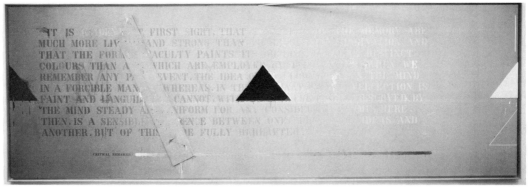

Romare Bearden

Romare Bearden is an artist who regularly incorporates into his art magazine photographs, bits of quilted cloth, and brightly painted flat shapes of color. Out of these and other materials he creates a nostalgic, glowing world of folk imagery and archetypal myth. Bearden's work is quite different from Arakawa's language-laden, conceptually oriented paintings; thematically it springs from personal biography and makes use of a fragmented, Cubistlike compositional structure.

Bearden is a black artist—perhaps the dean of black artists in America—and most of his work draws its inspiration and meaning from recollections of the rural south where he was born well over fifty years ago. Symbolism plays a major role in the thematic content of this artist's work; train images (illustration), for instance, constantly recur in many of his paintings—"journeying things" he calls them. "Trains are so much a part of Negro life," he observed. In the deep south blacks often lived alongside the railroad tracks (rents were cheaper there), found employment with the railroad, and during the great movement north rode them to what they hoped would be freedom and better lives. Train images are symbolic and metaphorical "journeying things" to Bearden, representing the physical and psychological means to leave the past—ridden with oppression and slavery—and "move on" to realize the potential for achievement in their lives. For over forty years painting has been Bearden's own "journeying thing"; but the road has been long and difficult. Only recently, at the age of fifty, has he been able to make a living through his art.

Faces are another important visual element in Bearden's work. From his earliest collages to recent exhibits they show up with consistent regularity. Black faces stare directly at us: old, young, beautiful, plain, street-wise city faces, and the innocent features of rural children coexist in the same paintings. Painful, anguished, poverty-stricken faces appear alongside cosmetically pretty faces from the pages of black fashion magazines. The individuals Bearden depicts vary in terms of physical beauty and economic situation but they have one powerful experience in common: they are all black people living in a white nation. The true source of Bearden's art springs from this common denominator of black experience but his paintings carefully avoid self-righteous dogma and lack the rancor of bitter, hard-core protest art. Expressions of torment and sadness can be seen in many of his works; but joy, beauty, and personal dignity reside there also. Naturally, Bearden has strong feelings about human rights, but he believes a greater contribution towards racial understanding can be made by revealing the pride and courage of black people's lives through the gentle and lyrical force of his art rather than blatant "social comment." In 1964 he wrote that he hoped to "establish a world through art in which the validity of my Negro experience could live and make its own logic." Thus Romare Bearden's paintings avoid narrow, stereotyped, media-induced images of American blacks—violent protest, anger, gross injustice—and safely allow us to drop our preconceptions and enter a world of commonly shared experience. A world of homes and families, a world that expresses the love of beauty we all feel, and a world that recognizes everyone's need for acceptance, compassion, and understanding. Thematically, Bearden's art stems from his personal experience of being a black man in America, but its directness and simplicity give it a universal appeal.

Of the Blues is the title of one of his most dynamic series of paintings to date. First shown at the Cordier & Ekstrom Gallery in 1975, these paintings document in freely painted, multilayered, and semihistorical style the jazz greats of the last fifty years. Austere compositions featuring Storyville bordellos with anonymous black musicians are juxtaposed with opulent images of "uptown" spots like the Savoy. As a young man growing up in Harlem during the twenties, Bearden got to meet and hear most of the great performers of the day: Fats Waller, Billie Holliday, the "Duke," Louis Armstrong, and Jelly Roll Morton. Even the

Romare Bearden, *Watching the Good Trains Go By* (Collection of Philip J. Schiller).

sound of their names rings with a musical note and syncopated rhythm. For sixteen years Bearden's studio, directly above the legendary Apollo Theater in Harlem, literally echoed with the music of these and many other great black musicians. Perhaps due to his strong personal attachment and love of the subject matter—jazz—this series represents some of his most effective and evocative work (illustration). Like the musicians he portrays, Bearden's paintings make music also; visual rhythms echo musical moods, from the cool, languid, silken sounds of a female blues singer to the brassy, swinging air of a Dixieland jazz band. "Art celebrates a victory," Bearden is fond of saying. This series celebrates the pride Bearden and black people all over the world feel about the contribution jazz has made to our culture.

Bearden himself is a gentle, somewhat shy, heavyset man who—to his great amusement—was often mistaken for Nikita Khrushchev in the early sixties when he strolled the streets of New York. Because of his light-skinned complexion many people, when they first meet him, assume he is white. But Bearden's ancestors first came to America some time in the early 1800s—by slave boat. Romare (he pronounces it "Rome-ery") Bearden was born in Charlotte, North Carolina, where his father's parents lived, but he grew up mostly in Harlem. Childhood summers were spent in Pittsburgh where he often stayed with his mother's mother. She was the owner of a boarding house near the steel mills, and he vividly recalls his grandmother's rubbing the new boarders (rural southern blacks who migrated north to work in the mills) with cocoa butter at night to soothe their irritated skins, scorched from the licking flames of the open-hearth furnaces. Bearden was fascinated with the mills and the

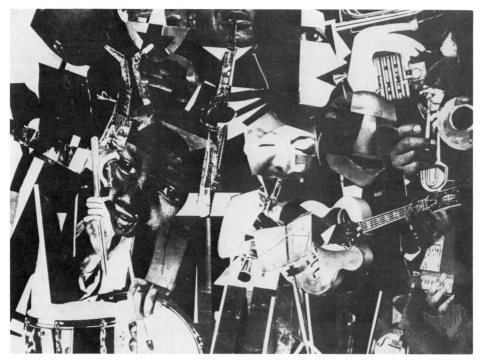

Romare Bearden, *Jazz Savoy—1930s.*

stories the workers told in the evenings about their childhoods in the south. The lure of this world was strong and by the time he was fifteen, Bearden decided to leave Manhattan and live with his grandmother in Pittsburgh.

Bessye, Bearden's mother, was an energetic, socially active woman of great political and social influence in Harlem. She managed several congressional campaigns and was one of the original organizers of the National Council of Negro Women. His father, Howard Bearden, worked for the New York City Department of Health as a sanitation inspector. He was a quiet, intelligent, sensitive man who remained very much in the family background. Bearden's mother was at the center of all family activity; judges, actors, writers, and musicians were always coming to their house to pay a visit. Even Duke Ellington was a family friend and later bought an oil painting from Bearden's first one-man show.

His parents had ambitious plans for their only son which included college and med-

ical school. Bearden had a gift for science and mathematics which he studied, first at Boston University and later at New York University. But plans for a medical career faded when he met E. Simms Campbell, a successful black cartoonist, at New York University. He then started doing political drawings for a black journal, *Afro-American.* Despite his bachelor's degree in science, cartooning was becoming more and more of a passion with him. Soon after he graduated from New York University, he enrolled at the Art Students League to study with George Grosz, a German artist whose art revolved around social and political satire. Grosz came to America when he sensed the impending upheaval in Europe and luckily got out before World War II started. Although his drawings and watercolors had a popular cartoon look, his knowledge of classical art and composition was significant. Grosz introduced Bearden to the work of Dutch artists like Vermeer and Rembrandt as well as European masters such as Holbein, Dürer, and Bruegel; after one semester of study with Grosz, Bearden's ideas about composition, drawing, and political cartooning drastically changed. His approach to art now was directed to the expressive possibilities of the fine arts rather than political drawings. After only one and a half years of study at the Art Students League, Bearden left and took a job working for the city as a caseworker in the Welfare Department.

Harlem in the late thirties was a cohesive and dynamic community; many of the black artists knew each other and met regularly at the house of a well-known artist, Charles Alston, to share ideas and dreams with a variety of performers, musicians, and intellectuals. The place was called "306"— after its address, 306 West 141st Street— and it remained open day and night to creative individuals of all races. Here Bearden met Langston Hughes, William Saroyan, Carl Van Vechten, John Hammond (the jazz scholar), and Garcia Lorca, the international poet. In those days, Bearden recalled, it was very easy to meet all kinds of interesting people.

Despite the lively social activity of Harlem during this era, it was hard to forget that, after all, a depression was going on. Because of this Bearden held on to his job with the Welfare Department and painted for a few hours in the evening. These were difficult times for him, not so much economically but emotionally and artistically. After his study at the Art Students League he was confused and unsure as to what specific direction he would pursue in painting. He no longer believed in political cartooning but lacked the confidence to venture beyond his early drawings and make a cohesive personal statement. Bearden was painfully frustrated and lost.

After leaving his studio with a friend one evening, a small, homely prostitute stopped them and offered her services. They were not interested; after some quick bargaining, her price dropped from $2 to nothing. "Just take me," she said, "I'm desperate." Bearden offered to help and talked to his mother about this woman—Ida—and together they soon found her a job. Part of her employment included coming to Bear-

den's studio every Saturday to clean. There was a big wooden easel in the studio and on it a piece of brown paper. Week after week Bearden stared at that piece of paper; he was incapable of making a mark. One Saturday Ida asked if the empty piece of paper was the same one she saw every week. Bearden shyly admitted he did not have any ideas so the paper just sat there. "Why don't you paint me?" she asked. The artist's mouth must have dropped because she quickly followed with, "I know what I look like, but when you can look and find what's beautiful in me, then you're going to be able to do something on that paper of yours." Bearden was struck by what she said and slowly began a series of works based on Ida. The aesthetic ice was broken; Bearden began to "unfreeze," and discovered the unique roots of *his* art in personal experience and the lives of people—black people.

Thus Bearden's mature work begins with *Ida* and a series of tempera paintings he did on brown paper that depict scenes from his southern childhood: a black sharecropper tilling his fields, a railway worker playing the mandolin after the sun has set. He realized at this time that his work would be to chronicle—much like Bruegel's detailed documents of Flemish life—the story of his people using the same formal elements that appear in classical works of art. Bruegel's detailed documents of Flemish life were historic models for Bearden. But the myth and ritual of the American Negro provided the on-going basis for his inspiration. Ida's portrait was the catalyst that enabled Bearden to transform his own experiences of black America into genre scenes of everyday life. Ida proved to be right; when he did see the beauty in her life, the struggle and anguish and triumph of his people, then he began to make statements that meant something to him.

By the forties Bearden's career was solidly launched and despite three years of active war service with the 372nd Infantry division, a Negro regiment, he had half a dozen important one-man shows by the end of the decade. Most of the work was done in somber earth colors and featured biblical and literary themes transposed into black terms. At this time he painted a series of works based on a poem of Garcia Lorca's, *Lament for Ignacio Sanchez Mejias*, and he also did a group called *The Passion of Christ*. Despite a sell-out show at the prestigious Kootz Gallery 9 (prices in the forties were not what they are today), Bearden had to go back to a job with the Department of Welfare. He listened to talk about Paris from various artist friends, and in late 1949 managed to get a leave of absence from his city job and moved to Paris for six months to live, paint, and soak up European culture. This old-world capital of culture was enormously appealing to him, and he admits to being seduced by its beauty and vitality. It was not a good time for artmaking, however; not a single painting was completed during his stay. Under the G. I. Bill of Rights Bearden studied philosophy at the Sorbonne and spent his free time visiting galleries and meeting friends at the cafes along Saint-Germain-des Prés. Late-night parties with French artists, writers, and intellectuals were also part of the scene in those days. Bearden met a few of the legendary greats from the Paris School, including Braque, Brancusi, and even Matisse, who could sometimes be seen strolling by the cafes causing flurries of excitement when he was recognized by the waiters. Leaving Paris was difficult; as soon as he got home, Bearden immediately applied for a Fulbright Grant that would hopefully send him back, but his application was turned down.

Discouraged by the failure of his scheme to return to Paris, Bearden diminished his painting efforts. Soon he stopped altogether. Music had always played a great role in his life, and musicians were among his closest friends. Perhaps the way to get back to Europe, he reasoned, was to write popular songs and have them recorded. Although Bearden had little formal training in music, he wholeheartedly set his mind to this task and over the years wrote about twenty money-making songs. *Seabreeze* was his biggest hit; some of the top names in jazz recorded it, and the Seagram Corporation employed it to sell a popular mixed drink.

Although he achieved a measure of success in the music business, these were not happy years for Bearden; the thought of painting again haunted him. Eventually, the sadness and frustration of leaving his art for the world of commercial music got to him. He recalled:

One night, I thought I was going to die. I called my father and said I thought I must have cancer of the stomach. But when I went to the doctor he examined me and said nothing was wrong. A couple of weeks later, I felt sure I was going to have a heart attack. Then one day, walking on the street, I suddenly felt I couldn't walk a step farther. The next thing I knew, I was in the hospital. I asked the nurse, "Where am I?" She said I was in the psychiatric ward at Bellevue. A doctor came by, and I asked him what had happened to me. "You blew a fuse," he said.[9]

After this frightening incident Bearden realized one thing very clearly: He just had to get back to painting. In 1954, at a Harlem benefit party for West Indies hurricane survivors, he met Nanette Rohan, whose parents came from the island of Saint Martin. Soon afterward they were married. Bearden credits Nanette with stabilizing his life and helping him build his confidence and return to painting.

In 1963, during the height of black activism in America, Bearden and several artist friends met in his Canal Street loft to discuss the political situation and particularly the plight of the black artist in America. They called themselves "The Spiral Group"—the name was supposed to suggest an upward movement, up and out—and soon rented a small exhibition space on Christopher Street in Greenwich Village. Soon regular shows by black artists who had difficulty finding a gallery in New York were organized and scheduled. At one of their weekly meetings Bearden suggested they all work on a composite collage along Negro themes. Bearden cut out a lot of photos from his wife's magazines—*Ebony, Look, Life, Harper's Bazaar*—and brought them to a Spiral meeting; no one seemed interested so he started to experiment on his own. He cut, pasted, and juxtaposed the magazine images and ads, sometimes painting over and around them with tempera and watercolors. What emerged were cubistic, fractured compositions from his real and imagined life. The pictures he chose all had something to do with his poetic vision of the black world: barnyard animals, country churches, valleys, locomotives, exotic jungle birds, black mothers holding babies, farmhands, guitars, African sculpture, and in almost every painting black faces of all kinds, staring directly at us. Before the Spiral period, Bearden had been working on brushy, abstract canvases, but when his dealer, Arne Ekstrom, came to the studio to pick out work for the next show, he was immediately attracted to this new work and decided to show the photocollages. After the opening many friends and artists felt that this new work was entirely his own, both aesthetically and emotionally. Bearden had "come home" with this series. Now

Romare Bearden, *Baptism* (courtesy of Cordier and Ekstrom Gallery).

and also discovered the visual means with which to express it.

Baptism (illustration), done in 1964, is a prime example of this breakthrough. Blue-green hills with a large yellow sun appear in the background, calling to mind archetypal children's drawings. A small white church with a prominent cross is in the distance to the left. Two figures in the foreground are half-immersed in a pale, cobalt-blue river which takes up a significant proportion of the composition. The fields of the valley are constructed from fragments of yellow-brown magazine reproductions. There is a simple, plain beauty to this rural baptismal scene. The colors, partly composed by paint, partly magazine reproduction, complement each other. Angular black hands of the participants punctuate the layout with staccato rhythms. The embracing and calming hands of the minister play against the upraised, excited hands of the immersed individual. A feeling of childlike grace and beauty pervades this picture, recalling another time and another place.

Memories (illustration), executed in 1970, reveals an entirely successful fusion of intense, saturated, painted color and detailed magazine photos that has become Bearden's trademark. From a distance it almost has the abstract, self-referential, geometric look of a Stuart Davis painting (Davis was an early influence of Bearden's) but upon closer inspection a circular yellow shape becomes a sun that lights a cubistic scene of touching sensibility. Within this tableau a beautiful, black female nude can be seen crowned with an African sculpture head. She stands quietly in an ultramarine pool of blue with a school of silver fish at her feet. Off in the background is a sharecropper's shack juxtaposed with photos of the Sphinx. To the far right, the outline of an intensely black crow visually activates an open area of pale grey-brown. Trees, pebbles, sky, water, and figure combine in a harmonious way to evoke a nostalgic mood and create a distinct sense of place. Bearden has given birth to a personal and unique world in these collages through his hard-won ability to organize color, shape, and thematic content into dynamic scenes.

Romare Bearden, *Memories* (Collection of Shorewood Press, New York).

Throughout most of Bearden's mature work there is a deep concern for classical themes in Western art such as *Susannah at the Bath, The Annunciation, Madonna and Child*, and the *Return of the Prodigal Son*. But all of these ancient myths are retold in terms of contemporary American Negro life. *The Prevalence of Ritual*, as a book about him is subtitled, is an aspect that Bearden incorporates into all of his paintings. The artist retells these archetypal myths — traditionally the beliefs of a white European society — but uses black people in the stories to signify their important relationship to the whole of humankind. Nowhere is this interweaving device used with more significance than in Bearden's recent series of paintings based on Homer's *Odyssey*. Constructed out of brightly colored sections of paper, touches of watercolor and gouache, they reveal black figures acting out, with a theatrical "prevalence of ritual," scenes from one of the earliest works of literature known to the Western world. In *The Return of Odysseus* (homage to Pinturicchio and Benin) (illustration), in a particularly Modernist gesture, Bearden connects an ancient African civilization to European origins. By relating these seemingly unrelated sources he takes us back in time and

reminds us of the role Africa played in the birth of Western culture. The Minoans, early inhabitants of Crete, were precursors to the classical Greeks, and it is widely acknowledged that Cretan civilization owes a great deal to the culture of North Africa, only 250 nautical miles across the Libyan Sea. In this way the spark of advanced civilization jumped across the Mediterranean, via Crete, to the cradle of European art, Greece. Consequently, the cultural roots of the West are not ethnically pure. Bearden reminds us, in this series, of the pluralistic sources of Western art in a civilization that usually perceives itself as isolated and untouched by other races and cultures. By setting black figures in a "white" drama, he reinforces the broad universality of "journey" myths as opposed to a presumed exclusivity.

In the *Odyssey* series Bearden sets up a visual system of brightly colored, tightly interlocking, two-dimensional shapes; within this gemoetric structure appear whitewashed buildings, palm-lined shores, Egyptian hieroglyphs, African artworks, and references to Japanese woodcuts. Bearden's Homeric world of ancient buildings, Mediterranean Sea, plotting gods, and emblematic black heroes transcends the original setting and historic period to touch us

Romare Bearden, *The Return of Odysseus* (homage to Pinturicchio and Benin) (courtesy of Cordier and Ekstrom Gallery).

in our own contemporary world. Homer's *Odyssey* is transformed by Bearden into a more specific and relevant translation than the one we may have studied in school. Classical myths encode information about life and living in such a way as to transcend the historical period in which they were written and to help us interpret the puzzle of our own personal, daily existences. Perhaps this ability to be reinterpreted in a variety of significant ways accounts for the durability of really great works of art. Certainly Bearden's work strives for and achieves an archetypal quality. His painting incorporates and celebrates the humble rituals and timeless elements of life through specific contemporary examples—a black family dwelling in a Harlem flat, a tenant farmer in overalls holding a rooster—and pictures them in heroic and mythic terms. Above all his work expresses a personal and essentially optimistic vision of life. Although he documents the particular joys, struggles, and triumphs of black people in his paintings, he does it in a way that allows

everyone to participate and perhaps discover the similarities in their own lives.

Bearden's success, aesthetic and financial, means a great deal to many black artists currently engaged in the difficult process of "finding themselves" as artists while they grapple with the problem of earning a living. Evenings, after his own studio work is through, Bearden devotes a lot of his spare time to helping young artists by offering advice and encouragement. The story of his long and eventful life, with its difficulties, struggles, and rewards is a source of inspiration to many friends and fellow artists. Bearden's life and art are both filled with optimism. Something at once simple and profound comes through in his paintings; they are songs, visual folk and jazz tunes that portray in pulsating rhythms, vivid colors, and moving imagery the poetic beauty of black life in America. Age-old symbols depict the transformation of despair into hope. Trains, open doorways, and birds in flight become the "journeying" symbols of escape—to freedom. "That's what art is about," Bearden keeps reaffirming. "It celebrates a victory."

Cy Twombly

Many painters active today grew up under the reign of Abstract Expressionism. Quite a few artists have since challenged the basic beliefs of this movement. Cy Twombly, however, is an artist who was deeply involved with Action painting during the early fifties and who has continued to evolve and revitalize this form of artmaking up to the present. His work has steadily developed personal themes and unique visual effects, but at the same time has remained closely in touch with some of the original sources of Abstract painting.

One important aspect of Abstract Expressionism that was often talked about by artists and critics of this school was the painter's "handwriting." This term refers to the particular calligraphic gestures and marks the Abstract artist uses to develop the painting's overall look or "image." In works that possessed neither realistic images nor geometric forms, the method of paint application—splashing, dribbling, staining, scrumbling, smearing—in effect became the artists' signatures and trademarks. This element of painting directly relates to one of the original sources of Abstract Expressionism—Surrealism. The Surrealists believed that by engaging in a form of gestural "automatic writing" they could unleash the creative power of the subconscious mind and attain a kind of visual magic, power, and truth in the process. Aspiring to a state of unconscious freedom and heightened perception, they sought through this "handwriting," to seize the creative moment and capture it on canvas. Far from conceiving of their work as "abstract"—that is, detached, remote, not concrete—they believed just the opposite: These automatic responses were very real, springing from untapped sources of authentic feeling and thought.

A strong relationship exists between the methodology of Action painting and another artform native to America—modern jazz. Both sought free-flowing, extemporaneous expressions of a deeply felt nature. Both wanted to break away from their respective restrictions: for painting, traditional composition and illusionistic images; for jazz, conventional melodic structure and prewritten scores. Of course these vanguard theories of "freedom" and "expression" in the hands of untalented, untrained artists and musicians were destined to produce chaotic, meaningless works of art. No amount of aesthetic philosophy can redeem incompetent work. Even though contemporary artists of this genre professed it was all about renouncing technique and "openly expressing yourself," most accomplished individuals had considerable training and experience in painting which they applied to their work, consciously or unconsciously.

Twombly's paintings and drawings are shrewdly deceptive; they appear at first to be haphazard collections of naively scrawled, randomly composed lines and marks without apparent grace or beauty. But behind outward appearances of disharmony and casualness, his works have a subtle and sophisticated sense of order that is both meditatively thoughtful and sensuously pleasurable. Twombly engages in a form of high-level, playful doodling; in his purely calligraphic work there is no image other than the image created by the act of painting, no symbolic messages, no sense of "logical" system, just a heightened and delightful sense of pleasurable activity and rediscovery of the primal "mark." And it is in the recognition and reiteration of the original mark—the line that verifies our existence—that Twombly as an artist achieves one of his goals: the reconnection of painting to writing which seeks to join sign to system.

Twombly's work takes us back to the origins of writing found in pictographic signs, symbols, and calligraphic marks. Even classical etymology links artmaking to writing; in ancient Greece the term *egraphen*, which means "written by," was always attached to the painter's signature. Cy Twombly's artworks are essentially based on gestural, autobiographic linear elements and express a highly developed graphic sensibility. His work forcefully embodies the Surrealist notion that drawing is the original source of all writing and visual art.

Considering Surrealism's obsession with archaic myth and Twombly's fixation with Surrealism, it comes as no surprise to learn of his passionate and abiding interest in classical literature. His fascination with Greco-Roman myth and history extends back to his secondary school days in Virginia. Classical antiquity with its rich interplay of fanciful legends and pragmatic philosophy operates as a creative springboard and catalyst in this artist's mature work. Through the use of literary quotations, Twombly is able to invoke the ancient past and consequently "return," through his art, to the ancient worlds of Rome and Greece.

Partly because of his love for antiquity, Twombly has made his permanent home, since 1957, in the historically rich city of Rome. From the eloquent ruins of the world-famed Coliseum to well-preserved medieval quarters along the Tiber, Rome often brings to mind thoughts of Apollo, Ovid, Leonardo, Dante, and Keats. This physically immense, culturally deep storehouse of 2000 years of Western thought has become Twombly's spiritual and artistic home.

After setting up a studio and adjusting to a European way of life and the hoard of riches there — Twombly has repeatedly referred to Rome as a "Luxurious City" — he produced an important painting, titled *Empire of Flora*, in 1961, that combined oil paint, crayon, and pencil on canvas. It is an exuberant, colorful composition of intense yellows, reds, flesh pink, and black paint that takes its theme from classical literature. According to Ovid, a relationship exists between destructive and creative forces which is expressed by the myth of heroic soldiers dying in the glory of battle and transforming into flowers. The mystery of death and beauty, dying and rebirth, is suggested by the frantic marks and lyrical color of this painting. A feeling of struggle is expressed by the primal way paint is applied to the canvas: Oil color is squeezed directly out of the tube, wet layers of paint are scratched through, and pencil lines and crayons streak the surface. An immense and powerful battle is taking place but the final outcome, according to Ovid, is not ignominious decay but rebirth and beauty.

Bay of Naples, 1961 (illustration) is very similar to *Empire of Flora* in terms of visual structure and use of materials. Both paintings represent an important change from earlier work which was primarily linear and sparse. In *Bay of Naples* the layering of elements is remarkably intense: thin and thick areas of paint, heavy crayon marks and faint pencil lines, and enigmatic diagrams. The large white field of the canvas highlights the strong red, yellow, and orange paint tones. An unlikely and creative synthesis is created out of Abstract Expressionism and concepts contained in classical literature — the gestural "New York School" of painting has been fused with the philosophical "School of Athens." Working in Italy also heightened Twombly's appreciation of baroque elements in art; in response his paintings grew larger and structurally complex. What energized and tied all of these aspects together was the artist's unique and highly effective calligraphic "handwriting."

The next distinctive phase in Twombly's work occurred in 1967 when he produced a widely exhibited series of paintings with a much narrower chromatic range — mainly greys and subtle whites — and a simplified calligraphic script reminiscent of the American artists Mark Tobey and Morris Graves. Perhaps Twombly was responding to the developing Minimalist movement in New York or had been influenced by Jasper Johns' overall hatching patterns. For whatever reasons, "cool," gestural restraint clearly distinguishes these works from the early emotive pieces. *Untitled*, 1967 (illustration), is an excellent example of the pared-down, diagrammatic approach Twombly used during this period. The cursive, white crayon marks on the grey housepaint-covered canvas bring to mind "penmanship" exercises on grade-school blackboards. All of the calligraphic elements that appear in the early works remain, transformed into a more formalized

Cy Twombly, *Bay of Naples, 1961* (photo by Geoffrey Clements, New York; copyright © Dia Art Foundation, 1979).

Cy Twombly, *Untitled*, 1967 (collection of Jasper Johns).

and restrained mode. The blackboard look is emphatic; perhaps Twombly felt the need to return to the simplicity and directness of a "school lesson" to reconfirm the original meaning of the handwritten mark.

This series is distinctive for the expression of movement, not just the recorded movement of the hand that made the marks, but the broader notion of movement as it relates to the natural world: the constant roll and pitch of the oceans, the movement, in our minds, of the earth in its orbit around the sun, and the normal motion of air as well as the violent action of storms. *Untitled, Captiva Island, Florida*, 1968 (illustration), is a pencil and collage drawing on rag paper that features a series of lassolike lines that roll across the surface left to right. At the top of this vertical composition is a facsimile of Leonardo's study for the "deluge," a cataclysmic drawing of remarkable power that could either be viewed as the creation

Cy Twombly, *Untitled, Captiva Island, Florida* (photo by Eric Pollitzer; copyright Cy Twombly/ VAGA, New York, 1989).

reviewed a show of Twombly's work and tersely wrote, "There isn't anything to the paintings." However, considering Judd's commitment to Minimal art (where less is more), this may have been a hidden compliment rather than criticism.

Drawing has always been an important medium for Twombly; many of his works on paper parallel and sometimes precede developments in painting. In recent years most of these pieces—which can be viewed as painterly drawings or draftsmanlike paintings—have taken on a significance that seems to rival and even surpass older paintings on canvas. This particular phenomenon is not just confined to Twombly, though: Many artists are discovering that working on paper opens up creative possibilities that traditional approaches with paint and canvas do not allow. To begin with, some of the historic pressure to "perform" is relieved; this might account for the spontaneity, freshness, and informality that so many contemporary drawings exhibit. Second, the reduced size invites an intimate relationship to the work which allows us to perceive it in a personal way. Finally, the open, simple nature of drawings reveals more clearly the way in which they are made. This allows us the opportunity to enter into a dialogue with the creative act they express.

Twombly readily admits to having a special "feeling" for paper; what he really likes is the open space of the creamy white surface—which, unlike in most paintings, is usually not completely covered. This pregnant "void" is particularly suited to focusing our attention on the expressive, cryptic markings that delineate the white space. These enormous open areas do more than contain and hold the marks; they inevitably take on a tangible presence of their own. In fact, these empty spaces are transformed by Twombly's carefully placed lines, marks, and graphic gestures into the most substantial matter of the whole picture. Paradoxically, the open planes of white space are the real prima materia of Twombly's work.

Curiously, many of Twombly's personal concerns, such as coded marks and artistic

or death of our planet and solar system. Twombly finds contemporary relevance in Leonardo's awesome study that graphically describes a catastrophic event of cosmic proportion and employs it in the context of his own visual work. The swirling lines of the reproduction play off in subtle counterpoint to the languid, cooler loops of the handdrawn marks. One amusing bibliographic footnote to this drawing hinges on the fact that Don Judd, a New York critic and artist, owns the piece. In 1964 Judd methodology, can be traced to the early fifties when he was drafted into the army and trained in the study of cryptology. During his tour of duty on the East Coast, Twombly had plenty of time to work with systems, think about concealment, and visit the art galleries in New York City. Confined to the barracks after "lights out," he would often, out of sheer frustration, make drawings in the dark so that his main connection to the work was tactile, not visual. Twombly feels that these early drawings helped him develop a certain physical relationship to his work that was important in later years. When he was finally discharged, he toured Europe and gravitated to the open, sun-drenched spaces of Italy and North Africa. Often he rented empty white rooms that looked out over the Mediterranean Sea in which to paint and draw. Although he soon returned to the tumultuous world of Manhattan and the developing saga of Abstract Expressionism, the airy, calm scenes of southern Europe never left his memory. Years later he was to synthesize these two powerful experiences — New York Action painting and "pure" space — into a major body of work.

In the spring of 1979, New York's Whitney Museum staged a major retrospective of Twombly's work. Along with the early paintings, a recent series of pieces on paper were shown that established him as a major force in the continuing evolution of "painterly" Abstract art. Many of these new works from the midseventies were multipaneled collage paintings on paper that seemed to tie together his visual and thematic concerns of the last twenty years: the early gestural brush strokes, the use of a simple repetitive mark, the "literary" and poetic titles visually incorporated within the work, and the use of central images and printed material.

Epithalamion III, 1976 (illustration), is a two-panel work of watercolor and pencil that utilizes all of these elements in a pictorial and verbal way. In a manner similar to the Leonardo "deluge" drawing, Twombly uses a postcard-sized reproduction of a Fredrick Church landscape painting; below this image a painterly reddish brown

passage appears that chromatically echoes the postcard print above. Between the two sections Twombly has printed the word *EPITHALAMION*, a generic term for ancient Greek nuptial poems. Epithalamion refers here to the marriage of nature and art, Church's painting and Twombly's art. To the right is a smaller panel — part of the same work — which reads like an abstract painting; there is no writing or collage on this section, but color relationships tie the panels together.

Idilli 1976 (illustration) uses some of the same elements and pictorial devices but operates with three panels, one large one on the left and two smaller ones, about half the size, on the right. No magazine reproduction is used but an intense green-black passage of freely brushed oil paint on paper is fixed to the top of the large, left panel. The usual, subtle, obliterated words and text appear throughout the first two panels. The right panel is empty except for the words, which read in part, *I AM THYRSIS*. Most of the language in this lyrical series manages to incorporate Twombly's classical references through an effective blend of visual pyrotechnics and verbal incantations.

The seventies was a popular time for returning to "basics" in art; artistic origins and sources were very much on the mind of many artists. Cy Twombly recently joined the ranks of artists like Romare Bearden to pay tribute to Homer, one of the earliest poets known to the West and a true classical "source." The piece *Fifty Days at Illiam* (illustration), a painterly installation exhibited in 1978, is an ambitious work composed of ten large white panels that loosely follows and is inspired by characters and events sung about in the *Iliad*. Twombly originally designed the huge painting — he definitely views it as one piece in ten sections — to be installed on both sides of a long corridor. Space limitations at Heiner Freidrich's New York gallery, where it was shown, required that nine panels be displayed against one wall, and the tenth panel at the far wall opposite the entrance. Consequently, this exhibit evolves along linear, narrative terms rather than environmentally surrounding

Cy Twombly, *Epithalamion III*
(photo by Eric Pollitzer; copyright
Cy Twombly/VAGA, New York,
1989).

Cy Twombly, *Idilli, 1976* (photos by
Geoffrey Clements, New York; © Dia
Art Foundation 1981).

CURRENTS

the viewer in the thick of battle as Twombly originally planned. Together, the ten panels describe the wrath of Achilles in a symbolic, painterly way, incorporating oil paint, crayon, and pencil on the vast, white surfaces that have become Twombly's trademark. When the show opened in New York, much to-do was made over the fact that this work was based on Alexander Pope's translation of Homer's epic tale. Twombly's fixation on the text (he feels the painting closely follows the text) reveals one of the problems in basing art on literary works: It is all too easy to make a painting a dull and meaningless rendering of a literary fact. Although the artist may have been intimately involved with the interpretation and meaning of this classic, we are primarily and legitimately concerned with the way the painting reads visually.

One of the central and recurrent images in *Fifty Days at Illiam* is Achilles' shield. It appears consistently throughout the ten panels that depict the ten-year Trojan–Greek war and is subtly transformed into many visual varients and forms. Twombly places the image of the shield at the center of the work's symbolic and narrative core. On various panels certain phrases from the Pope translation occur along with the names of various military protagonists. In one of the sections titled *The Heroes of the Achaeans*, Twombly lists the names in his distinctive, terse, calligraphic hand. Accompanying this high-brow literary graffiti are peculiar shapes and pictorial symbols that comprise his unique painterly "image."

Despite the historically rich, physically ambitious aspects of this installation piece, it seems to lack the visual and cognitive cohesiveness of Twombly's recent collage paintings on paper. There is something elegant, primal, and convincing about these physically modest works on paper that invoke the name of an ancient poet, god, or hero, and with a few deft and accomplished strokes of the brush conjure visual fantasies of sea and land gardens. By contrast, *Fifty Days at Illiam* was a Cecil B. De Mille extravaganza — heroic, grandiose, and perhaps slightly obtuse — that sacrificed, to a certain extent, the intellectual intimacy, human scale, and thematic clarity found in Twombly's smaller work.

One recent piece that exhibits these qualities is *Aristaeus Mourning the Loss of His Bees*, completed in 1973. It is a small work on paper that makes elegent use of simple materials such as paint, crayon, and pencil. *Aristaeus Mourning the Loss of His Bees* artistically invokes this relatively obscure Greek god by writing his name amid a cloudlike passage of lightly tinted blue-green paint and matching crayon lines. A fragile, suggestive feeling of wind, air, and sky sweeps through this 27-by-39-inch painting. Near the top of the painting the title *Aristaeus Mourning the Loss of His Bees* is written simply and touchingly across the surface. Aristaeus was worshiped in antiquity as the first to cultivate bees, as well as the developer of vineyards and olive groves; he was the protector of herdsmen and hunters and possessed the arts of healing

Cy Twombly, *Fifty Days at Illiam* (photo by Jon Abbott, New York; © Dia Art Foundation, 1979).

and prophecy. Without directly illustrating these mythological events, Twombly breathes poetic life into archaic myth and gets us to think about, once again, the demigods of the classical past who, through their exploits and adventures, explained the workings and mysteries of the temporal world to humanity.

In the final analysis, all of Twombly's work takes us on an unpredictably sensuous, expressionistic voyage from past to present, fact to fiction, and perception to concept. In all of his paintings and drawings the spirit and body of the original "mark" emerges to record unique impressions of space and time. Abstract Expressionism—a particularly modern idiom—is linked to the timeless thought and literature of Ovid, Virgil, and Homer. Twombly's accomplished hand and crayon conjure up these ancient bards; and in a pure arcadia of poetic fantasy, they come to life locked in the calligraphic writing, repetitious gestures, and passages of lyrically flowing paint.

Susan Rothenberg

In 1979 New York's Whitney Museum mounted an exhibit of "new-trend" paintings that caused a great stir in the art world. Titled "New Image Painting," this exhibit heralded the emergence of a group of artists who incorporated some form of representational imagery. What distinguished this group of *new* image makers from Realists of the past was their interest in combining and reconciling what heretofore had been antagonistic aesthetic directions: abstraction and representational imagery. Along with the work of Nicholas Africano, Jennifer Bartlett, Neil Jenny, and six other artists, "New Image Painting" featured the work of Susan Rothenberg.

From the mid-1970s Rothenberg had become increasingly well known for her large, flat, almost diagrammatic paintings of horses that skillfully balanced and combined elements of abstraction with greatly simplified, yet easily comprehensible, images. In her earliest paintings of this series the image of the horse emerges subtly from a ground of brushy earth colors and assumes the haunting presence of a ghostlike form — neither blatantly becoming an illustration of a horse nor lapsing into complete abstraction.

The shape of the horse image in *Flanders* (illustration), done in 1976, takes its form from the feathered edges of massed paint. Even though Rothenberg is representing an image (her earlier paintings were completely abstract and minimal) her concerns are still in large part compositional: the use of simple shapes, the relation of shape to edge, and the presence of linear and geometric elements. *Flanders* clearly makes use of many of the formal design elements Rothenberg had used in her earlier self-referential work — but with the important inclusion of an image. Fitting snuggly within the rectangular canvas, the image of the horse is symbolic of something primal yet half remembered. Mysteriously appearing on both sides of the canvas are vertical bars that visually enclose, and to a certain extent contain, the image of this historically important animal.

Rothenberg's visual and thematic treat-ment of this potentially romantic image leaves us questioning its meaning. Obviously the image of the horse is intended for our contemplation, but it languishes curiously in a neutral thematic and pictorial space. Unlike many Realist painters who draw us into their images, Rothenberg distances herself, and us, from the image by emphasizing abstract elements. We are forced to notice formal qualities of the painting — colors, paint texture, composition — as well as the image itself. The combining of abstract and imagistic features characterizes most of the New Image painters.

In an interview Rothenberg explained some of her feelings about using the horse as a primary image:

The way the horse image appeared in my paintings was not an intellectual procedure. Most of my work is not run through a rational part of my brain. It comes from a place in me that I don't choose to examine. I just let it come. I don't have any special affection for horses. A terrific cypress will do it for me too. But I knew that the horse is a powerful, recognizable thing, and that it would take care of my need for an image. For years I didn't give much thought

Susan Rothenberg, *Flanders*, 1976 (photo by Zindman/Fremont; courtesy of Sperone Westwater Gallery).

to why I was using a horse. I just thought about wholes and parts, figures and space.[10]

At about the same time that Rothenberg was painting her large emblematic horse images (1978), she also began a series of small 9-by-9-inch studies that incorporated—in the same simplified style—heads and hands. Because of their reduced scale these images evoked feelings of intimacy and vulnerability. Soon Rothenberg enlarged these images to 10 feet in height and found that in the process "... they became very confrontational." *Rose*, 1980 (illustration), reveals childlike images of an overlapping hand and face executed in thickly outlined black-blue, and red paint set against a heavily textured white ground. Like the horse, the hand is a primal image that goes back to cave drawings of prehistoric times. Along with simple, linear profile drawings of hunting animals—often portrayed overlapped—the outlined shape of a hand is a constantly recurring theme of early cave art.

By visually connecting these two parts of the body Rothenberg simultaneously presents us with something utterly direct, yet exceedingly complex in its psychological overtones. Confronting this painting in person is quite a different experience than seeing it reduced in an illustration to a 3-by-5-inch surface of grey tones. Towering far above a person's height, it achieves—through this dramatic scale and the physicality of paint—a feeling of power few contemporary paintings can equal.

Rose—by comparison with the subtle harmonies of *Flanders*—exhibits more aggressive visual and psychological elements. The lines in this painting boldly thicken to become tangible forms in themselves and can be read as independent shapes. Negative and positive forms are created by the interplay of brusque dark lines and the ponderous white enclosed forms.

Rothenberg draws on an elementary form of color coding in *Rose* to differentiate the two images. Black lines edged with a blue-purple color make up the hands, while a brilliant red hue forms the image of the head. The juxtaposition of these primary colors dramatically enhances the dualistic imagery. Along with her horse series, this painting of body parts appears to raise more questions about meaning and interpretation than it answers.

In 1981 a temporary change of scene from Manhattan to nearby Long Island led Rothenberg in a new direction. While renting a house on a picturesque creek near the ocean, Rothenberg became fascinated with visual elements of the landscape just outside her door. She recalled, "There were boats parked out front. I started painting boats just as a vehicle to learn how to use oil paint after a dozen years of acrylic." *Reflections*, 1981 (illustration), represents a departure from her earlier austere canvases. Here Rothenberg expands her repertoire with the rich surfaces of oil paint,

Susan Rothenberg, *Rose*, 1980 (photo by Roy M. Elkind; courtesy of Sperone Westwater Gallery).

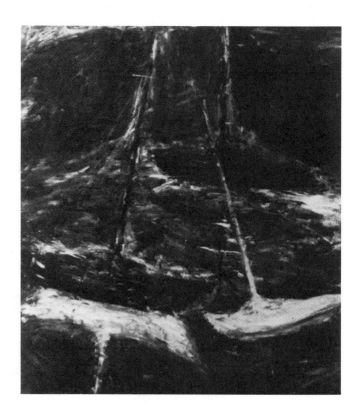

complex color modulation, and new compositional modes. "The boat became a symbol to me about the freedom I was feeling," she said. "Sailboats are beautiful—they're light and they depend on wind. They suggest qualities of light and atmospheric conditions. They started to lead me down a different kind of space now. Shadow and movement. Depth and resonance . . . At first I was horrified—what is this Neo-Impressionism?"

Rothenberg's uneasiness with aspects of traditional painting is typical of a generation of painters schooled in Modernist abstraction but anxious to discover their own way. Rather than dogmatically rejecting the past or blindly pursuing the new, artists like Rothenberg are interested in pursuing a path of personal artmaking that can bridge movements and eras and reconcile the seemingly irreconcilable.

Rothenberg's recent paintings continue to further explore mysterious imagery and use richly textured oil-painted surfaces. *Mist from the Chest*, 1983 (illustration), is a half-comic, half-sinister depiction of a man floating on his back in the water. His elongated neck emerges cartoon-style from beneath the water like a submarine's periscope; perched on top of this appendage is the man's head which quizzically stares at the escaping steam coming from his chest. Because of her early experience as a dancer, the concept of the body and its orientation to the world remain important ideas to Rothenberg. Like many of her simple, primal images, *Mist from the Chest* defies easy interpretation. This may be part of Rothenberg's strategy. A lack of thematic certainty may direct the viewer toward noticing visual elements of the painting—surface qualities, spatial relationships, and the interaction of color. Rothenberg seems more interested in creating open-ended dialogues about meaning that suggest a variety of possibilities rather than one interpretation. Her images appear to develop and emerge from the act of painting rather than being planned beforehand. As Rothenberg admits, intellectual rumination does not figure greatly in her work—feeling and intuition play a far greater role. Summing

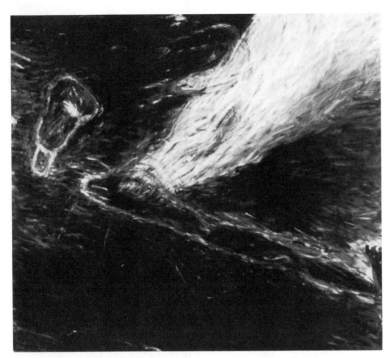

Susan Rothenberg, *Mist from the Chest,* 1983 (courtesy of Sperone West-water Gallery).

up what she is trying to communicate through her paintings, Rothenberg recently stated, "It's trying to get at something very basic—you could call it primitive desire."

Anselm Kiefer

Two decades ago American supremacy in the visual arts went for the most part unchallenged. Since World War II it had been taken for granted, not only here but in other parts of the world, that American artists were leaders in terms of aesthetic originality, relevance, and influence. Despite the fact that European artists of an older generation continued to make their presence felt on our shores, most had established themselves well before the outbreak of the war. Many artists born in Europe in the 1940s grew up in an environment of economic deprivation and cultural uncertainty. Generous support systems enjoyed by American artists were unknown to their

European counterparts. But all of this has changed. Since the miraculous economic recovery of once war-torn countries such as Germany and Italy, conditions abroad now favor a strong artistic renaissance.

In the early 1980s European artists—led by the Germans and Italians—began to make a significant impact on the American cultural front. Major galleries and museums scrambled to get on the bandwagon and hailed the work of these cultural emissaries as a fresh vision for the new decade. America's monopoly on artistic innovation and cultural relevance had lost its hold and a wave of new European art arrived on our shores with much fanfare and publicity.

One important characteristic of most of the new European art was its bold use of imagery and literary allusion. America's hunger for images after twenty-five years of self-referential Abstract art no doubt helped win acceptance for this kind of art. Not only were these new European paintings laden with images—they told compelling stories and were symbolically potent.

By comparison, American art of the recent past was more analytic in style and "cooler" in feeling. Much of this new painting also expressed an awareness of Europe's unique cultural history and the role art could play in the expression of national pride. With elements like this at work it was no wonder critics and public alike responded so favorably to this new work.

Anselm Kiefer, along with Georg Baselitz, Markus Lüpertz, and Jörg Immendorff, is considered to be one of the most distinguished representatives of Germany's new painting—much of it done in an Expressionistic style. His work, like that of many of his contemporaries, appears to be directed toward a search for spiritual values that could be relevant to his native country. Kiefer's images return time and again to the heroic past and folk myths of Germany's ancient traditions.

Kiefer, an intensely private person who refuses to be photographed and interviewed, lives quietly with his family in a large converted schoolhouse in the heavily wooded region between Frankfurt and Stuttgart. Unlike many of his contemporaries (artists born during or just after the war) who chose to live in industrial loft sections of large cities like Berlin and Düsseldorf, Kiefer prefers the quiet and contemplative life of the countryside. Not far from his home is a region of Germany referred to as the Nibelungenstrasse, or Road of the Nibelungs. The Nibelungs were an evil family in German medieval mythology who possessed a magic hoard of bewitched gold. For Kiefer, the German countryside that surrounds his home is bristling with legendary material, and more often than not it forms the thematic basis of his art.

Perhaps more effectively than many of his equally well-known German colleagues, Kiefer manages to synthesize Abstract painterly elements and narrative structures that relate to Germany's cultural traditions and recent historical past. Kiefer's broadly brushed passages of paint may evoke memories of American Abstract Expressionism, but thematically they probe, through the use of metaphoric landscapes, the spiritual soul of Germany's "Fatherland." *Margarethe*, 1981 (illustration), offers an example of how Kiefer manages to achieve a meaningful balance between painterly concerns and perceptions of the social and cultural history of Germany. This painting—one of a series—was inspired by Paul Celan's poem *Dein Goldenes Haar Margarete*, which describes the German destruction of Jewish culture. In the poem, Margarete, an aryan blond, is juxtaposed with a dark-haired Jewish counterpart named Shulamite. Kiefer makes visual use of the symbolic dichotomy between Margarete and Shulamite through the contrast of yellow and black. One of the most striking features of *Margarethe* is the way in which the artist has made use of actual straw to represent blond hair; this physical material is glued to the canvas and functions as elements of color, texture, and symbol. Set against a pale blue ground (perhaps signifying hope) the yellow straw, and the simulation of dark, scorched earth in *Margarethe* combine to evoke a complex emotional response to the political and religious history of twentieth-century Germany.

The Meistersinger (illustration), also done in 1981, makes similar use of texturally rich paint, straw, and written language inscribed directly onto the canvas. The light grey-green fingers of paint that descend from the top of this painting visually echo the thin reeds of straw that push up tenuously from the dense earth colors in the circle below. According to German mythology there exist thirteen meistersingers, or members of an ancient musical guild. In this painting Kiefer has sequentially numbered each of these figures within the canvas. Like *Margarethe*, *The Meistersinger* offers us the same provocative blend of abstract elements and literary symbols.

Sand of the March, 1982 (illustration), presents us with the specter of an immense and foreboding landscape. Dark furrows lead our eyes in a slow and majestic curve from the bottom right to the horizon line above. Written boldly in this stretch of open sky is the legend "Märkischer Land," the name of a region in Germany that was once the central core of the Prussian state. To-

Anselm Kiefer, *Margarethe*, 1981 (courtesy Marian Goodman Gallery).

day, in a divided country, this area lies between Berlin and East Germany.

Because of this painting's immense size —about 10 feet high and 18 feet across— it assumes a visual presence entirely in keeping with Kiefer's recurring themes of the importance of Germany's history and lands. Although many elements of the painting suggest a vague, symbolic landscape, Kiefer has written the names of towns and places in the Märkischer region on sheets of paper which he has collaged onto the canvas. As the names move vertically toward the horizon line they proportionally become smaller, reinforcing the pictorial space of this landscape painting.

To further emphasize the symbolic meaning of this painterly meditation about German history, Kiefer literally mixes the sand of his homeland with his pigments— upon close inspection the paint on the canvas is visually transmuted into earth and land. *Sand of the March* intricately and successfully blends visual elements of Abstract painting with the symbolic evocation of actual places in a region that once comprised Germany's political and cultural heartland.

Kiefer's simple but effective color scheme does much to reinforce the somber thematic mood of this painting. Dark yellows, a hint of Mars red, pale creams, blacks and greys create a pervasive mood of foreboding and dread. *Sand of the March* becomes an open-ended set upon which our thoughts and memories of recent history can be played out in a variety of ways. Rather than making fixed ideological statements with his paintings, Kiefer presents us with images that become visual catalysts for our own opinions; how we read them depends on our past experiences and points of view.

In a recent series of paintings, Kiefer focuses on another aspect of German history: the cultural role of government ar-

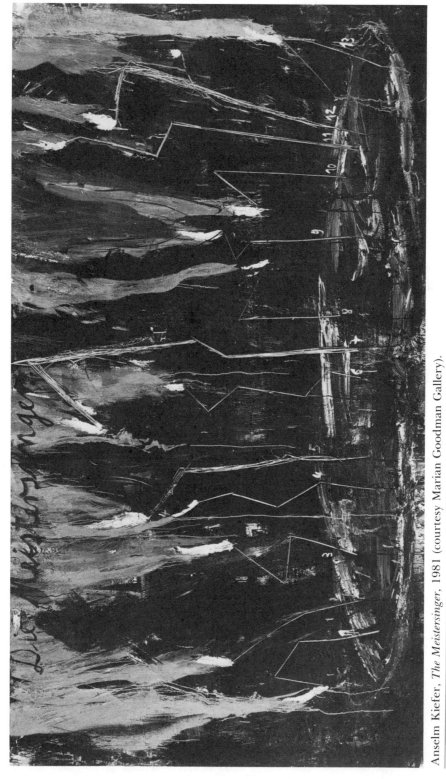

Anselm Kiefer, *The Meistersinger*, 1981 (courtesy Marian Goodman Gallery).

Anselm Kiefer, *Sand of the March*, 1982 (courtesy Marian Goodman Gallery).

chitecture during the war years in Germany. Over the past decade there has been growing interest on the part of art historians and critics in the architectural program of the Third Reich. Exemplified by the buildings of Albert Speer, Hitler's chief designer and builder, this architecture can be characterized by an oppressive, dwarfing scale and rigid formality. In *Interior, 1981* (illustration), Kiefer uses pictorial elements of this fascist architectural style to raise questions about the relationship between political power and artistic freedom, ethics, and cultural responsibility. Speer was not the only artist of his generation to be seduced by easy access to political power. Questions about the moral integrity of government policies and the artist's responsibility to challenge political decisions are issues which still haunt us today. One need not go back to Germany in 1940 to confront these political dilemmas. Witness the Vietnam era in America—a period of heated debate during which dissenting, socially conscious artists were pitted against government war policy.

Interior reveals the vast inner courtyard of an immense building constructed in a vaguely Neoclassic style. The scale and proportion of this building, unlike humanistic, classical architecture, are disorienting and ominous. Vast halls, without the comforting presence of one human figure, stretch enormous distances. Kiefer's painting presents us with the image of a building in which we might easily become lost. Although obviously referring to fascist building programs of the 1940s, this painting makes us question our present-day official architectural styles and what they express in terms of political power and social control. Inevitably all of Kiefer's images are imbued with an introspective, meditative quality that more often than not makes us think about and question the political realities that inspired them.

Sean Scully

Working in counterpoint to a body of art based on the appropriation of popular im-

---

Anselm Kiefer, *Interior, 1981* (courtesy Marian Goodman Gallery).

agery and Neo-Expressionist modes, a tight cadre of mature and emerging abstract painters have been making their presence felt in the art world. Influenced by a variety of historic sources—Cubism, Suprematism, Minimalism—these artists have been busily at work expanding the boundaries of an earlier, classic abstraction and making abstract painting relevant to the present time. Many of these artists, in fact, believe that the current focus on figurative, narrative, and Neo-Expressionist issues is healthy in the long-run for nonrepresentational art. After a long, unbroken ascendency into the realm of the near-sacred, Abstract art now finds itself involved in a fruitful process of introspection and transformation.

Sean Scully, one of the staunchest defenders of Abstract painting—a style closely associated with America—ironically was born in Ireland in 1945. Since the late 1970s this artist has become the leading practitioner of a rejuvenated, "born-again" brand of Abstract art. Scully believes passionately in the relevance of this kind of artmaking and feels that the specificity of contemporary narrative art, with its heavily laden symbolic program, ". . . does not allow the painting to express things far bigger." Attracted to the spiritual qualities that he believes paintings can express through color, light, scale, and the organization of form, Scully states "I'm interested in art that addresses itself to our highest aspirations, that's why I can't do figurative painting—I think figurative painting's ultimately trivial now. It's all humanism and no form."

But Scully is quick to point out that he admires certain Neo-Expressionist artists for their abundant energy and the way they use materials and scale. Abstract art had become too self-referential and hermetic Scully admits, and he is thankful that new representational painting has led to an interest in ". . . restoring subject matter and guts to painting, and relating it to life—which is what it needs."

At the age of four Scully's family fled from the poverty of Dublin and sought better economic opportunity in London. Starting with a job as printer's apprentice at the age of fifteen, Scully held a variety of temporary positions in London until he enrolled at the Croyden College of Art when he was twenty years old. Scully recalls that his prior work-experience lent a sense of urgency and focus to his studies at the art school. Other students, most of them right out of high school, did not have Scully's perspective on the realities of life. "They hadn't experienced getting up in the dark at six o'clock every morning, driving to work and working for twelve hours," Scully observed.

While at Croyden, Scully's painting evolved through many of the typical stages of student work: tightly painted figures, large-scale brushy landscapes and the beginnings of an abstract decorative use of striping that would eventually become his hallmark as a mature artist. Scully traces the origin of his infatuation with the stripe to a summer camping trip in Morocco. There he saw native tents made of brightly colored panels juxtaposed with one another, creating unusually striking optical arrangements. The memory of that North African trip with its riotous compositions of multihued stripes is still vivid in Scully's mind.

While studying art at Newcastle University, where he received his degree in 1971, Scully began painting a series of acrylic canvases that were composed of densely packed grids. This preoccupation with horizontal and vertical pictorial elements—which in a modified way continues to the present—was inspired by the work of Dutch artist Piet Mondrian and the contemporary American artist Sol Lewitt.

Winning an exchange fellowship to Harvard University in 1972 enabled Scully to study and travel in the United States. Here he experienced an openness and generosity toward his work that he felt was absent in England. In 1975 he was able to return to America on a Harkness Fellowship, which allowed him to live and work in New York City for two years. After his grant was over, Scully taught at Princeton University from 1977 until 1983, the same year he became an American citizen.

Hidden Drawing No. 3 (illustration) was completed in 1975 during Scully's first year in Manhattan. It is a visually dense painting constructed of layers of masking tape and acrylic paint. Within a square format, the space is divided into nine uniform geometric areas reminiscent of Ad Reinhardt's mysteriously dark cruciform image. The interlaced color bands which form the basis of the painting are created out of a subdued but evocative palette of greys and toned-down primary colors. Although *Hidden Drawing No. 3* owes a debt to Minimalist art of the 1970s, it also establishes the compositional basis for many of Scully's mature visual themes.

By 1981, after having gone through a series of predominantly black paintings, Scully started working in oil rather than acrylic paint. At this time the stripes became broader and took on vibrant hues of Indian red, sienna, and an intense blue-grey. *Enough*, 1981 (illustration), has juxtaposed panels of vertical stripes with horizontal bars. Dynamic spatial effects are achieved through these contrasting elements. Panels with thicker stripes appear visually closer; while patterns composed of thin stripes recede into the distance. The rich textural paint qualities of this painting also distinguish it from earlier work in acrylic such as *Hidden Drawing*. Also, rather than masking his stripe edges with tape (as he had previously done) Scully was now hand-painting the edges, which created subtle and visually interesting effects.

1981 was an important transitional year for this artist. It was about this time that Scully began to use shallow three-dimensional panels in his work, a device which lent unusual sculptural qualities to his painting. This important new development came about through chance. For quite some time, Scully has made small personal paintings as a counterpoint to his more public, large-scale works. In 1980 he made a black and white vertical painting on a 10-inch-wide plank of wood. Feeling it was too long, Scully cut off a small piece and, in an experimental mood, placed it on top of the main section. This small painting became *Solomon*; within a year Scully had embarked on a whole series of broadly striped, bas-relief paintings.

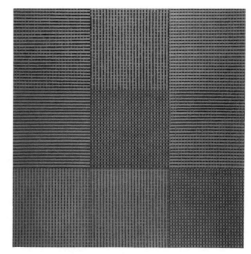

Sean Scully, *Hidden Drawing No. 3*, 1975 (courtesy of the Mayor Rowan Gallery).

Along with these changes in materials and structures came a heightened use of color reminiscent of his Fauvist-inspired canvases of art-school days. Intense oranges, yellows, blues, and greens—only slightly muted by the addition of earth tones—paraded in syncopated rhythms across his paintings. Scully has managed to incorporate many of his past motifs and

Sean Scully, *Enough*, 1981 (Collection of the Mellon Bank, Pittsburgh).

sensibilities within an enlarged framework of attenuated color and three-dimensional forms.

Flyer, 1985 (colorplate), synthesizes many of Scully's concerns: complex rhythms, expressive scale shifts, vertical and horizontal opposition, and intense color relationships form a cohesive unity. What makes *Flyer* such a compelling arrangement is its ability to reconcile opposites harmoniously — complexity exists within an eloquent simplicity, and dark, light-absorbing colors contrast beautifully with passages of reflective white. Between some color bands the faint outline of previously painted passages emerges (Scully constantly adjusts the colors by overpainting). In large part this revelation of the painting process is what distinguishes Scully's work from previous geometric painters who worked in strictly flat, hard-edged Minimalist modes. *Flyer* updates elements associated with the early 1970s (such as geometric abstraction) by making use of expressive paint-handling techniques. In the process Scully's recent paintings are both reassuringly familiar and refreshingly new.

Not one to remain fixed, Scully has recently added even a few more elements to his visual repertoire. Along with a subtle reduction of visual activity, he has introduced diagonal stripes. "A diagonal can go in any direction; you gain movement," Scully observes. In *Desire*, 1986 (illustration), the abrupt shift of rhythm by the addition of a diagonal passage is quite apparent. Scully's masterful composition creates many interesting spatial effects, yet the result is anything but a formalist display of visual pyrotechnics — *Desire* evokes feelings and sensibilities one rarely encounters in geometric abstraction.

The use of black is a prominent unifying feature of this painting. Basically, only two other colors appear in *Desire* — an earthy yellow and a sky blue. Subtle variances of color appear, however, within these broad passages and visually enliven the painting. Traces of dull red, dark green, and vestiges of white hint at the process of painting and overpainting. Within this simple arrangement of forms and colors there is much variety — paint shifts from thin to thick, glossy to dull. The overall effect is anything but simple.

Scully acknowledges that he does not know what direction his work will explore in the future — whether it will become more purified or more complex. But he does express a desire to keep his work aggressively explorative and transitional.

The issue of meaning in abstraction is an important concern for Scully. While admitting that figurative painting thematically involves people with larger issues of life, Scully wants to expand the experiential

Sean Scully, *Desire* (Collection of Colonia Insurance, Cologne, West Germany; photo by Robert E. Mates).

base of Abstract art so that it too becomes as interesting to people as the best representational art. "That's the main focus for me," he reflects, "and that's the central issue right now for abstraction."

David Salle

Few artists emerging in recent times have received anywhere near the volume of critical commentary and public attention that David Salle has garnered over the past decade. And with a major retrospective at New York's prestigious Whitney Museum in early 1987, he appears assured of a prominent place in the evolving history of the contemporary avant-garde.

To be sure, not all of the critical response to his work has been entirely favorable. Much of his work is confounding even to critics with minds nimble enough to attempt to explain fashionable aesthetic theories such as "appropriation" and "simulation." Salle juxtaposes and combines such a wide and unusual range of images, textures, materials, and visual devices that any attempt to logically interpret his paintings leaves us temporarily short-circuited. By embracing such a broad range of images and painting modes, Salle's work ends up meaning nothing in particular, and this in itself is an important aspect of his intentions. Searching for specific meanings in Salle's paintings is a frustrating task and likely to end in defeat. What heightens this frustration is that while Salle seems to encourage interpretation by consistently using recognizable images that hint at narrative structures, he places them next to comparatively meaningless geometric patterns or biomorphic forms. Actual objects are also attached to his canvases but apparently at random. Contrary to logic, thin fields of color wash across some canvases; other sections of the same painting might contain thickly painted passages that remind us of Abstract art of the 1950s. Salle's barrage of visual signs and symbols does not stop. Blue-chip images from the annals of art history are placed next to cartoonish funhouse faces, and sleazy images of women taken from soft-core porno magazines are juxtaposed

with brightly colored commercial textiles. All of these images, textures, and materials, however, are carefully controlled and organized — they are anything but haphazardly thrown together. Considering the care with which the various elements are manipulated, they slyly hint at shrouded meanings.

Try as we might we are unable to make any progress at interpreting this material — images and styles of painting do not line up in any discernible order. Eventually, we are left with the disconcerting feeling that something is wrong with us for not getting the message. Despite repeated attempts at comprehension, Salle's kaleidoscopic fragments lock us out at the same time they urge us forward. His paintings seem to have plundered some late twentieth-century image bank and transformed these signs into visual symbols that titillate our expectations and play with our dreams. From the beginning Salle sets us up for a fall. His paintings rely on the fact that we are constantly in search of personal meaning when we scan the images presented to us everyday by the mass media. Compared to the blatantly manipulative (and reassuring) images of advertising and the entertainment industry, Salle's images confuse and misdirect us — they become the diabolic doubles of popular culture, insidiously undermining the meaning of these familiar images from within.

Salle comments on his work and its relationship to popular culture:

If the work is about anything, it is about a certain kind of representation which is more often found in popular culture than high culture because high culture is about exalted means of presentation — in my mind — and popular culture is about denying access to the means of presentation, to the mechanics of presentation. That's why popular culture is more mysterious than high culture, because it's more covert. In all my images there's a notion of complicity and covertness that makes you think about popular culture.[11]

Salle's images slice through the strata of commonly seen imagery with the precision of a surgeon's scalpel. This artist manipulates styles of Modernist art the way other artists manipulate colors and forms. What makes these paintings particularly fascinating to our era (which seems to be questioning the foundations of Modernism) is the way in which they seem to celebrate and/or mock the values and standards of twentieth-century art.

Some of Salle's early paintings appear to be engaged in a vigorous dialectic between various aesthetic tenets of contemporary artmaking. *Autopsy,* 1981 (illustration), pits representational imagery against the hermetic meanings of abstract color and shape relationships. Salle presents us with two very different artistic sensibilities in this painting — representation and abstraction. By showing both modes side by side he puts us in a critically difficult position. Which style is more truthful and relevant? There is no answer to this question and the two approaches effectively cancel each other out. The title suggests that this is a postmortem investigation that seeks, through dissection of the corpses, to reveal the cause of death. At the same time that Salle questions the validity of representation and abstraction, he uses them to reveal something about themselves. On the left side, geometric, hard-edged abstraction is represented by means of rectangles painted in flat black, blue, white, and flesh tones. On the right side, a photographically generated image

of a nude woman wearing outlandish paper cones on her head and over her breasts confronts us with a tragicomic image of human degradation and the turning of sex into a commodity. For the purposes of his critique Salle brings together images which represent high culture and popular culture. The geometric left side typifies the kind of art which has enjoyed much critical acclaim in the world of the serious collector and the museum — this is the paradigm of serious, high culture. The image on the right side is symbolic of popular imagery — it parodies the sexual enticements used by media advertisers to sell everything imaginable. Unlike the glamorous images of professional models who luxuriate in fantasy settings, Salle's woman is set in the shabby surroundings of a rumpled bed in an ordinary apartment. Presenting these images side by side, Salle makes us aware of the cultural overtones they generate and of how visual symbols are manipulated in our society. Salle's main concerns do not lie with the images themselves — he is interested in the complexities of meanings that surround these combinations.

Salle's work has become increasingly complex over the past few years. *Poverty Is No Disgrace* (illustration) was completed in 1982 only a year after *Autopsy.* The painting

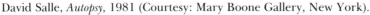

David Salle, *Autopsy,* 1981 (Courtesy: Mary Boone Gallery, New York).

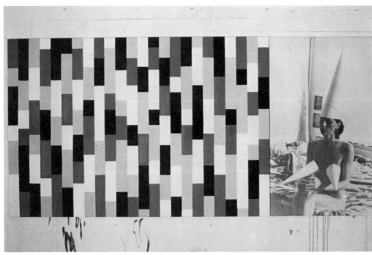

David Salle, *Poverty Is No Disgrace,* 1982 (Courtesy: Mary Boone Gallery, New York).

is composed of three separate sections and, because of their irregular shapes, this work assumes sculptural qualities. Flanked by panels containing abstract shapes and an academic nude, the center section contains a multitude of overlapping images, painting styles, and three-dimensional forms. The relative calm and clarity of the flanking panels provides a perfect foil for the complexities of the center section. In this panel a crowd scene drawn with thick dark lines shares the pictorial space with a smeared passage of paint that obliterates some of the figures in the background. Over this darkened passage of "abstract" paint a woman's head is painted in linear style looking at the struggling mass of humanity. Fixed to the lower portion of this panel is an actual piece of a chair.

Many aspects of this painting relate to modern film technique—montage in particular. As our eyes cross the canvas reading images, passages of paint, shapes, and three-dimensional forms, we begin to assemble a cognitive image based on our perceptions of the parts. Each viewer reads this painting differently and assimilates it in personal ways. Film and media experiences in general have much to do with Salle's aesthetic sensibility. Driving down the street or watching television presents us with an enormous amount of kaleidoscopic im-

agery, much of it contradictory in nature. Rarely do we question the meanings of these surreal juxtapositions—they have become too much a part of twentieth-century life to see, much less to question. Salle presents these experiences to us in a highly codified form and within a different setting—the world of the museum and high culture.

My Head, 1984 (illustration), is even more compositionally and conceptually complex than *Poverty Is No Disgrace.* One of the most striking and puzzling elements of this painting is the use of raw plywood and protruding dowels tipped with blue. The lower portion of this large-scale work is made up of five discrete monochromatic paintings illustrating tabletop sculptures of various abstract and semiabstract Modernist styles. Superimposed in these panels are sketchy figure drawings of a seventeenth-century artist—rolled drawings under his arm—juxtaposed with a provocatively posed female figure.

Thematically, *My Head* largely deals with two of Salle's central issues—art history and sexual imagery. References to modern sculpture, romantic images of the artist, and provocatively posed female figures merge and blend with each other. When Salle first moved to New York from California, where he attended art school, he worked in the design department of a midtown Manhat-

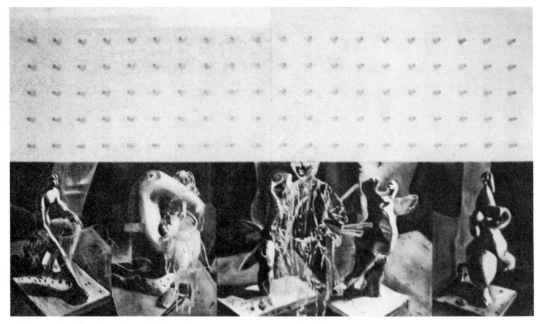

David Salle, *My Head,* 1984 (Courtesy: Mary Boone Gallery, New York).

tan magazine publisher of soft-core pornography. No doubt this exposure to the sex-for-sale industry and his behind-the-scenes knowledge of how sexually titillating images are presented had an influence on his work. Seen in the context of this painting, the grid of dowels suggests both a pervasive phallic presence and the elementary, repetitive forms of Minimalist art. This kind of ambiguity is basic to Salle's work. Through these transformative elements he suggests that our society is a culture of hybridization whereby things are so completely blended together that they cease to be single entities or to have single meanings.

Contrary to some of the most sacred doctrines of the Modernist era, Salle refuses to produce work of a consistent iconographic nature. Unlike the repetitive, trademark styles of a few decades ago, Salle draws from a seemingly bottomless well of images. He is consistent, however, in the way in which he juxtaposes symbols of high culture with those of popular imagery. *Making the Bed,* 1985 (illustration), presents us with a collection of images that, in terms

of deciphering narrative meaning, is bewildering and inconclusive. The top section of this painting is lifted, or "appropriated," from the annals of Western art history: seventeenth-century sailing ships make their way through stormy waters. Below, (as if perhaps set symbolically below these violent waves), an erotic image of a self-absorbed woman is posed bending over a couch. Superimposed over both of these macroimages we find a variety of images and details that stylistically represent many aspects of modern painting: a bas-relief sculpture of an ear makes direct reference to an art work by Jasper Johns, small sketches of figures are painted over the image of the woman, and an "abstract" passage of paint runs vertically across the top and bottom panels, linking them together. Incongruously overlapping the ships of war is a drawing of what appears to be a biomorphic light fixture. Many details of this painting are quite understandable viewed in isolation; seen together they offer enormous resistance to formal interpretation and classical unity. Yet all of these individual voices do somehow form larger patterns of multifaceted thought and present us with haunting visual arrangements both familiar and strange.

Ultimately, Salle's work addresses two of the most powerful realities that now confront contemporary artmaking. The first is that the concept of Modernism in art (once so clearly understood) appears to be becoming more elusive and questionable. The second is that, at the same time, the influence of the media on our psyche appears to be increasing. Responding to these forces, Salle's paintings chart and explore an aesthetic territory that lies beyond the paths of classic Modernism—to what, for lack of a better term, might be called Postmodernism.

The exponential growth of artmaking in the late sixties and seventies left in its wake a broadened vocabulary of new forms, concepts, images, and ideologies. Many artists once strongly identified with major movements continued to evolve and develop individual aesthetic identities. Simultaneously, a younger generation of creative individuals reexamined personal sources for painting and found their "voices" without the distractive influence of powerful movements. Old polarities between avant-garde and traditional, or abstraction and realism, all but faded from the scene by the midseventies. Stylistic coexistence was the new watchword for a new era of artmaking.

The advent of the eighties was for the most part a period of reconciliation and synthesis after decades of artistic advancement and refinement. Artists seemed to signal for time-out while they took stock of the situation and determined the directions they were to follow. The art of painting— seemingly forgotten during the heady, conceptual maneuvers of the seventies— returned with a vengeance. Historic styles long thought played out, were dusted off

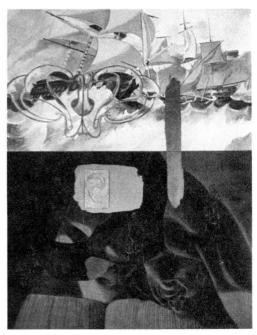

David Salle, *Making the Bed*, 1985 (Courtesy: Mary Boone Gallery, New York).

and resurrected. Neo-Expressionism made the biggest splash in this world of reinvention; not since the Abstract Expressionist era of the 1950s had so many angst-ridden, paint-laden canvases been produced. But the pendulum of style does not remain fixed for long nowadays. By the mideighties new forms of abstract and geometric art were enjoying a renaissance.

Debate continues about the meaning and relevance of Modernist thought as we near the end of the twentieth century—some view it as over, and others see in recent developments a discontinuous extension of this movement. One classical aspect of the Modern era remains unchanged. The vision of artistic freedom and invention dreamed of by Modern Art's pioneers still seems very much alive in the work of many artists today.

3 Printmaking

Art in the Age of Mechanical Reproduction

Historically, printmaking has always been a "commercial" medium: a mechanical means to reproduce images or forms economically. In this sense its origins extend all the way back to before the Guttenberg Bible; if Oriental wood-block prints and stone rubbings are considered, the beginnings of what we think of as an essentially modern process start around A.D. 100.

With the advent of modern technology various forms of printmaking have become more sophisticated, popular, and widespread. The basic underlying principle still remains simple: An ordinary potato can be carved to make a relief stamp and with it an aesthetically satisfying print.

Over the last fifty years a bewildering variety of new technical means have been developed and made available to the graphic artist for the purpose of duplicating an image; they include: high-speed offset lithography; xerography; and many photographically allied processes. Even though there is great validity to these new tools (no doubt they can enrich an artist's technical repertoire), most contemporary artists make use of four traditional mediums. This quartet—woodcuts, etchings, lithographs, and silk screens—accounts for an overwhelming majority of the graphic output today.

Because of the confusing proliferation of duplicative methods, terminology, and works themselves, it may be helpful to carefully define just what is meant by an "original" print (an apparent contradiction in terms). Most of the printed images with which we come in contact in our daily lives are photomechanical reproductions, produced on commercial rotary offset presses for mass distribution. The text and illustrations in this book, magazine photographs, and commercial posters of master paintings are all examples of this type of printing.

By contrast, fine-art prints are generally worked and printed by hand in much smaller, individually numbered editions. The main reason for this seemingly anachronistic procedure is to achieve better control over the process and produce effects that might approach those of original artworks—pen and ink drawings, for example. For this reason the centuries-old technique of etching, which can closely resemble an ink line on paper, might well suit a modern sensibility.

The direct intervention of the artist's hand at all stages of the print's production is a primary criterion nowadays for the definition of an "original" graphic work. In the late fifties the Print Council of America published certain guidelines about printmaking and with some clarification of terms they are still valid today:

An original print is a work of graphic art, the general requirements of which are: 1. The artist alone has made the image in or upon the plate, stone, woodblock or other material for the purpose of creating a work of graphic art. 2. The impression is made directly from that original picture by the artist or pursuant to his direction. 3. The finished print is approved by the artist.[1]

When photomechanical imagery began to attract a significant number of artists in the early sixties — Rauschenberg's brilliant use of photosources and processes comes to mind — some purists adhered to strict historical guidelines and did not consider these prints to be "original" because the artist did not hand draw the image. Today this is a dead issue; many of the most successful prints of the past fifteen years have incorporated, either wholly or in part, photomechanically reproduced images. What counts is the artist's ability to understand the printmaking process and achieve results.

Several decades ago American printmaking received a solid boost when Stanley William Hayter — an English artist who operated the most successful printmaking workshop in Paris — moved his operation to New York in the forties. Hayter brought with him a legacy of Northern European graphic traditions that reached back to Dürer and Rembrandt. But he was also extremely interested in contemporary printmaking; his Paris workshop, Atelier 17, counted among its patrons Picasso and Miro. Thanks to Hayter a new generation of young painters, including Jackson Pollock, were able to try their hand at etching and lithography; but most of these artists made only a handful of prints. It must be remembered that due to the pioneering nature of their imagery and lack of public interest in avant-garde art there was little demand for their work. Within twenty years this situation dramatically changed. By the midsixties many popular artists had waiting lists for paintings as yet unproduced; printmaking was an artform that, aesthetically and commercially, satisfied this demand for art. Several new workshops devoted to developing a strong American graphic style and tradition joined forces with established artists and moved printmaking out of the backwaters of the art world into the mainstream.

Although prints have unique visual qualities that cannot be duplicated by one-of-a-kind artworks, they strongly relate to painting and drawing; in fact many of the visual problems and concerns of these two modes are expressed and extended through graphic work.

Four outstanding contemporary painters, working in today's most significant reproductive mediums, are discussed in depth here: James Dine, an accomplished etcher; Jasper Johns, known for his remarkably expressive lithographs; Pat Steir who works in a variety of graphic media; and Francesco Clement, an Italian artist known for his remarkable woodcuts that resemble watercolors.

Dine's personal connection to printmaking is a particularly interesting story. In the late sixties — weary from the pressures and restrictions of fame — he rediscovered in traditional modes of draftmanship, and their relationship to etching in particular, a way out of an aesthetic bind in which he found himself. Printmaking became more than a way to quickly produce a body of marketable work: It became a means of salvation. Through it, he regained a better sense of what he wanted to accomplish as an artist and in the process learned how to incorporate personally meaningful sources into his work.

Etching is a relatively straightforward means of reproducing linear drawings; for this reason Dine felt it was an appropriate medium for his interest in classical drawing. Etching falls within the larger category of intaglio, which refers to prints made from incised metal plates. An older intaglio form, called engraving, makes use of hand-cut grooves; etching, however, employs the corrosive action of certain acids to "bite" the zinc or copper plate. A thin, acid-resistant coating is first painted on the metal surface. Various hand tools (burins) are used to scratch through the coated plate and to expose the bare metal to the corrosive action of the acid. Shallow depressions, or grooves, which hold the ink are left on the face of the printing plate. The whole process can be repeated after the print is proofed; in this way, the "drawing" can be slowly built up to a finished state.

To make an impression, ink is first forced into the shallow lines, and the shiny surface is carefully wiped clean with several changes of stiff cloth; the ink in the subsurface areas remains intact. Dampened paper is placed on the face of the plate and run through a flatbed roller press under enormous pressure. The soft printmaking paper is squeezed into the shallow grooves and picks up the ink. Other techniques of etching such as aquatint, soft ground, and lift ground produce different visual effects, but the general principle remains the same.

Lithography — a relatively recent process developed in Germany during the early 1800s — is another popular form of printmaking. Although Jasper Johns has distinguished himself in a variety of graphic mediums, he has perhaps contributed most notably to this one. Because of its extremely wide tonal range and adaptability to painterly images, it is used more than any other medium.

Lithography, which means reproduced from stone, is accomplished by drawing with a grease pencil or oily ink on the surface of a thick slab of Bavarian limestone. The stone is then etched with a weak solution of nitric acid which fixes the image and is then treated with gum arabic to desensitize the stone to other grease marks. A large, rubber roller selectively deposits ink on the surface as it passes over the wet limestone. Because the properties of the stone have been altered by the chemical process it has undergone, ink is *accepted* by the grease-covered areas and *rejected* by the clear limestone. Slightly dampened paper is placed on the inked stone and run through the press. Under great pressure a bladelike scraper forces the paper against the limestone and transfers the ink onto it. In essence, lithography is simply a chemical process that is based on the antipathy of grease and water.

The two basic materials for drawing on a stone are grease pencils that produce chalklike effects and Tusche, a greasy black liquid that can make watercolorlike wash drawings. Although the marks on the stone appear black, they can be printed in any color desired. Through careful print registration and multiple printing, complex color effects can be achieved. Jasper Johns has pushed lithography to new aesthetic heights by combining various media and techniques in a remarkably sophisticated way. Many of his prints go through eleven or twelve different press runs to attain visual effects that approach the richness of his paintings.

Silk screen was, and still is to a certain extent, a popular reproductive medium. During the sixties, when simple, "hard-edge" graphic imagery appeared frequently in painting, screen printing was often used to express this look graphically. Robert Rauschenberg and Andy Warhol (see Chapter 2) utilized this medium to produce one-of-a-kind works as well as multiples.

In fact, silk screen is nothing more than a sophisticated stenciling process. A fine nylon or metal screen (silk is rarely used today) is tightly stretched over a wooden frame. Stencils blocking out specific sections are fixed to the mesh, and paint is squeezed across the screen, forcing the pigment through the open areas onto the paper. As with lithography, a separate "run" is necessary for each color. During the last twenty-five years, sophisticated photostencil processes have been developed that enable the screen print to reproduce photographic imagery with ease. Although it does not enjoy the great popularity it had in the sixties, it continues to be a popular process and is often used in combination with other graphic modes.

As mentioned earlier, the woodcut, or relief print, is one of the oldest forms of printing known. But, until contemporary artists such as Francesco Clemente adapted this ancient process to fit current aesthetic demands, it had largely fallen into disuse by modern artists. Perhaps the utter simplicity and directness of the procedure proved confounding to popular tastes. Most woodcuts possess a vital, intense visual quality that is reflected by the materials and tools used in their making. A plank of wood is cut, carved, and drawn upon with various sharp knives and special chisels; it is a subtractive process, only the untouched wood surface will print. After the block is inked with a roller, paper is placed on top, and the back of the paper is rubbed with a curved wooden instrument to transfer the image to the print. The cut-away sections hold no ink and leave no marks.

Contemporary woodcuts make use of special materials such as plywood, translucent inks, powered cutting tools, and mechanized presses, but the simplicity and directness are still evident. Recent interest in Neo-Expressionism has sparked renewed interest in this medium by many young artists. Visually, the woodcut lends itself quite well to brusque and forceful art-making sensibilities.

Modern graphics, because of their relatively low cost, reasonable size (they seem to physically fit most homes and offices, unlike many contemporary paintings), and sophisticated design enjoy unprecedented popularity today. Thanks to time-honored printmaking processes, evolved throughout the centuries and refined today in special workshops, many individuals and public institutions can own and enjoy the same original images. Perhaps another reason behind its prevalence is the fact that there is something intrinsically beautiful and essentially modern about the idea of multiplicity itself. What is even more intriguing is the knowledge that the seeds for this contemporary artform were planted so long ago in archaic processes and ancient traditions.

James Dine

James Dine arrived in New York in the early sixties, just out of a midwestern art school, and within a few years skyrocketed to fame as a prominent painter in the developing Pop-art movement. Soon, however, the intense career pressure and narrow aesthetic limits of this avant-garde group proved disconcerting and restrictive. Dine remembers being afraid to follow his natural inclinations—to paint a realistic figure, for example—for fear of being labeled reactionary and retrograde. Peer disapproval and censure were inhibiting factors. His psychological and artistic health required getting away from the oppressive New York art scene. Dine, his wife Nancy,

and their three children set out for Europe on an extended vacation.

Away from the pressure to "do something new," Dine was able to relax, enjoy himself more, and begin to study European art and traditional draftsmanship. Through these sources he was able to reinvest his art with meaningful new themes and in large part turned to printmaking as a means with which to express them. Although his influences were historic works, the prodigious body of prints and drawings he has made over the last fourteen years have been anything but dry and academic. Today he can be looked upon as a highly inventive artist who uses traditional techniques and materials to express personal ideas about objects, people, and their mysterious and complex relationships.

Since the late sixties Dine has been steadily producing fine-art editions of prints that encompass a broad but coherent series of subjects and themes: his well-known multicolored "hearts"; vast assortments of ordinary hardware-store handtools such as saws, pliers, hammers, and drills; articles of men's clothing such as ties and bathrobes (a particularly meaningful and obsessive image for Dine); and, more recently, a series of psychologically penetrating portraits.

Although the figure is one of his main thematic concerns, today all of Dine's work can be viewed as metaphorical portraits: The informal cartoonlike hearts are really warm and loving symbols for his wife Nancy; neckties and empty bathrobes are autobiographical references to himself; even the vast tool series can be seen in self-referential anthropomorphic terms—the wrench becomes an extension of the hand and arm, and the hairy paintbrush refers to the artist's own beard.

To a great extent Dine always was a figurative artist. In the old days, he obliquely referred to the figure in his Pop-art assemblages; today, he openly works in this genre. By shifting from abstract to imagistic themes, however, he is not making any judgmental statement about the merits of one approach over the other; he is more concerned with personal meaning in his art than with following the dictates of any camp or movement.

A broad feature of his recent work is its overt human dimension. Many of these prints are based on people he personally knows or literary artists whose work he admires: Dine's wife Nancy; the romantic French poets Rimbaud and Flaubert. Archetypal figures such as "a nurse" and "the swimmer" and scores of self-portraits obsessively drawn over the last ten years serve as examples. They range in mood and technical scope from simple linear etchings to complex prints that employ thirteen colors (each color requiring a separate plate), and several printing processes such as screen prints, lithography, and etching. Although Dine started out as a painter and still produces many fine pastel and oil paintings, his prints seem to embody, more forcefully than any other medium perhaps, the essence of his artistic purpose.

In 1960, the stirrings of a new movement dubbed Pop art by many critics were being felt in the art world. At this time, Jim Dine produced two painterly theatrical events at the Reuben Gallery: *Car Crash* and *Jim Dine's Vaudeville*. *Car Crash* contained none of the comic overtones found in many of the other Happenings; in fact it was almost gruesome. Dine was motivated by the strong emotional feelings he felt when a good friend of his died in an automobile accident. This close connection of personal experience to artistic themes typifies the artist's approach. Many of his early constructions featured personal belongings, such as his shoes and dress suit, that were physically attached to the paintings. The use of everyday objects in this way quickly earned him a place in the widening circle of so-called Pop or New Realist artists.

Although many elements of his work related to the Pop school, Dine never really considered himself a card-carrying member of the group. Pop art was known for its "cool" and detached approach; Dine was thematically, emotionally, and passionately committed to his art. "Pop is concerned with exteriors," he told John Gordon of the Whitney Museum, "I'm concerned with *in-teriors*. When I use objects, I see them as a vocabulary of feelings. I can spend a lot of time with objects, and they leave me as satisfied as a good meal. I don't think Pop artists feel that way. . . . And I think my work is very autobiographical. What I try to do in my work is explore myself in physical terms—to explain something in terms of my own sensibilities."

Dine believes that his growing up in Cincinnati, the son of a midwestern family of German Jews, had much to do with forming this particular sensibility. Dine's father ran a hardware store in town and was generally unsympathetic to the arts; his mother, however, came from an unusual, artistically active family. She enthusiastically supported her son's involvement with drawing and painting and encouraged him to continue. Dine attributes the artistic interest of his mother's side of the family to their ethnic heritage: ". . . they're Jews in the Midwest," he observed. "It's quite strange to be a Jew in the Midwest—at least, it was for their generation . . . but they had this history of finesse—one grew up with some culture."

Dine's grandfather—on his mother's side—came from Poland and also ran a local hardware store; he was a big, strong, amiable man who loved to work with tools on various odd jobs and projects. Dine had free access to all of these implements and loved to play with them as a child. At an early age, he developed a strong affection for the visual look and physical feel of these simple hand tools. "From the time I was very small I found the display of tools in his store very satisfying," Dine wrote in a biographical statement. "It wasn't or isn't the craftsmanship that interests me, but the juxtaposition of tools to ground or air or the way a piece of galvanized pipe rolls down a flight of gray enamel steps."

From the ages of nine to eighteen, Dine worked in his father's hardware store, which he found pleasant but boring. He enjoyed being surrounded by utilitarian objects that were strangely disconnected from their future life: gleaming white sinks and toilets, odd sections of multicolored ceramic tile, sparkling rows of chrome towel rods. He remembers being transfixed by the dazzling chromatic spectrum of commercial paint charts. The haunting specter of all of these functional items was to emerge years later in his painterly assemblages that incorporated actual objects—a lawnmower, chromed shower head, and a roller-type applicator of house paint, for example.

His mother's death when he was fifteen threw him into a state of emotional turmoil. Not too long afterward, when his father remarried, Dine chose not to live with him and his new stepmother and moved to his grandparents' house. These were emotionally difficult years; three evenings a week were spent studying painting at the Cincinnati Art Academy's extension program. Art became a release from the pain and anxiety he experienced after his mother's death and helped relieve the confusion and hostility he felt towards his father and stepmother. To compound difficulties Dine was also struggling with a reading disability. "You see," he told the art critic John Gruen in an interview, "I'm left-handed, and there's a screw loose in my head in terms of a motor thing. So I have trouble reading, and literature has meant more to me than anything, except for music. The fact is, both music and literature really mean more to me than painting." Over the years Dine has actively pursued an interest in literature by collaborating with poets like Ron Padget and Robert Creely, by designing the sets for a production of Shakespeare's *A Midsummer Night's Dream*, even occasionally by writing poetry himself.

After graduating from high school, Dine spent one year at the University of Cincinnati, disliked it, and left for the Boston Museum School, another disappointment. Soon he was back in his native state studying painting at Ohio University in the town of Athens. During this period of his life he remembers reading and almost memorizing the pictures of well-known New York School artists that regularly appeared in magazines like *Art News*. The paintings of Franz Klein, Robert Motherwell, and de Kooning became indelibly etched in his mind. Manhattan was his mecca; somehow he had to get there. In 1958 he and his young wife Nancy—they met while studying at Ohio University—packed up their few possessions and moved to New York, no job prospects in sight, but determined to live in an area where exciting art was being done. Dine's enthusiastic confidence paid off; almost immediately, he found a job teaching art at a school on Long Island. Although this quiet, suburban area was not the center of activity in which he had hoped to live, it was close enough to enable him to come to New York City regularly.

The artistic excitement of Manhattan in the early sixties was heady: Jasper Johns and Robert Rauschenberg galvanized the art scene with their bold departures from party-line Abstract Expressionism. Other, younger artists like Red Grooms and Claes Oldenburg were originating performance events called Happenings that would lead to even further departures from traditional painting. The most important rallying cry of these artists was "make it New"; like the whispered word of well-meaning advice offered to the young man in a film of that era, *The Graduate*—"plastics"—newness was not just a desirable artistic attribute, it was a religious obligation.

Allan Kaprow, at that time a young art professor at Rutgers University, influenced Dine's early New York period profoundly. Kaprow was a persuasive, charming, and articulate individual who zealously championed the view that the essence of contemporary art was rooted in dynamic change, not in the static qualities of the painting medium. "Painting is dead," was his message; visual events like Happenings and performances were more meaningful. Kaprow wrote an article in *Art News* called "The Legacy of Jackson Pollock," which claimed that Pollock's last work (in which he "danced" around the canvas and employed materials other than paint) predicted the future of art to be in movement, change, and the use of materials other than canvas and paint. The environment would be our "canvas" and real things our "paint," Kaprow suggested.

At this time Dine was twenty-two years old. Today he acknowledges that he was very impressionable and was swept up by the prevailing avant-garde spirit of the time. His first Happening attracted a great deal of media attention. Dine, however, felt uneasy and slightly guilty about all the attention at the time. He was aware of the many talented artists in New York who had been working for many years without the slightest bit of recognition — "and here is this young schmuck who comes up and pours paint over himself (in a Happening) and it's meant to be profound. I was just uncomfortable."

Dine's discomfort quickly turned to severe anxiety. Good things were happening with his artistic career; Martha Jackson, an art dealer, gave him a regular monthly stipend, which enabled him to quit teaching and devote himself fully to art, but he withdrew more and more from the outside world and rarely left the studio. His life consisted primarily of his work and his family. Finally, the inevitable happened: A personal breakdown occurred which led to his ongoing involvement with psychoanalysis that continues today. Once a week Dine leaves his Putney Vermont Farm, where he lives and works, to drive to New York for therapy sessions.

The artist credits his great productivity of the midsixties to a greater awareness of deep-seated feelings and anxieties about his family, work, and the New York art scene. Between the years 1962 to 1966 Dine had three one-person exhibits at Sidney Janis and one show at the Martha Jackson Gallery. By 1966 Dine was established as one of the prime movers in the New Realist or Pop-art movement. But he was dissatisfied with his status in the New York art world and had arrived at a personal impasse with his work. Even though he had freely participated in Happenings and actively constructed assemblages out of familiar objects from his childhood — sinks, paintbrushes, shower heads — he felt that he was essentially interested in traditional painting and had scant knowledge of its history or craft.

Consequently, between the years 1966 to 1969, Dine stopped painting entirely. Money was no problem for him at this time, so he took his family to Europe, where they settled in London and lived for five years. Editions Electo, an English fine-arts publishing house, commissioned Dine to do a suite of prints while he was there; "they were lousy," he felt; but working in the printmaking medium freed him from the pressure of making historically important paintings. They opened up new possibilities for him, emotionally and artistically. London was a dream city for Dine; he walked its ancient streets, wrote poetry, played tennis every day, went out often in the evening (this was the Beatles era), and most importantly looked at traditional European painting with a fresh eye and open mind. Dine recalls with some amusement today how long it took him to work up the courage to enter the Louvre. He was terrified to confront the masterpieces he had neglected studying for so many years. Many of his peers had rejected these historic masterpieces as being invalid for our present time; Dine felt he had been conditioned in the early sixties to wholeheartedly embrace the new and discard the old.

By living as an expatriot in Europe, Dine bought time for himself away from New York's commercial pressure-cooker art scene. London provided a psychologically safe home for him to begin to rethink his life and the future of his work. After a pleasant round of social engagements, the sheer luxury of waking up late and experiencing European customs, Dine slowly began to work again. He designed and helped paint the set for a stage version of Oscar Wilde's *Dorian Gray*, made some paintings to be used in an Italian film, and, most significantly, began to realize the wonderful potential of printmaking. Because of its directness, etching was a particularly enticing medium for Dine; through it he was able to explore a wide range of possibilities and employ traditional aspects of art such as realistic imagery and classical draftsmanship.

During this period of his life, Dine turned away from the restrictions of the New York School and taught himself how to draw in a more articulate and expressive way. Through the medium of the print he was able to establish a richer, more symbolic imagery based on a new vocabulary of expressive drawing skills.

Dine's *2 Hearts (The Donut)*, 1970 (illustration), is a print that makes ample use of his enlarged visual repertoire. Dine did not leave behind the conceptual rigor of his New Realist assemblages; he was able to enrich this earlier body of work with new graphic possibilities. The work *2 Hearts (The Donut)* is a two-part lithograph of great complexity and visual richness. Twelve stones and eight sensitized aluminum plates were used to print this eleven-color edition of seventeen prints. Central to each of the two sections that make up the diptych is Dine's by now ubiquitous heart. These two shapes—commonly symbolizing love and the seat of our emotions—along with many watercolor splashes and washes are identical in both panels, having been printed from the same stone or plate. Subtle and pronounced differences occur between the left and right sections. Sometimes Dine changes the color of the ink in a passage from a dark bold tone to a delicate pale wash. There is a forward, courageous, and disarming quality to this piece. By working with a potentially trite and mawkishly sentimental image like the heart, Dine runs the risk of pandering to popular taste. But the treatment is too rich and complex to be

Jim Dine, *2 Hearts (The Donut)* (photo courtesy of Universal Limited Art Editions, Inc.).

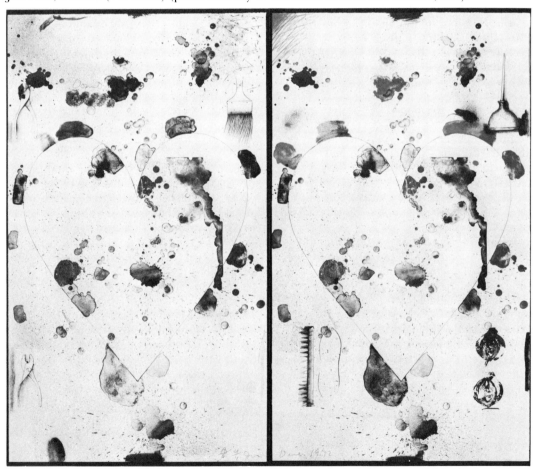

defined in such terms. Along with the lyric colors, and touching thematic sensibility, there is an element of strange and anxious beauty contained in his renderings of common hand tools that surround the hearts. Pliers, paintbrush, oil can, and scrub brush metaphorically suggest many relationships between the work-a-day world and the inner recesses of the heart: Pliers can grip and firmly hold as well as painfully pinch; house painters' brushes are instruments of renewal; oil cans are used to unlock and free frozen components; and brushes can cleanse and purify. All of these implements surround and orbit the hearts, touching and extending the meaning of this ancient symbol of love and emotional commitment. Dine uses images of hearts and tools as a "vocabulary of feelings" that direct us towards interior shadings of the psyche.

Concurrent with his "heart" prints, the artist began to explore the visually expressive possibilities of the human face. In 1971 he began a comprehensive series of selfportraits. These, in turn, prefaced an important suite of prints that narrated the life of Arthur Rimbaud, the nineteenth-century French symbolist poet. As he has mentioned, Dine is passionate about literature. He is fascinated with romantic French writers like Flaubert, Appolinare, and, particularly, Rimbaud, whose short, meteoric life and untimely death became the motif for a series of etchings titled, *Rimbaud, Alchemy.*

This printerly obsession with Rimbaud's life began in earnest when Dine found an old copy of *Historia*, a French literary magazine, at a Parisian bookstall in 1970; a striking four-color illustration of the poet on the cover—dreamily staring into space—became the basis for a whole series of prints. *Historia* was the first work produced from this source; it combined silk screening, lithography, and etching with hand painting of watercolor after the prints were editioned. Several years later the Petersburg Press in New York published *Rimbaud, Alchemy,* a six-print series.

Over the years Dine read everything he could find about Rimbaud and became fascinated with the drama of his life as well as his literary output. One of the key themes found throughout the work of this important poet was the idea of symbolic transformation. Dine incorporated this concept of change into the printmaking process by physically altering the zinc plate from one print to another. For instance, he would physically cut and reetch the first etching plate to produce a second series of prints, and so on. As many as five or six separate etching editions might be produced from one original zinc or copper plate.

The *Rimbaud, Alchemy* series, published in 1973, illustrates this particular process. *Rimbaud, Cool Impudence on His Part* (illustration), the first in the series, is printed from an 11- by 17-inch copper plate; the poet's youthful face appears on the bottom center of the print surrounded by a generous amount of empty space and flowing, energetic, lines emerging from his hair.

The next print, *Rimbaud, Alchemy on Japanese Paper* (illustration), is probably the most beautiful and evocative print of the series. A less youthful, world-weary, brooding image of the poet stares out at us. Rimbaud's head has been cut out of the large copper plate and has been reworked to produce an effect of aging and transformation. This second state—now measuring only 5 by 7 inches—is printed in the center of a special 20- by 16-inch, handmade Japanese Chiri paper. The heavily etched, dense black areas of eye sockets, hair, and shoulders contrast strikingly with the delicate, brown-flecked surface of the Japan paper.

The third through sixth prints—all, sequentially printed by altering the original plate—are titled: *Rimbaud Wounded in Brussels; Rimbaud, the Coffee Exporter; Rimbaud at Harar in 1883*; and finally *Rimbaud, Dead at Marseilles.* The first print of this group is an even more closely cropped portrait of the poet with the word *RIMBAUD* written across his forehead; the next in the series is cut down further and ominously darkened. *Rimbaud at Harar in 1883*, the fifth plate, has been altered by scraping and burnishing to bring out piercing highlights in the eyes. The final print, *Rimbaud, Dead at Marseilles* (illustration), features embossed white circles where his eyes were

Jim Dine, *Rimbaud, Cool Impudence on His Part,* etching (photo courtesy of Petersburg Press, © 1973 Jim Dine).

Jim Dine, *Rimbaud, Alchemy on Japanese Paper,* etching (photo courtesy of Petersburg Press, © 1973 Jim Dine).

Jim Dine, *Rimbaud, Dead at Marseilles,* etching (photo courtesy of Petersburg Press, © 1973 Jim Dine).

drawn. Holes were drilled into the copper, carefully following the outline of Rimbaud's eyes. When the plate was run through the press, the paper was squeezed into these inkless depressions and emerged stark white. The overall effect of this print is that of a powerful, haunting, deathlike image of Rimbaud staring vacantly at us from beyond the grave.

Over the years Dine has exploited certain themes and images such as hearts and portraits of French poets. Another important symbolic image that keeps appearing is the empty bathrobe. For about twenty years Dine has been making sculptures, paintings, drawings, and prints based on this mysterious and enigmatic image. The artist attributes the source of this autobiographical metaphor to a clothing ad that appeared in *The New York Times* in 1969. The figure had been airbrushed out, leaving only the free-standing robe. In the mid-sixties Dine wanted to do a self-portrait but felt that if he did a figurative painting he would be labeled a conservative (one of the dirtiest words in the avant-garde vocabulary at the time). So he used the empty robe as a symbolic stand-in for himself.

Red Etching Robe, 1976 (illustration), and two similar prints (reworked in the same manner as *Rimbaud, Alchemy*) clearly reveal Dine's long-standing obsession with this mysterious image. The history of his personal involvement with this ghost garment is fascinating. Over the past decade Dine has made many assemblages, drawings, paintings, sculptures, and prints of the robe; in fact it has become one of his most prominent visual trademarks. A recent 1980 exhibit at the Pace Gallery features a series of paintings on this theme collectively titled *4 Robes Existing in a Vale of Tears*. These canvases and *Red Etching Robe* seem to evoke in a strangely beautiful and visually lush way new perceptions about this old symbol. Dine confirmed this feeling in an interview with John Gruen in *Art News*; the artist remarked, "The robes have become much more mysterious than they used to be, and that's because I understand them more."

Red Etching Robe is a bicolor etching printed from two 35- by 23-inch copper

Jim Dine, *Red Etching Robe* (courtesy of the Pace Gallery, New York).

plates. A red aquatint establishes the general outline and vibrant color of the upright robe, which is portrayed standing with folded sleeves that suggest an invisible figure with hands placed on the hips. The second plate employs deeply etched black lines that define details of the robe — folds, lapels, and a hanging belt — and create a somber background. Two other editions of this print were created by reetching the plate and changing the color of the inks. Although the three prints vary, they are all distinguished by their bold, frontal imagery and concise visual organization.

The World (for Anne Waldman) (illustration), printed in the early seventies, beautifully connects visual symbol to conceptual idea. Like the robe, the image of the heart has persisted in the artist's visual repertoire far longer than he had originally anticipated. Returning to the heart, for Dine, is like renewing an acquaintance with an old friend. Through this popular, common symbol he can express his enthusiasm and admiration for friends such as the poet Anne Waldman. One of the problems connected with cartoon imagery is the risk that the artwork itself becomes obvious and trite; Dine, however, usually transcends the kitsch superficiality with which this sign is associated and uses it to symbolically represent the passion he feels for his work, his friends, and his wife Nancy. Ultimately the image of the heart expresses the artist's keen romantic sensibility. As his young son proudly explained to a friend who questioned a whole studio-barn full of heart embellished prints and drawings: "My Daddy is in love."

Specifically, this print acknowledges Dine's long-term friendship with the writer Anne Waldman, a prominent figure in the New York School of Poetry. Various multicolored hearts, along with collaged images of vegetables and flowers, are distributed over the surface of this complex eleven-color screen- and block-printed work. Labels, written in pencil, are placed above each of the hearts and plant images. The mythical "World" for Anne Waldman is represented by a turnip. The hearts were titled with specific geographic locations: Köln, Chez Max, Vermont, Syracuse, Oslo, and Ohio. Dine often enjoys making "multiples" (works that are close but not exactly alike) by collaging, hand tinting, and directly writing on the print rather than by producing mechanically exact reproductions. He also loves the idea of working and

Jim Dine, *The World (for Anne Waldman),* color lithograph with screen and block printing (photo courtesy of Petersburg Press, © 1971 Jim Dine).

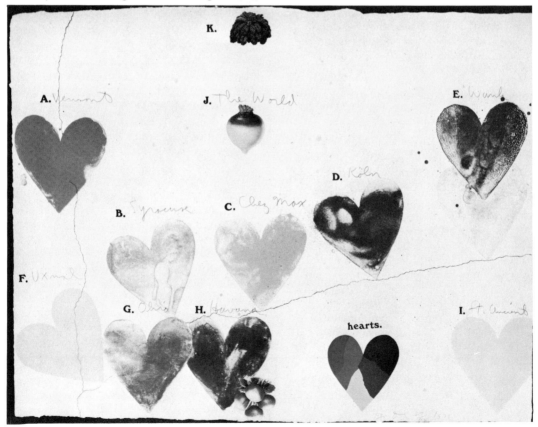

reworking a plate, pulling editions from various "states." Rembrandt also loved to work this way; his many print variants offer revealing insight into the way his highly creative mind worked.

8 Sheets from an Undefined Novel is the title of a suite of Dine's hand-colored etchings published in 1976 by Pyramid Arts, Ltd., in Tampa, Florida. Dine presents us with eight novelistic characters of his own invention, all of whom seem to be in search of a unifying plot. The artist supplies us with all of the necessary ingredients for a slightly sinister, romantic, vaguely nineteenth-century novel. He also challenges us to come up with a storyline that ties this varied group of individuals together. The cast of elegantly drawn characters includes a wistful *Sufi Baker*; a pensive, tie-clad *Die-Maker*; a perverse, seminude *Nurse*; a *Russian Poetess*, stylishly thin and manishly attired; a sensuous, naked *Swimmer*; and *A Fancy Lady*, drawn in the nude, with her legs provocatively parted. *The Leaning Man* (illustration) is the last and most detailed print of the series. It represents Dine, the master artist, as a self-contained, aloof figure who remains oblivious to the vanity and pretensions of the characters (like Alfred Hitchcock, Dine, the printmaker, appears in his own script).

There is an unusual, visually active, organic look to the prints that comes about because of the particular materials and process used in this series. Instead of employing mechanically perfect printmakers's plates, Dine bought a roll of roofer's copper and cut the etching plates out of this rough material. The bruised and scratched surface of the unpolished metal lends an interesting background of random marks and patterns to the prints.

Another specialized printmaking technique contributed to their unique "look." Preliminary drawings of each character were made on 30- by 40-inch paper; the best studies were then photographed and reduced to 20 by 30 inches. Dine made a tracing of the photograph and transferred the drawing to the plate by retracing it on a pressure-sensitive asphalt ground. Wherever the tracing stylus made contact with the plate, the softened ground was lifted by the paper, exposing the bare copper to the acid. Instead of scratching through a hard layer of asphalt resist and creating a sharp pen-and-inklike line in the print, the soft-ground method looked more like a pencil or chalk drawing. After various proofs were pulled and the print was brought to a finished state, the soft copper plates were steel plated to enable them to survive the editioning process.

This group represents a high point in Dine's printmaking career. In it he is able to incorporate and develop many of his past concerns — such as fictive biography and visual symbol — in a remarkably inventive way. Thematically, compositionally, and technically these prints effectively reflect Dine's abiding love of literary sources in art.

The unique qualities of this series very much relate to Dine's present living and working situation. Since his return to America in 1971, he has quietly pursued a personal artistic vision — based largely on

Jim Dine, *The Leaning Man* (courtesy of the Pace Gallery, New York).

classical drawing styles — with a steadfast vigor. His Vermont retreat provided a necessary separation from the fluctuating standards and fickle demands of the New York art world. Dine used this psychological distance to reexamine and redefine *his* artistic needs, not the marketplace's, and to come up with a personally meaningful method of artmaking.

Today, as far as he is concerned, James Dine is disinterested in "movements" of any kind and does not consider himself part of any group or "scene" — either figurative or New Realist; he has developed a body of work over the past decade that is based on literary sources, personal beliefs, and traditional methods of artmaking.

Dine's career crisis in the late sixties was the catalyst for growth and change — an opportunity for him to look at himself more clearly. As he relates it:

I stopped everything and went back, as it were, to draw the figure. . . . I read about a man who had open heart surgery just recently and I thought about it. He described how it's done. I think they give curare as part of the anesthetic, which "stops" the body. That's the way I felt. I put a stop to my so-called art and tried to sit with myself and look harder. I taught myself to look, to develop the tools to eventually make something that would be more meaningful — not absurdly more meaningful, but more meaningful to the viewer and to me as an art experience.[2]

Jasper Johns

In 1960 Tanya Grosman wrote a letter to Jasper Johns, a young artist just establishing himself as an important Post-Abstract Expressionistic painter, inviting him to make lithographic prints at her recently formed Universal Limited Art Editions (ULAE). In those days the facilities were modest — Mrs. Grosman's garage in Long Island housed the work studio and press — but the eventual results of this collaboration on the American fine-art print market were auspicious.

For the last twenty years Johns has produced such an abundant array of striking prints, mainly lithographic, that if all of his paintings and drawings were destroyed, he would still be considered a major artist on the basis of his graphic work alone.

The first print made at ULAE was a 22-inch-square black-and-white lithograph called *Target.* Johns simply made use of the mysterious, by now well-known, archery-like target as the print's central image. The aesthetic importance of this iconographic symbol was profound at the time. This target, an intrinsically flat object — a map of sorts — led the way out of a confounding dilemma in which the painting world found itself years ago. The art critic Joseph Young noted that both Pollock and de Kooning, the reigning monarchs of Abstract Expressionism, seemed to favor a return to illusionistic subject matter in the early fifties. Both artists hinted through their work that making viable paintings entirely out of the act of painting was becoming more and more untenable. The problem for young painters — if they read and believed the handwriting on the wall — was in which direction to turn. In the early fifties, Pollock supposedly freed painting from the constraints of the easel with his huge splatter works; at the same time de Kooning was creating canvases of great presence and tangibilities without the use of recognizable imagery. But by 1956 Pollock had died in a tragic automobile accident, and de Kooning was fervently painting the figure again — with heavy overtones of Action painting, it must be added. Did these events signal a reactionary return to old themes, like the figure or landscape, painters wondered? The seemingly unshakable doctrines of the Abstract school — like expressiveness, romantic notions of protean struggle, and an emphasis on the *act* of painting — were being questioned. Soon the debate over whether to work imagistically or nonimagistically, subconsciously or with deliberate control became heated.

About this time Jasper Johns moved to New York from his native state of North Carolina. Within a relatively short period he began making paintings of objects that iconographically resolved some of these dilemmas. Johns filled his studio with canvases that pictured common archery targets and images of the American flag. Both subjects are intrinsically flat but exist as tangible objects or things. On a two-dimensional surface Johns painted "figures" that were visually flat in appearance. The art critic Joseph Young believes that by pursuing this course the artist joined the hermetic, inner expressiveness of Action painting with an external need for tangible imagery, thus opening painting to a wide variety of new thematic sources and compositional options.

In an interview with Walter Hopps, Johns explained: "The target seemed to me to occupy a certain kind of relationship to seeing the way we see and to things in the world which we see, and this is the same kind of relationship that the flag had. . . . They're both things which are seen and not looked at, not examined, and they both have clearly defined areas which could be measured and transferred to canvas."

Johns' meteoric career in painting was solidly launched by these curious and enigmatic targets. A painting of a large green target (now in the collection of the Museum of Modern Art) was exhibited at the Jewish Museum where the art dealer Leo Castelli happened to see it. Intrigued with its cool presence and frontality, he contacted Johns, saw an overwhelming collection of similar pieces at his studio, and offered him a show at his new gallery. The rest is history.

Since his much heralded debut at Castelli's, Johns has continued to produce some of the most arresting and enigmatic icons of the twentieth century: Light bulbs, rows of alphabets, ordinary flashlights, even common coat hangers are transformed by this quiet introspective artist into simple, eloquent, visual statements that have much to do—thematically and visually— with our era and culture. These simple paintings, drawings, and sculptures tease our minds. Their utter simplicity activates our imaginations and makes us see in a way we have never seen before. In a sense the subject is not entirely the commonplace object but the act of looking—and thinking.

Ironically, the utter simplicity and directness of Johns' early work spawned an enormous amount of speculative, complex, and sometimes intellectually convoluted articles in many of the professional art journals, including: "Infraiconography of Jasper Johns"; "Jasper Johns' Ambiguity: Exploring the Hermeneutical Implications"; "Cruciformality"; "Aesthetic of Indifference"; and "Passionless Subjects Passionately Painted," to name only a few. These learned essays plummet the murky metaphysical depths of Johns' exquisitely simple, intrinsically flat, "object paintings" that by their very being seem to defy this kind of interpretation and symbolic reading. The baffling mystery of their meaning seems to be locked *in* and not under their barb of commonness. Leo Steinberg, one of the earliest and most perceptive critics to write about these quizzical target, number, and flag paintings realized that the puzzle provided its own answer. "Seeing them becomes thinking," he observed. They compel us to look at them closely, to dwell upon them, and to slowly turn them over in our minds; attempts to figure them out logically fail. These paintings represent and illustrate an exquisite paradox: They simultaneously exist as both solid objects and intangible illusions.

"The early things to me were very strongly objects," Johns said. "Then it occurs that, well, any painting is an object. . . . I don't know how to describe the sense alterations that I went through in doing this in thinking and seeing. But I thought how then to make an object which is not so easily defined as an object, and how to add space and still keep it an object painting."

Johns' contribution to late twentieth-century art seems to have been to lead the way beyond the limitations of Abstract Expressionism. By injecting a new illusionism and subject matter into the mainstream of modern painting in the late fifties, he opened it up to many of the pluralistic and individual approaches we see today.

Johns is an intensely personal and private individual, who generally refuses to explain or talk about the meaning of his work. Paradox and contradiction abound in his life as well as his art. In quotations from the late fifties he claimed to have no ideas about the meaning of his work and added that he believed the painter's business is to make paintings and not worry about what they might mean. Johns, however, is a complex and thoughtful person; one must be careful not to interpret him literally. It appears that his reticent attitude towards "meaning" stems, in part, from a reaction against the Abstract Expressionists who ascribed profound significance to the most inconsequential gestures; in counterpoint Johns shrugs off the idea that there is any specific meaning to his deliberate and carefully thought out compositions. Another factor was his early fame. Having to deal with the Machiavellian workings of the art world taught him to guard against the onslaught of both well-meaning and hostile critics.

In 1961 he constructed a small bas-relief sculpture out of sculp-metal, called *The Critic Sees*, which comments quite facetiously on the dilemma of criticism and critics. Behind the rims of a pair of glasses are two open mouths instead of eyes. The sculpture was prompted by an encounter Johns had with a professional critic. The artist had tried to answer a question raised by the reviewer, but the critic refused to listen and launched into his own explanation of what the work was about. *The Critic Sees* is a witty comment about the dangers of talking without listening and looking without seeing. It is a subtle reminder that the ultimate meaning of Johns' art is locked in its visual references rather than in verbal explanation — seeing "the things the mind already knows."

Like the work, Johns himself is a curious and unique blend of contradictions. Cool and reserved with acquaintances, open and generous to good friends, he is at once an artist of great deliberation who also maintains that the "accidental" plays a major role in his life, a provincial southerner who captivated the sophisticated art world of New York by storm, and a passionately thoughtful individual who consistently denies the importance of intellectual content in his work. Emotionally, he is both haughty and whimsically naive.

John Cage, a composer and good friend of Johns, wrote about a conversation that took place between Johns and a society lady in the Museum of Modern Art's sculpture garden. Walking with Johns amid the Henry Moore and Matisse bronzes, she asked, "Jasper, you must be from Southern Aristocracy." He replied, "No, Jean, I'm just trash." She replied, "It's hard to understand how anyone who's trash could be as nice as you are."

Johns was born in the rural community of Augusta, Georgia, on May 15, 1930. When he was quite young, his parents separated and he went to live with his grandfather in Allendale, South Carolina. For years he shuttled back and forth between various aunts and uncles, eventually returning to live with his grandparents. Moving from place to place was an unsettling experience for Johns; each new household took some getting used to. When he entered the third grade, his mother sent for him and that year he lived in Columbia, South Carolina, with her and his stepfather. His longest stay was with Aunt Gladys — a total of six years. She lived at a lakeside community called "The Corner" and operated a one-room school called "Climax" that Johns attended. When Johns was a teenager, he returned to live with his mother and eventually finished high school in Sumpter.

He was always interested in art; his grandmother was a painter and he grew up with the notion that artists were important and useful members of the community. After only a year and a half of art study at the University of South Carolina, Johns grew dissatisfied with the training and longed for associations with real artists that he felt were available only in New York. In 1949 Johns left for Manhattan, where he lived and worked for about a year before being drafted into the Army. When he returned two years later, he resumed painting and supported himself by working at bookstores and, occasionally, designing window displays for Tiffany's. Outside of his one and a half years of formal training, Johns is a self-taught artist. When he returned to New York from his Army tour, he enrolled in a degree program at Hunter College. After one day he quit. Courses in Drawing, French, and English Literature — this last class opened with a lecture on

Beowulf—seemed far removed from his concerns at the time. Johns was impatient to pursue his painting and needed a certain isolation in his life to focus his thoughts. At the age of twenty-four, after working for several years without the influence of very many other artists, he came to a crisis in his life and destroyed almost the entire output of these early years. Only four privately owned pieces escaped destruction. In many ways these surviving artworks were prophetic: One construction shows a plaster cast of a head with a collaged map above it (similar in content and structure to *Target with Four Faces* now in the permanent collection of the Museum of Modern Art); another piece employed a series of numbered toy piano keys reminiscent of his *0–9* series.

Johns attributes the destruction of his work to a spiritual awakening of sorts: "Before, whenever anybody asked me what I did, I said I was going to become an artist. Finally, I decided that I could be going to become an artist forever, all my life. I decided to stop *becoming* and to be an artist." From that point on Johns refocused his efforts and within four years produced the well-known series of targets, flags, and number paintings that won great critical acclaim at the Castelli Gallery.

After producing a mature body of work, Johns discovered that the printmaking medium, with its slow, deliberate, and tedious demands, was a perfect match for his temperament. Throughout the sixties, primarily at Tanya Grosman's, he sharpened and refined his graphic techniques and image-making skills with familiar themes such as the number series and targets. Johns even recycled three-dimensional ideas into printmaking forms with his famous *Ale Cans* print.

The story of the Ballantine ale can sculpture is an interesting one and by now part of the mythology of contemporary art. Leo Castelli had such spectacular success with Post-Abstract Expressionist art in the late fifties—which threatened and challenged the modern masters of Action painting—that de Kooning remarked of the gallery dealer, "You could give that S.O.B. two empty beer cans and he could sell them. . . ." Johns got wind of this remark and thought it was a wonderful idea for a sculpture. Johns made two life-size plaster models of the ale cans—he was drinking ale at the time and felt de Kooning would not mind the minor substitution—and reworked them by hand. These plaster models were then cast in bronze and painted to simulate the originals in label design and color. They are not physically identical, however; one can is slightly taller and has two punctured openings (after all, 1960 had not witnessed the miracle of the easy-open, pop-top can). The companion can is sealed, the contents presumably intact.

Johns beautifully recreated this pop icon in a print called *Ale Cans*, 1964, published at Grosman's workshop in an edition number of thirty-one. This two-dimensional lithograph closely follows the sculpture in structure, and feeling. The ale cans are placed in the center towards the bottom of the composition on a velvety black background created by a dense thicket of heavily drawn black lines. They sit there radiating a quiet, mysterious, self-contained beauty. Even the format of the print seems to be about containing and being contained. A rectangular black shape is drawn within a creamy white field of paper; a small pedestal quietly sits inside the darkened frame; and on top of the shallow stand are two glowing, metallic green and gold cans of Ballantine ale. Like the sculptures, one can is opened; one is closed. The simple containers become trapped within an elaborate web of graphic lines and marks suggesting visual complexity and elusive meanings.

This print is one of the most valuable and sought after graphic images produced in the last twenty years. Tanya Grosman, the director of the printmaking studio that published it, even has a pet name for this piece: She refers to it as *Peter and Paul* in deference to its translucent radiance and puzzling duality. To her it is nothing less than a contemporary icon—disarmingly simple yet mysteriously eloquent. *Ale Cans* is the joining of technical mastery (the lustrous black ground heightens the luminosity of the pale green and gold cans), economical execution, and thematic clarity. If any of the thirty one copies of this print were to appear on the market today, it would command a price in excess of $15,000: an enormous increase from its modest publication price of $175 in 1964.

Until her death in the early eighties, Tanya Grosman worked closely with Johns on a number of important printmaking projects. What linked them was their mutual concern for uncompromising excellence. ULAE was dedicated to producing the finest prints possible, aesthetically and technically; Johns was interested in challenging, and in his words "complicating," the already demanding process of lithography. Over the last twenty years scores of high-quality print houses have sprung up throughout America — from San Francisco to New York, Chicago, Illinois, to Tampa, Florida — but in the late fifties ULAE and a few others were lone pioneers in this developing field.

Jasper Johns, *Target* (photo courtesy of Universal Limited Art Editions, Inc.).

Mrs. Grosman recalled vividly how Johns tried to rethink and push the formidable technical restrictions of lithography on his first prints. "I left the stones with Jasper and after a time he said I could come to collect them. There were two: a target and a zero. The target was all right, but the other one! He explained that the zero had to be printed, and then the drawing partially erased, and a numeral one printed, and than that would be erased. . . . I didn't know if it was even possible to make these changes, but I said we would try."

True to character Johns loved the slow, deliberate, protracted process of delayed gratification that lithography embodied. *Target* (illustration), a relatively straightforward lithographic crayon drawing of his now-familiar image, was printed immediately; *0–9* turned out to be an extensive series of prints that pushed the medium and took a total of three years to complete. Obtaining the proper paper accounted for much of this delay. Mrs. Grosman realized

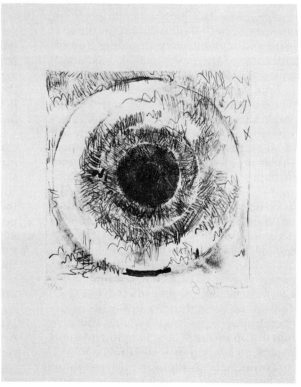

the important role paper plays in the success of a print. Unlike canvas, a support completely covered by the paint, drawings and prints usually make use of substantial areas of unworked paper as part of the design. Several years went by before Grosman had accumulated a supply of paper that was appropriate to Johns' complex graphic ideas and rigorous technical demands.

0–9, 1960 to 1963, actually turned out to be ten editions, not one. The most remarkable aspect of this series, from a production point of view, was that they were all produced from one original stone by grinding down the printing surface until the previous image was partially erased but not entirely removed. Through this operation Johns created an evolving series of images that embodied change and transformation not only thematically but quite literally.

On the first state of the stone he drew, in two rows, the numbers 0 through 9 with lithographic crayon and tusche, a special greased-based ink. Below the small series of ciphers he drew one much larger number starting with 0 (illustration) in the first print and sequentially proceeding to 9 (illustration) This basic format was maintained throughout the series.

Everything about this group suggests the theme of infinite variation and subtle change within carefully determined parameters— even the choice of papers. Johns made use of an extensive stock of handmade papers Grosman gleaned from various sources throughout the world. Each of the ten variations was printed on three kinds of paper, in different colored inks. The first group was pulled on a soft, creamy off-white with black ink; another with a special gray on eggshell paper; the third group was printed in vivid colors on a smooth, brilliantly white paper. It is easy to see why Johns fell in love with the lithographic process: The slow, demanding procedure naturally fit his deliberate and inquisitive sensibility and gave him the opportunity, by changing inks and papers, to easily transform an image.

Through the printmaking process Johns has been able to express his interest in variation and change to a degree that would be impossible in any other medium. Graphics naturally lends itself to variation. Five or six years after the completion of his first print, Johns had been acknowledged as a master in that field; by 1970 he had produced well over 100 different prints.

As an artist Johns is fascinated with the technical problems and limitations inherent to printmaking and sees opportunity within those constraints. His work thrives on and profits from the possibilities of doing something with a medium that others have deemed too difficult or even impossible.

In 1968 he went to the Los Angeles print workshop of Gemini to work with Ken Tyler, a young master printer who was also interested in exploring new processes. During the late sixties, Tyler was developing daring methods of printmaking that made extensive use of sophisticated technology. Like the city of Los Angeles itself, Gemini championed a new, modern, streamlined vision of contemporary graphics; Johns' work there reflected this boldness.

The *Figures in Color* series, executed at Gemini, clearly illustrates this point; it is quite different in look than the *0–9* edition printed at Grosman's studio. Some of the aggressiveness of Los Angeles, as well as the liberal nature of the late sixties, is reflected in the Gemini group. Like the earlier number series, each of the ten prints was produced from one source—in this case, a lithographic stone with an aluminum plate. But *Figures in Color* differs substantially from *0–9* in compositional structure and chromatic range. First of all, the series makes use of one large numeral per print (illustration), rather than include all nine digits in each composition. Second, the colors are quite vivid and fade one into the other—a result of the split-font inking techniques. Large-diameter gelatin rollers are used in lithography to pick up ink from a mixing slab and deposit it on the stone. With this process, instead of a single color, two or three contrasting inks are placed on the slab in parallel bands and are blended with the roller until they gradually and evenly fade from one to the other. The

Andy Warhol, *Marilyn* (photo by Eric Pollitzer; courtesy of Leo Castelli Gallery, New York).

Richard Estes, *Escalator*.

Shusaku Arakawa, *Courbet's Canvas* (courtesy of Ronald Feldman Fine Arts, Inc.).

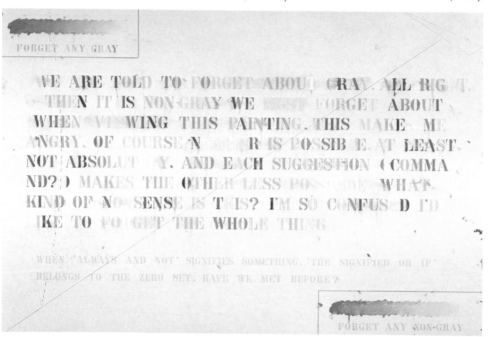

Sean Scully, *Flyer* (Collection of the Corcoran Gallery of Art, Washington, D.C.).

Pat Steir, *The Brueghel Series*
(A Vanitas of Style)
(courtesy of Michael Klein, Inc.).

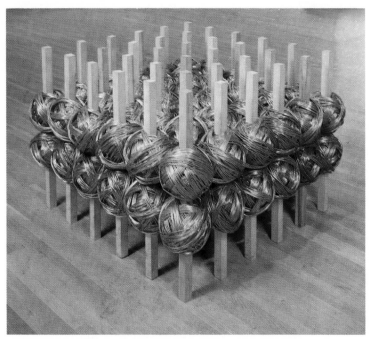

Jackie Winsor, *#2 Copper* (Collection: H. H. K. Foundation
for Contemporary Art, Inc., Milwaukee;
photo courtesy of Paula Cooper Gallery, New York).

Lucas Samaras, *Chair Transformation
No. 25B* (photo courtesy of the Pace
Gallery, New York).

Jasper Johns, *0–9* (first piece "0") (photo courtesy of Universal Limited Art Editions, Inc.).

Jasper Johns, *0–9* (last piece "9") (photo courtesy of Universal Limited Art Editions, Inc.).

Jasper Johns, *Figure 0* (Color Numeral Series) (courtesy of GEMINI, G.E.L., Los Angeles, California).

roller then transfers these split colors to the lithographic plate. Johns makes use of this transitional device, moving from one color to the next, following the progression of the color wheel, in order to reinforce some of the ideas expressed by the series: variation and sequential change along with a reassuring sense of continuity.

In 1970, Ken Tyler and Johns had the opportunity to make full use of modern technology to extend and enlarge the historical concept of relief printing well into the twentieth century. Instead of squeezing dampened paper into the depressions of a chemically etched plate, they enlisted the use of an enormously powerful hydraulic press to "print" on thin sheets of lead from hardened steel dies. Five unusual and striking bas-relief prints by Johns were published by Gemini via this unusual method.

For each of the five embossed prints, Johns first made a full-scale, shallow, three-dimensional mock-up. This was sent out to an industrial tool and die shop that fabricated the master dies from his model. Thin sheets of lead were cut from a large roll, placed between the male and female mold, and formed under intense, even pressure.

In each of the prints Johns isolates and places in the center of the lustrous field of natural lead one commonplace shallow-relief object: *High School Days* (illustration) shows a man's shoe with a small mirror (real) mounted on the tip; *The Critic Smiles* is a profile of a toothbrush with teeth instead of bristles; *Flag*, a textured version of the American flag; *Light Bulb*, a naked bulb suspended from an old-fashioned twisted cord; and *Bread*, a hand-painted slice of commercially baked white bread.

All of these ordinary objects assume an uncommon presence because of the compositional format and unusual printmaking materials used. *High School Days*, once one breaks the code and figures out the meaning of the mirror, is one of the most humorous reliefs in the series. The shoe is beautifully detailed but quite ordinary, except for the presence of a small circular mirror with the legend "high school days"

Jasper Johns, *High School Days* (courtesy of GEMINI, G.E.L., Los Angeles, California).

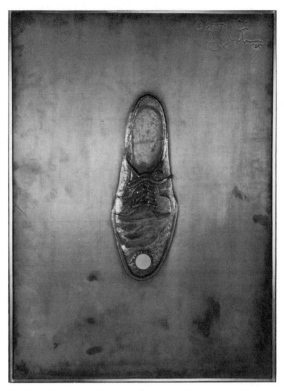

inscribed along its edge. Johns evokes, in a witty and nostalgic way, the myth-laden years of adolescence with this particular piece. Supposedly such an apparatus was used by boys to enable them to look discreetly up the skirts of girls.

Fragments—According to What is another series of prints that illustrates Johns' shift toward a "cooler," more controlled approach. Perhaps working with the clean, machinelike surfaces of the lead reliefs suggested other "looks" to the artist. In any case, this group clearly departs from the loose, gestural, painterly look of the ULAE era. *Fragments—According to What* are, for the most part, clean, analytical, sectionlike compositions that visually appear to have been derived from larger works. One device that appears in all of these prints is the split-font inking technique seen earlier in *Figures in Color*. Dense, fully saturated dark areas gradually fade into delicate, transparent tones.

Richard Field, a curator who specializes

in Johns' graphic work, observed that *Fragments—According to What*

are diagrams of devices, systems, actions, and words, all implying shifts of focus, change, and motion. They exist in an environment of bendable space, largely established by the precisely modulated bands of blended ink. The total effect of this precision is what I have called transparency. Although the viewer is aware of the textures and flatness of the lithographs, he nevertheless has some feeling of looking into a further space. This carefully controlled transparency strengthens the illusionistic potential of the entire image, and is really the framing language with which the artist and the spectator go about their analyses.[3]

The six prints in *Fragments—According to What* are thematically based on an important mixed-media painting Johns completed in 1964, titled *According to What* (illustration). *Leg and Chair* (illustration) is a print that refers to the section of the painting that incorporates an actual chair and three dimensional leg cast into the composition. Johns had this detail of the

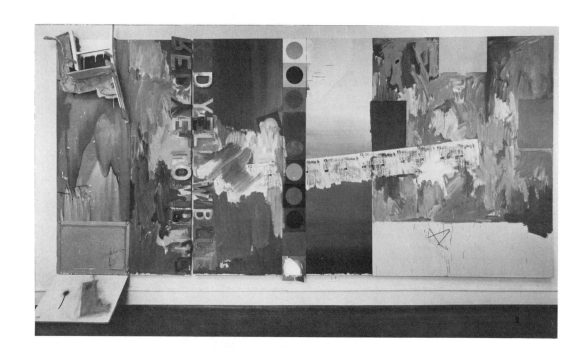

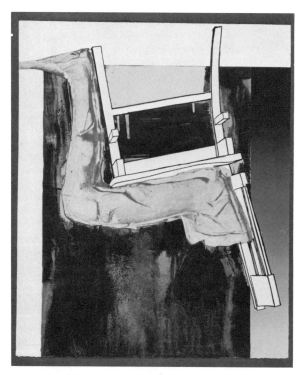

ABOVE: Jasper Johns, *According to What* (photograph by Rudolph Burckhardt; courtesy of Leo Castelli Gallery, New York).

LEFT: Jasper Johns, *Leg and Chair* (Malcolm Lubliner Photography; courtesy of GEMINI, G.E.L., Los Angeles, California).

painting photographed and prepared a photolithographic plate from the negative; in the finished print various washes and crayon drawings merge with the photo image to create a unifying effect.

Bent Blue (First State) selects as its "detail" an area in the painting that incorporates twisted three-dimensional letters that spell B-L-U-E. This lithograph — made entirely of a washlike brush drawing — incorporates this particular section of the canvas.

Bent Blue (Second State) (illustration) is one of the outstanding pieces in the whole series. It makes use of the "first-state" image but adds a deep and spatially open field of color that fades, with a deliberate slowness, from dark blue-black at the bottom to a transparent gray at the top. In the righthand corner is a faint, torn fragment of newspaper with a dark blue X "canceling" it and contradicting the illusionistic space set up by the blue-gray field. Johns juxtaposes many different elements in this particular print: the loose, complex, or-

ganic washes of the letter forms; the skylike translucency and depth of the open space; and the canceled printed section. In a sense each print in this work is unique; many different sections of newspaper were used to individually produce each print in the edition. By sandwiching the newspaper section between the plate and the dampened paper, the newspaper ink was transferred to the print by the pressure of the press.

Hinged Canvas, *Bent "U"*, and *Bent Stencil* are the remaining prints in the *Fragments — According to What* series. Like the previous works, they specifically refer to three sections of *According to What*: the small stretched canvas placed on the large painting with its back to us, the three-dimensional letter "U" twisted out of shape; and the warped stencil attached to a painted passage that alludes to traffic lights.

In response to criticism that Johns merely repeats old motifs rather than develop new images, it must be pointed out that he usu-

Jasper Johns, *Bent Blue (Second State)* (Malcolm Lubliner Photography; courtesy of GEMINI, G.E.L., Los Angeles, California).

ally manages to breathe new life into familiar themes. *Fragments—According to What* is a good example of the transformative power of Johns' aesthetic (some observers fail to take into account how important this creative recycling is to his work). One of the reasons for Johns' using preexisting imagery—like his flag and target series—is to free himself from confusing decisions about *what* to paint so he can concentrate on *how* he is painting it. Michael Crichton's engaging catalogue essay on Johns points out several similarities between this contemporary artist and Edouard Manet, one of the founding fathers of Modernism. Both painters made use of preexisting imagery: Manet literally borrowed from and repainted many of the historic compositions of old masters and even contemporaries like Delacroix. Another aspect of Manet's painting that relates to Johns' work is the French artist's interest in flatness. One of his most famous paintings, *A Bar at the Folies Bergères*, shows a barmaid standing in front of a mirror—a surface physically flat but illusionistically deep.

Jasper Johns and Samuel Beckett, the contemporary playwright, are both modern existential artists; Johns is to painting as Beckett is to drama. The language and theatrical form of Beckett's classic play *Waiting for Godot* may express and reflect the nature of twentieth-century life more succinctly than any other modern play. The language is sparse, laconic, obtuse, and perhaps brilliant in its perceptions of modern-day hope and despair. Contradiction and conflict energize its fragile plot: It is both funny and tragic, cautiously hopeful and innately pessimistic. In many ways Johns' work parallels Beckett's. No answers are provided by *Waiting for Godot* to puzzling questions such as who is this mysterious figure and when, if ever, will he appear? Beckett suggests that these questions—like many questions of life and living—can never be answered. Like Johns and his mentor Marcel Duchamp, Beckett maintains that there can be no solution to life and art because no problem exists. Leo Steinberg's perceptive remark about Jasper Johns' work could just as well apply to

Beckett's plays: "Seeing them becomes thinking."

Given the similarity of sensibility between these two artists, it was more than appropriate that they should collaborate on a special project. Vera Lindsay was the catalyst in this venture, first getting in touch with Beckett's literary agent and then Johns'. Both artists knew and respected the other's work but were cautious about the details of the collaboration. Beckett did not want his work to be "illustrated" and in fact, believed that it could not be done effectively; Johns concurred. The last thing he wanted to do was to illustrate a literary work, no matter how good it was. When both artists met in Paris to discuss the idea, they immediately liked each other; both felt strongly, however, about maintaining the independence of their work. Beckett suggested using an early draft of *Waiting for Godot*, but Johns rejected the idea; he wanted new, unpublished work. Finally, Beckett submitted five never-before-printed prose pieces written between 1960 and 1974 that proved acceptable to Johns.

This uniquely collaborative work of art between two outstanding artists from separate disciplines was bilingually titled *Foirades/Fizzles*. Both the visual and verbal contributions are fiercely independent, yet mutually harmonious overtones are generated by placing them together. Johns' etched images are clear and open, yet hermetic and closed to facile interpretation. Beckett's text embodies negation and doubt; any position is soon contradicted: "I do the impossible, it is not possible otherwise." The language is dense and impenetrable, much like the tangles and textures in an early Johns' painting that puzzle the eye and mind.

Both the text and the images constantly refer to the self yet seem autobiographically distant. Consider the way each artist begins the book. Johns opens dramatically with a delicate gray, aquatint self-portrait created by coating his face with a "sugar-lift" medium and pressing it directly to the plate with a sideways rocking motion. Directly opposite this print, Beckett's text begins with (the writing is in French with an

English translation). "J'ai renoncé avant de naître, ce n'est pas possible autrement, il fallait cependant que ca naisse." ("I gave up before birth, it is not possible otherwise, but birth there had to be.")

True to Johns' methodology much of the thematic imagery is derived from past work, particularly *Untitled*, a large four-paneled painting made in 1972 that combines planks of wood with wax body casts, flagstonelike shapes, and a distinctive "cross-hatching" pattern he began to use at this time.

The visual structure of the book has been very carefully worked out: Johns designed it with the help of Mark Lancaster. Most of the etchings appear to the left of the text; versions of Johns' familiar numerals introduce each prose piece which first appears in French, then in English. Throughout the book a wide variety of Johnsian images appear, alone and alongside the letterset printed type.

It almost seems inappropriate to refer to this remarkable collaboration as a "book." The physical presence and overall effect of this magnum opus are stunning. *Foirades/ Fizzles* overwhelmingly becomes more of an exquisite object, a package, a unique container, and holder of verbal and visual ideas than it does a book in any usual sense of the term. Enclosed within its soft, white liner case fabricated by Rudolf Rieser in Cologne, the whole work becomes a pristine Mimimalist sculpture of sorts. And the credits of this book read like an international *Who's Who* in the fine-arts graphic world. The Crommelynck brothers of Paris — unequaled masters of etching — outdid themselves and achieved intaglio effects rarely seen in modern times. Their prints range from simple line drawings to lithographiclike etchings and extended the medium beyond its normal limits. Typesetting was done by Fequet and Baudier of Paris. Even the paper was specially made by Richard de Bas, in Auvergne, watermarked with Beckett's initials and Johns' signature.

Once the elegant linen cover of *Foirades/ Fizzles* is opened, a rich experience of images, textures, and verbal dialogues begins to unfold. The front endpaper is a four-color — orange, green, purple, and white — print with Johns' recently developed "hatching pattern." The origin of this image is typically Johnsian; it first showed up in a painting called *The Barber's Tree* made in 1975. Johns "borrowed" this pattern from a photograph in *National Geographic* magazine that showed a Mexican barber painting the tree outside his shop so that it resembled a red and white, candy-cane-striped pole.

Jasper Johns, words from the *Foirades/Fizzles* book (photo courtesy of Petersburg Press, © Jasper Johns & Petersburg Press S.A., 1976).

Jasper Johns, "Torse" from the *Foirades/Fizzles* book (photo courtesy of Petersburg Press, © Jasper Johns & Petersburg Press S.A., 1976).

The fourth etching in the book (illustration) is a double-page spread that, in a painterly fashion, verbally lists in two columns various parts of the body — English on the left, French on the right: Buttocks/Fesses; Knee/Genou; Foot/Pied, and so on. This print along with others reinforces the self-referential concept of the "body" that runs through both artists' work.

Above all, Johns realized that the overall relationship of the images and stories — their scale, placement, and sequential order — was of primary importance. The enormous amount of trial proofs generated by the artist attests to this overriding concern. Many of the early prints were printed from large plates that enabled the artist to work and rework an image. Originally, Johns had intended to visually inset the etched images within the printed text, like a contemporary magazine layout, but decided instead on a structure of juxtapositional counterpoint. The etched drawings on the left page were played against the pristine texture of the typography to the right. *Torse* (illustration) is a fine example of the interplay between image and text. This image of the torso was photographically transferred to the plate by screen process, etched, and then reworked by hand. Drawn into the bottom of the plate are the stenciled letters which spell out T-O-R-S-E (torso in French); a subtle watercolor like aquatint distinguishes this print. Duality as a thematic and structural theme characterizes the whole book and this print in particular: The light and airy texture of the type contrasts with a dark foreboding density in the etching; image and name stand side by side; a few pages further, a similar plate is recycled, reversed, with the torso in white and the text and label in English.

The book ends visually, as it begins, with impressions of the artist's body — this time his hand and foot. *HandFootSockFloor* is a plate that makes use of sugar-lift aquatints to record the artist's imprint as part of his body literally touch the etching plate. There is much that is intimately personal, vulnerable, and touching to this gesture. Its simple directness takes us back to neolithic times when individuals — similar, no doubt, to us in many ways — left their hand prints behind as mute evidence of their passage through the narrow walls of a cave and the broader avenues of life.

As an artist Johns loves to balance many opposing qualities of thought: The literalness of visual patterns with the symbolic meaning of figurative images; aspects of prehistoric art with sophisticated twentieth-century processes and materials; and a conceptual use of language coupled with sensuously beautiful visual effects.

Foirades/Fizzles represents only a recent chapter in the long, dynamic, and ongoing career of this painter, sculptor, and graphic artist. By refusing to do the expected, Johns continually reinvests himself with renewed sight and insight. Through this generative process he continually surprises himself and us. New meanings and new perceptions constantly surge through his work; simple images present themselves in complex terms and everything becomes far more than it seems.

Pat Steir

For six months out of the year Pat Steir, an American painter and printmaker, works in a top-floor studio in Amsterdam located just across the canal from a house that Rembrandt lived in centuries ago. For many contemporary artists this connection to a past master's life might hold great charm; for Pat Steir it holds the key to much of the meaning of her art. Steir lives in Amsterdam for a variety of reasons (for one, her husband is a Dutch publisher), and through her connection with this ancient city she has developed a fascination with the issue of style in the visual arts. Amsterdam is a fertile place for the study of art — its museums hold a vast array of paintings, drawings, and prints; and within a short train ride even more museums are available for study. Steir passionately believes that "All art making is research, selection, a combination of thinking and intuition, a connection between history and humanity."

For all of her current interest in history, Steir has not always felt this way. She admits attending art school in such a "self-absorbed daze" that the concept of art history meant little to her. At this stage of her education Steir was laboring under the premise that to be an artist she needed only to "be herself." The idea of originality was almost a religious belief. Jungian thought was an important influence at this time — universal symbols were expected to emerge from the artist's unconscious. Steir's struggle during this period, at least in terms of how she defined it, was to get in touch with her own innate images and discover her power as an artist.

1969 was an important year for Steir. This was the year she painted *Bird*, a piece she considered her first mature piece of art. Things were opening up to her in many significant ways. "Until now," she remembered, "I had hardly ever gone to galleries and museums. My friends were literary. I wasn't really involved with the art world. Literature and music were my main interest. I was completely isolated from the art world. With *Bird* and the paintings that followed after it, I decided that I was a painter after all, so I left my job in publishing (Steir designed book jackets) and began to teach painting instead."

In the early 1970s, Jack Lemon, owner of Landfall Press in Chicago, invited Steir to work at his printmaking facility. Except for a course in printmaking at Pratt Institute in the early 1960s, Steir had no previous experience with this technically complex process of artmaking. *Wish #2, Breadfruit,* 1974 (illustration), was one of the prints that emerged from this early collaboration. In part this image was based on a painting Steir completed a year earlier in 1973 titled *Breadfruit.* This painting makes compositional use of a large black square that almost fills the canvas. Steir's print employs many of the same visual elements as her painting: brush strokes, an energized, scribbled mark, and an image of a flower.

But other visual elements play an important role in this print and distinguish it from her painting style — particularly, the color scales that flank the center section

Pat Steir, *Wish #2, Breadfruit*, 1974 (courtesy of Michael Klein, Inc.).

and the thin horizontal lines that run through the composition. The use of color scales can, in part, be attributed to Steir's experience as an art director in the publishing industry. When proof of printed matter came to her office for approval, it often had barlike color scales along the border to attest to color accuracy in the reproduction process. The fine lines in the print can be traced to Steir's developing friendship with Agnes Martin, a Minimalist painter who made extensive use of pencil lines and grids in her work.

The reference to the tropical breadfruit plant has an interesting literary origin. Steir had been reading Darwin and had become infatuated with his enthusiastic belief that widespread cultivation of this plant would put an end to starvation in many places of the world. Every element in this print can be related in some way to personal aspects of the artist's life. Even the iris in the print's lower right makes an autobiographical reference. Steir's first name is Iris, although she has preferred to use her middle name since early childhood (Iris seemed too exotic for the northern New Jersey towns she grew up in).

The early seventies were a prolific period for Steir; during this period she was able to combine and synthesize a variety of visual and thematic elements developing a distinct and mature style in the process. By 1975, however, Steir found that she had become less interested in art as a vehicle for research into other subjects, and she became absorbed in the study of classical Western painting—Courbet, Rubens, the Breughels, Van Gogh, and Rembrandt. What fascinated her in particular was the issue of style and what visual elements made one approach to painting different from another. About this time she remembers seeing an early self-portrait by Rembrandt that was particularly eye-opening. In this painting Rembrandt's face was rendered in dark shadow; his hair, silhouetted against bright sunlight, was painted in a decidedly modern, scribbled style. "It took my breath away," Steir remembered, "because there it was, everything we call contemporary. Yet it belonged completely to its own moment. That is what gave it its humanity, its insight."

After investigating a wide range of historic painting styles Steir concluded that little substantive difference exists between various art historical periods. Elements of all styles are evident in all paintings if you look close enough, she maintains. The "alphabet," or colors, remain the same but the "handwriting" (the method of paint application) changes. Differences exist, Steir believes, in the artist's use of scale and space. For example, "small black and white strokes that make up detail—that's Rembrandt. Expand them and it's Kline." From her research, Steir has come to believe that the issue of styles in the history of art is more a continuation of shared concerns rather than an expression of divergent interests.

Using this theory of the unification of styles, Steir began work on a large-scale painting in 1982 called *The Brueghel Series (A Vanitas of Style)* (colorplate). This work is more than a monumental painting—it is a visual manifesto and meditation on the meaning of style in painting. *The Brueghel Series* measures almost 20 feet high by over 15 feet across; it is made up of 64 painted sections, each canvas measuring roughly 28 by 23 inches. Every individual section is painted in the style of one of the great masters in Western art history—from Sandro Botticelli (1444–1510) to Jackson Pollock (1912–1956). Stylistically, one could hardly think of a greater range of work. Steir studied all of these artists and tried to determine what was quintessential about their styles. Each panel in *The Brueghel Series* represents her efforts to paint as if she were that master artist.

While *The Brueghel Series* was in preliminary planning stages Steir had been on the lookout for a painting that could be used as the template or central image for her exploration of painting styles. One day on a trip to a museum in Rotterdam, Steir bought a poster that attracted her. It was a large reproduction of a vase of flowers painted by Jan Brueghel the Elder. On the train ride back to Amsterdam she studied the image excitedly and thought that this could be the image she was looking for— an image to express her vision of the synthesis of painting styles. Later, after folding and cutting this poster in her studio to see if the image would be suitable, Steir began

to work on a series that was to take her two years to complete.

Almost every style of Western art is documented in this ambitious painting—from the Renaissance to contemporary art. Viewing it is like assembling the pieces of a gigantic, complex puzzle. Steir spent a good deal of time deciding which artists would share adjacent panels. After many trials the painting began to take form. Interesting conceptual relationships were suggested by juxtaposed artists: De Kooning and Manet meet at panels near the center. A geometric Mondrian and an expressionistic Pollock are placed near the borders of the composition. Considering the difficulties of arrangement and inherent diversity, *The Breughel Series* holds together remarkably well.

While Steir was analyzing and incorporating these diverse painting styles into one monumental work, she began to find that her attitudes toward history were undergoing profound changes:

To struggle to understand history is so much more touching than self-expression. Working my way through this project was an incredibly moving experience. I found that there were strange, very intimate things to be learned—which painters were left-handed, for instance. But all that is private, between me and the work. Or me and the painters I dealt with. The public meaning of The Brueghel Series *is different. To try to paint the history of painting is to reflect the history of one's learning to perceive, to be in the world. And to show how the culture learned the same things, though of course Western culture didn't have to learn to be in the world. It created our world, so* The Brueghel Series *is about that too—the story of a culture inventing itself.*[4]

Another concept that informs Steir's work is her belief that the more closely we look at a painting or print, the more the barriers between realism and abstraction disappear. *Form, Illusion, Myth* (illustration) is a lithographic print in three sections that reflects this point of view. Reading the three panels that make up this work from left to right, we are presented with a cinematic treatment of a still-life. These three sections correspond, in terms of film, to a medium shot, close-up, and ultra-close-up. In the print on the far left we see a pitcher containing

Pat Steir, *Form, Illusion, Myth* (courtesy of Michael Klein, Inc.).

chrysanthemums rendered in a Realist style. In the center section the flower fills the frame and is semiabstracted. Finally, in the piece on the far right, the image of the flower dissolves into a maze of abstract patterns, lines, and marks in a close-up view. Steir has chosen the chrysanthemum for the didactic exploration of "Realist" and "Abstract" style because this flower is the traditional symbol of oriental scholar/painters who, through their art, sought to break down barriers between the viewer and the illusionistic space of the painting.

Recently Steir has been working on a series of monoprints and etchings that feature portraits of herself as a historic artist such as Van Gogh, Rembrandt, and Courbet. Each print is done in a style appropriate to the work of that artist. These conceptually convoluted portraits are intriguing on a variety of levels. Doing a self-portrait in the distinctive style of a recognizable famous artist is a risky endeavor at best. Whose artistic personality will dominate? Does it even matter today? What does this say about one of the hallmarks of the Modern era — originality? These questions bring the issue of borrowing images and styles (more fashionably known as "appropriation") to the forefront. Perhaps this term is *the* buzzword of the eighties. But, while other young artists have made their careers on the basis of copying modern masters verbatim (Sherrie Levine for instance), Steir has established hers on the basis of stylistic analysis, which she calls "quotation." Putting semantic differences aside, the issue appears to be not whether one appropriates (with or without the pertinent philosophical theories) or quotes, but how successful the work is visually and conceptually. Steir — who believes the term "appropriation" has a negative connotation — manages in these monoprint self-portraits to come up with graphically beautiful images, whose power lies equally in their visual presence and conceptual origins. *Self as Picasso as a Young Man #1* (illustration) economically captures the feeling of an early Picasso drawing with its lyrical brushwork and subdued "Blue Period" color.

Because the monoprint technique is a

Pat Steir, *Self as Picasso as a Young Man #1* (courtesy Crown Point Press, San Francisco and New York).

combination of two processes — printmaking and painting — it is conceptually well suited to the nature of these dualistic prints. To make one of these monoprints, Steir paints directly on the etching plate with inks. Some of the plates in this series were blank and some of them — as was the case with *Self as Picasso* — were etched with soft ground lines that roughly established the form of the head. After the inks are painted on with a brush, the plate is run through the press much as a standard etching would be. Although this process makes use of some printmaking techniques, each print is unique and cannot be duplicated exactly.

Self as Picasso fits easily into Steir's views of art and life: It is part print, part unique image, part Picasso, and all Pat Steir. The sensibilities of synthesis and historic awareness that characterize her work are also motivating factors found at large in the world of art. New visions are being created out of familiar source material, and what seems fresh and relevant appears to be developing in large part out of important concerns that are remembered from the past.

Francesco Clemente

Like their German counterparts, postwar Italian artists address concerns that arise out of their consciousness of European history and traditions. Rather than focusing on angst-ridden subject matter and boldly Expressionistic painting styles as the Germans are likely to do, recent Italian artists assume more introspective, wry, and aesthetically mutable stances. For the Italians, art is not something locked away in a museum to be visited once or twice a year, it is a part of everyday life. Sandro Chia, one of Italy's prominent New Wave artists who hails from Florence, recalls as a boy playing soccer in the Piazza Santo Spirito, and during breaks visiting the adjacent church to admire Filippino Lippi's magnificent altarpiece. Walking with friends, playing, or on his way home from school, Chia might pass Ghiberti's bronze Baptistery doors, or look up and see Brunelleschi's imposing dome. Most Italian artists—whether they grew up in the north or south—could recount similar stories. American children by comparison were raised with the Flintstones, TV sit-coms, and talking dolls. And, while the United States has many fine museums, Italy itself *is* a museum. Italian artists carry this cultural legacy around with them the way we carry around memories of rock-and-roll songs and Hollywood films.

Contemporary printmaking in Italy is particularly well informed by an awareness of the history and stylistic heritage of this medium. For artists such as Mimmo Paladino, Sandro Chia, and Francesco Clemente, printmaking is not seen primarily as a means of reproducing images developed through painting, but as a rich medium that resonates with its own inherent visual capabilities and historical references.

In the early eighties, Francesco Clemente, along with Enzo Cucchi and Sandro Chia (the trio were known waggishly as the "three Cs"), took up residence in New York City and proceeded to garner much publicity for their lively approach to artmaking. Coming at a time when there was a resurgence of interest in narrative art and the figure, their fortunes and popularity skyrocketed. These artists were not like the usual expatriots who made their way to America. They seemed to move through the world—adapting different life-styles to suit different geographical situations—with an ease and comfort that were enviable. These citizens of the world divided their time between New York lofts, family residences in Italy, and studios in exotic countries like India. German artists appear to be wedded to their native soil and haunted by their political past—few of them leave Germany for any period of time to work in foreign countries. The Italians, however, have attained the ability to travel freely, setting up studios whereever they might stay for a few months. It is not as if they have rejected Italy; rather, they seem to carry with them the essence of Italy—an attitude that enables them to learn and profit from a variety of cultures and ways of life.

Also what some contemporary Italian artists bring with them is a classical education—the kind of schooling that until the student unrest of 1968 meant learning Greek and Latin, not as electives but as core requirements of the curriculum. Because of this formal education, Clemente is the most serious reader and scholar of the group. The range of literature that captures his imagination is formidable; one day he may be fascinated with Ezra Pound's pro-Fascist speeches. The next day he can be found making his way happily through thirty volumes of the Loeb Classical Library. But Clemente is not a traditional scholar in any real sense. Both in his art and in his readings, the past is looked upon as a rich treasure trove to be picked through, borrowed from where necessary, and discarded when irrelevant. Refering to the classical courses that schools abandoned in the late sixties, Clemente said: "What ever was good about 1968 was lost. But then, you cannot capitalize on good things. You must use them, spend them, let them go. They will be born again, and again, and go away again and again." No doubt coming from a country as deeply layered with historic epochs as Italy contributes to this perspective of cycles and rebirth.

Because Clemente is so steeped in his cultural origins, he seems to have an inner security which allows him to flaunt traditional conventions. For instance, despite his love of reading and classical literature, he has no bookshelves. For that matter he has no furniture except for a few chairs and a wooden table. "When my wife and daughter were coming from Italy (to his New York loft), I knew they would expect some furniture," Clemente recalled. "This is the first real home we ever had, after all. But I couldn't think what to get, so I bought the myna bird over there, and I let him go free, and he furnished the room all by himself."

Clemente's prints are as varied and inventive as his attitudes about life and living. Not surprisingly, Clemente is a born adventurer. When he was in his twenties he literally traveled around the world learning new artistic techniques and ways of using materials. Often the traditions of the country Clemente settles in for a while profoundly affects his art. Consequently, the range of media used by this artist is remarkably wide; Clemente has worked in mosaic, fresco, watercolor, pastel, books, woodcuts, etchings, and drawing, as well as painting.

Although Clemente's 1984 self-portrait (illustration) looks remarkably like a watercolor, it is really a woodcut done in the ancient Japanese woodblock tradition. As Westerners, we expect a contemporary woodcut to be rough-hewn and vigorous like the work of the early twentieth-century German Expressionists. But Clemente loves to do art that displaces our normal expectations. Working in Japan with master printer Todashi Toda, Clemente was able to achieve, through numerous superimposed impressions, the effect of wet pigment spreading and soaking onto the paper. This print represents the multicultural fusion of influences and sources that are typical of Clemente's work. As a self-portrait the subject matter of this print is unmistakably European in origin, yet the technical means of execution are clearly Japanese. Clemente's art delights in cultural displacement. By using these juxtapositions in his work he calls attention to the fact that our world has become increasingly multinational in terms of economic and social interactions.

The one constant in Clemente's prints is the repetitive use of his own image—they are all to some extent self-portraits. In print after print Clemente stares out at us, clothed, naked, active, passive, and usually surrounded by a variety of iconographic props that lend his work a vague air of mysticism. *Self-Portrait #4* (illustration) is an etching that shows Clemente staring directly at us with a bemused expression. His left hand holds a fork that menaces a snake curled into a circle and swallowing its tail. Forming the circular backdrop for the snake is a symbolic representation of the earth suspended in the sky.

Clemente assumes a variety of roles throughout this series of self-portraits. But there is nothing dishonest about his assumption of various identities. Clemente freely invents himself and alters his person

Francesco Clemente, self-portrait (woodcut), 1984 (courtesy Crown Point Press, San Francisco and New York).

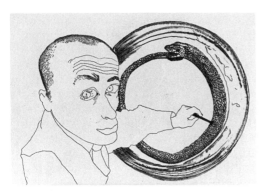

Francesco Clemente, *Self-Portrait #4* (courtesy Crown Point Press, San Francisco and New York).

to fit each artmaking experience. In a sense this is what we all do everyday as we go to work, relax with our friends, or spend time by ourselves. Each activity calls for a different persona. All of Clemente's aesthetic masks reveal some aspect of his personality and allow us a glimpse of the mental images that form the life of his imagination. Clemente's prints make use of representational imagery — not to show us the world through an illusionistic window — but to give form to his fertile imaginary experiences.

Central to Clemente's imagery is the idea of mythology. He is in love with the conceptual notion that legends and half-forgotten cultural beliefs are still a subliminal part of our lives — however buried and distant they may be. Clemente explains that he wants to ". . . just travel through mythology. . . . I don't know about any mythology, but I fall back again and again on images which belong to some mythology. The way it goes is just not to know about

any, and to have faith in the possibility of the tradition of art to give truthfulness to any image you come across."

This Side Up (illustration) is an etching that features an image of the artist as a male caryatid, or mythological Greek priestess from the temple of Diana at Caryae. Clemente draws himself in this print as a hermaphrodite: male and female anatomical features allude to masculine and feminine personality traits in all of us. Many features of this image are strangely disorienting. We are not sure whether the figure shown here is lying down, or whether it looks like this only because the print has been tilted on its side. An arrow to the far left points upward indicating — in no uncertain terms — which way the print should be displayed. In Clemente's prints, images are often presented to us in a cool and dispassionate way. Clemente recognizes the power of images and cautions us about making too much of their "meaning." "When you look to any image in art," he says, "you should be able to be detached and feel a kind of irony. . . . Art is not a religion, it's not a dogmatic thing, but images are very dogmatic sometimes. If you enlarge something, if you take a picture, then enlarge it and put it in an empty room, it would be a terribly hypnotic presence. I don't like that hypnotic presence. I like images to leave a kind of detachment in the person who looks at them."

In one sense this commentary appears to be contradictory — particularly since Clemente loads his prints with symbolic references. Yet the self-canceling nature of all of these mysterious images lends credence to his statement. There is a detached

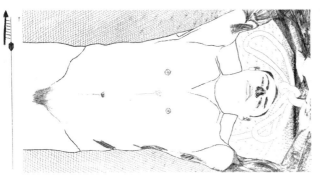

Francesco Clemente, *This Side Up* (photo by Colin C. McRae; courtesy Crown Point Press, San Francisco and New York).

CURRENTS

quality to these prints. Clemente presents these visual allusions to us in a half-serious, half-comic tone. Truth, he seems to say, is no longer a verifiable certainty — believe what you will. Ultimately, Clemente's belief is pragmatism — he uses whatever works.

Paradoxically this artist sees himself as part mystic visionary and part twentieth-century artist burdened by the realities of artistic success and the struggle of making his way in a world devoid of traditional belief and value systems.

4 Form, Environment, Process

The Context for Contemporary Sculpture

The development of a contemporary sculptural aesthetic in America in many ways parallels the path painting followed. In the twenties and thirties both painting and sculpture went to one of two extremes. Some artists were greatly influenced by classical European styles, and produced dry, academic work. Others turned their backs on this "foreign" circle and cultivated an overtly naive, provincial look. But, whereas painting had a healthy body of avant-garde explorers such as Arthur Dove, Georgia O'Keeffe, and Stuart Davis to serve as role models, sculpture had few, if any. Until the arrival of bronze caster Jacques Lipchitz from Paris in the forties, three-dimensional work in metal consisted of uninspired portraiture and dreary park statuary. Gatson Lachaise, another Parisian immigrant, also made significant contributions to sculpture, but because of his favorite subject matter—large, erotic, female figures—he was politely ignored by a puritanical, insecure art audience.

Elie Nadelman was a notable exception; his sculptural work parallels many of the early avant-garde painters mentioned in terms of original Modernist experimentation. Aspects of American folk sculpture —such as cigar-store Indians and ships' figureheads—were subtly incorporated into his elegantly carved polychromal figures. But generally, sculptural activity took a less innovative and low-key turn until David Smith's mature, aggressive, large-scale metal pieces triumphantly emerged in the late forties.

Just as modern American painting has as its native hero, Jackson Pollock (who brought a Western "openness" with him to the cosmopolitan East), so contemporary sculpture can point with pride to David Smith—from the heartland of America, Decatur, Illinois.

Smith (as his name might indicate) was descended from a long line of metalworkers. One of his relatives was the first blacksmith to set up shop in the Decatur area when the westward land movement first settled the prairie. Like Pollock, Smith went back East to study at the Art Students League in Manhattan; Jan Matulka, a Czech artist teaching there, was a great influence, introducing him to Cubism and Surrealism.

Although Smith was primarily trained as a painter, between 1930 and 1933 he became more aware of the possibilities of sculpture. The change came about through the process of collage: As Smith described it, his painting developed into three-dimensional forms raised *off* the surface of the canvas. Gradually, the canvas became the base, and the painting became the sculpture.

Brute physicality impressed him. Smith claimed he liked working with industrial metals and techniques because they freed him from the historical constraints of traditional materials. The primitive struggle of bending and shaping large pieces of iron and steel greatly appealed to Smith's rough, independent personality. During his youth he had worked as a riveter at the Studebaker plant in South Bend, Indiana. It was there he developed a strong feeling for the power and raw beauty of heavy industry as well as a thorough knowledge of metal craft. In the early thirties—while his paintings were becoming more three-dimensional—he was impressed with the reproductions he saw of Picasso's Cubist sculpture and Julio Gonzalez's welded iron pieces.

During the forties Smith stated publicly that the aim of his sculptural work was to give people access to contemporary symbols—symbols of an industrial era. Although his mature sculpture was never narrative or illustrative, it never became totally "abstract" or cerebral in feeling: References to human feelings and values were laced throughout his works.

In fact, strong moral outrage led him to produce a series of pieces titled *Medals for Dishonor* between 1937 and 1940. These were bronze plaques depicting the injustices of modern society. Later Smith believed them to be among his weaker works because of their heavy-handed self-righteousness.

There was always a strong conceptual framework to Smith's work but not the type of solipsistic rumination found in "Conceptual art" of the late sixties. His work expressed a strong emotional commitment to the life of the mind and an overriding concern for human values. Towards the end of his all too short life—he died at 39—Smith had produced the *Cubi* series (illustration)—one of the most original and forceful sculptural groups ever seen in America.

In the early sixties this body of work galvanized American sculpture and established a firm footing upon which it could build. These bold works issued a challenge that was to stimulate other artists like Donald Judd, Tony Smith (no relation), Mark di Suvero, and most significantly Robert Morris.

The environmental impact of Smith's Bolton's Landing studio on these artists was significant and revelatory: The fields were covered with his large, cubular, stainless steel structures; the blue sky, hills, and grass seemed to be part of the work. They were dynamic, forceful constructions that ap-

David Smith, *Cubi XXIII* (photo by David Smith; David Smith Papers, on deposit at the Archives of American Art, Smithsonian Institution).

peared to be suspended in a delicate, precipitous balance: poetic symbols of industrial beauty and form. This original artist enthusiastically and optimistically embraced the industrial age. Clearly a modern romantic, he saw the machine as the embodiment and extension of our body and the artworks produced by it as an expression of our civilization's soul.

Smith had created a new syntax of sculptural forms and fabrication processes but, just as importantly, he created a new spatial context in which to view three-dimensional work: The environmental space beyond museum walls. People could now meet this work on new terms, decent terms, face to face and functioning in the outside world.

When David Smith covered the field around his New York State studio with this energetic outpouring of large-scale artwork, he was pointing the way towards a sculptural renaissance of great significance. This work was directed towards the establishment of a new dialogue between the viewer and the work of art, and exchange that demanded more from the observer but also promised more. In this way he directed our attention to the sculptural *process* not merely the *product*.

Robert Morris is perhaps the single most important sculptor to emerge immediately after the death of Smith. The refined precisionist forms he executed in the midsixties manifested the spirit of Smith's work without directly imitating his "look" which was largely informed by Abstract Expressionism. Metaphorically, Morris drew on the *Cubi* series for direction: He detached and isolated one of the many-faceted cubular forms and presented single images that confront us on new spatial terms. All signs of struggle with the materials—used to such great effect by Smith—were missing in Morris' early works. These constructions are distinguished by their architectural overtones. Most importantly, like architecture, they greatly affect the way our bodies respond to them. Even the materials used are related to architectural fabrication: aluminum, fiber glass, plywood, and sheet metal. Through these simple and industrial materials, Morris, like Smith, created

an art of spatial experience rather than sculptural illusion.

Jackie Winsor is a sculptor who, in a sense, synthesized the antithetical aspects of Smith and Morris (organic structure and refined geometricity) into one body of work. Her sculpture is both geometric and organic, expressive and controlled, complex and simple. It functions within broad spatial contexts as well as a self-contained "Object." Aspects of primitive art mingle with sophisticated futuristic images in the work of this unique artist.

Lucas Samaras' art functions outside of any clearly identified mainstream movements. Yet his work has curiously anticipated many of the concerns of the past two decades—body art, individualism, narrative art, and pattern painting. Because Samaras was born in Greece and came to America as a young boy, his sculptural assemblages reflect a Byzantine sensibility which results in his use of exotic materials and sensuous surfaces. At heart Samaras is a mixed-media artist who has worked with the widest range of sculptural materials imaginable.

Jonathan Borofsky's sculptural work also defies classification. Influenced by conceptual work of the late sixties, Borofsky has managed to develop an aesthetic that, while maintaining a concept-oriented base, reaches out toward populist concerns. While he challenges popular standards of art-making, he creates accessible art that engages the public attention.

Tony Cragg is representative of the English school of sculpture that is largely based on the use and transformation of found objects. No doubt England's economic woes of the past fifteen years have fostered this direction; by making use of the discarded material of an ailing industrial society, Cragg's work makes a statement about the relationship of consumerism in the West to developments in late twentieth-century aesthetics.

Robert Morris

Robert Morris is a sculptor who has been working with a variety of sculptural materials and new spatial forms since the early sixties. Together with Carle Andre, Don Judd, and Sol LeWitt, he was one of the founding fathers of a movement that came to be known as *Minimalism*. Critics also applied the terms *Cool Art*, *Post-Geometric Structures*, *ABC Art* and *Primary Structures* to this reductionist, carefully thought-out artform.

Unlike the Abstract Expressionists who expressed an "I can't-explain-it-I-just-do-it" attitude, many of these early Primary Structure artists were verbally articulate and university trained (often in art history and philosophy). In fact, Don Judd, a leading artist of this movement, was a professional writer and critic before he started making art. Consequently many of these individuals were well-versed in sophisticated concepts and aesthetic philosophy, and were keenly aware of Modernist art history and its precedents. Generally, critical theory usually follows artistic developments — "explaining" them, as it were. But with the Minimalists the reverse held true: A priori intellectual considerations and speculation in large part gave birth to this movement.

In 1965, even though no major exhibit of this radical, geometric art had been staged, lengthy articles establishing the "theory" of Minimalism appeared in major journals such as *Artforum*, *Arts*, and *Art in America*. At this time, only a few group shows and one or two solo exhibits of this new work had been held.

Today the radical edge of these pieces has been dulled by the passage of time, broad exposure, and museum acceptance, but their initial effect was great controversy. Recent criticism accuses the movement of being overly dependent on extravisual concepts and solipsistic discourse. Although this may hold true for some work, the best of it, particularly Morris', admirably stands the test of time.

One of the reasons Robert Morris' work continues to generate excitement is its ability to constantly pose relevant questions and seek new answers to sculptural problems. Besides his Primary Structures of the early sixties he had distinguished himself in other arenas of artistic investigation: sculptural performance, "process art" (a mode of working seemingly at odds with Minimalist concerns), site-specific environmental sculpture, and, recently, a return to sculptural "objects" that reconcile theoretical concepts with traditional forms. This ability to continually challenge himself and question the spatial syntax of modern sculpture provides the basis for an engaging and lively body of work. body of work.

Like much of the art of the seventies, Morris' work during the past ten years has been intimately involved with sculptural space, place, and a concern for the ongoing process rather than traditional concerns of three-dimensional form. Even the early "Minimal" arrangements of cubular forms must be considered within the context of what they do to and with the space in which they are exhibited. If they are plucked from this spatial matrix and looked at in isolation, they yield little pleasure and stimulation. Traditional criticism made little headway with the utterly straightforward, literal forms used in Morris' early work. Direct, descriptive analysis proves disappointing; simple cubes, rectangular slabs, and elemental wedges yield little material for waxing poetic in a romantic vein. Some viewers concluded that since these works utilized a restrained visual language, they expressed minimal aesthetic meaning. Actually, these works reveal fine subtleties that people often miss because of their expectation of what a sculptural work should be. Morris' early art raises questions about the role *visual* perception plays in our evaluation of contemporary works of art. Many modes of perception and thought are activated by these works and the context in which we view them. Morris wants the viewer to expand his or her role to include experiences of a participatory nature. He is interested in how we *act* around these works and how we relate to a new three-dimensional situation. Rather than only ask us to look at them, he invites us to experience them in a variety of complex ways: through movement, walking around and between them; through touching their smooth industrial surfaces; through responding to them intellectually; through measuring them literally and figuratively; and perhaps most importantly through considering them within the matrix of the space they create and inhabit. Seen in this broad context they become three-dimensional mirrors reflecting our movements, feelings, and physical presence. Their "coolness" acts as a catalyst urging us to move beyond passive looking into active participation.

By means of these simple, yet spatially complex, works Morris removes the veil of illusion that usually surrounds traditional sculpture. He wants us to see how they are made — no tricks, just simple, straightforward construction techniques. They do possess mystery; but the secret lies not so much in how they are made but in what they might mean and what they become in our presence. Our minds turn them over and examine not only the deceptively simple forms but the entire space they become immersed in and activate.

Robert Morris is also a manipulator of time. One of the first sculptural pieces he ever made was a 9-inch walnut cube called *Box with the Sound of Its Own Making*, 1961 (illustration). During the construction of this early piece, he recorded the sounds of the work process: measuring with a pencil and rule, sawing the wood, and nailing the sides together. Later, when the work was exhibited, a small tape recorder was placed inside the cube and for a duration of three hours would play back the sound of the entire process. Morris seems to be imprisoned inside the box like a contemporary Houdini. But instead of attempting to escape like the past master of showmanship and illusion, this contemporarty artist seems bent on his own containment. Although obviously finished, the box introduces information that contradicts the visual evidence at hand. A physically dematerialized Morris — made manifest through recorded sound waves — continues to measure, hammer, and saw, in contradiction to the presence of the completed form. Through the use of a box and tape recorder the artist expresses an interest in the redefinition of sculptural space as an interior experience. The sealed but vocal cube — manifesting an inner and outer volume — echoes our psychological and physical experience of the piece. Also, by sharing with us the creative activity (the sounds of its fabrication),

Robert Morris, *Box with the Sound of Its Own Making* (copyright Robert Morris/VAGA, New York, 1989).

Morris puts us in touch with a primary and sometimes overlooked aspect of art: It is *made* by human hand and does not exist in nature outside of the context of human thought and activity.

During the next three years other boxes appeared that contained similar conceptual and perceptual information. The *I Box*, completed in 1962, is a shallow, 12-inch by 18-inch wall construction that features a large letter *I* fixed to the surface and is hinged on the left side. A small drawer knob is placed to the right. When we open this door, a frontal, full-length picture of the artist wearing only a silly grin appears. Morris juxtaposes a literal use of language with a photographic representation of himself. The letter *I* can also be read like a Chinese ideogram, as representing the configuration of an upright human figure. The alphabetic door leads to an illusionistic representation of the artist in a primal state of being: no clothes, no hat, no shoes, no umbrella.

Perhaps the most enigmatic and humorous of the early boxes is an untitled work, measuring 13 by 7½ by 13½ inches, that depicts a padlocked box with the inscription, "Leave key on hook inside cabinet." The function of the cabinet seems to be totally defeated by the apparent contradiction of the statement. Upon more careful thought, we realize that the inscription does not specifically refer to the key needed to unlock the padlocked door. As in carefully devised works of art, a *key* may be necessary to unlock a puzzle. So much of Western art has relied on visual illusion to fool and deceive us; but Morris makes use of verbal allusion to enlarge our vision and make us "see" the complex nature of contemporary art. This artwork tantalizes us and becomes an exquisite object of desire: The desire to know the truth about an impenetrable mystery.

In 1965, at the Green Gallery in Manhattan, Morris first exhibited a group of sculptures that employed mirrors to remarkable effect. It was the start of an obsession with reflective sculptural surfaces that continued for eighteen years and produced scores of varied artworks. Along with other sculptures in the Green Gallery exhibit was a grouping of four cubes measuring 36 inches by 36 inches, standing exactly 6 feet apart. Morris had constructed other simple forms before; what set this work apart and signified an interesting shift from other primary sculpture was the mirrored surfaces of these cubes. The artist discovered a mirrored form of plexiglass that could almost seamlessly cover a plywood cube, dematerializing its geometric solidity and presence in one swift stroke. These works expressed an exquisite paradox: Physically they were still cubular forms but illusionistically they had dematerialized to reflect the environment in which they were placed. Morris is extremely fond of this paradox; much of his work directly relates to themes of contradiction and reconciliation.

He loves the parable of Japanese swordmaking prior to the eleventh century. Until this time an ideal sabre was a contradiction in terms. If hard metal—the kind that would hold a sharp edge—was used, the sword was relatively brittle and might fracture in battle. Tempered flexibility, on the other hand, was achieved at the expense of a sturdy, sharp edge. Until Japanese metal workers figured out a way to bond two thin outer layers of hard steel to a flexible, soft inner core, the ideal sword was unattainable.

Inspired by the struggle of early Japanese metallurgists, Morris bonded thin mirrors onto solid surfaces to achieve similar results in reconciling the irreconcilable. *Untitled*, 1965 (illustration), is an excellent example of his early mirrored sculpture. To try to describe the way these sculptures look is an act of folly; one ends up describing the room or landscape in which they are situated. Placed on a wooden floor, as they are in this illustration, they almost disappear. Our perception of the cubes is elusive; they are not entirely present and not entirely gone. These sculptures are active rather than passive; they invite kinesthetic play; walking around and between them sets them and us in motion. Sculpturally they are fabricated out of one of the most insubstantial surfaces we know: the mirror.

In a preface to a retrospective exhibit titled *Mirror works: 1961–78*, Morris wrote about his infatuation with the phenomena of mirrored surfaces:

In the beginning I was ambivalent about its fraudulent space, its blatant illusionism. Later its very suspiciousness seemed a virtue. I came to like its hovering connotation of abject narcissism, its reek of the cheaply decorative, its status as a kind of disco-degenerate category. Mirrors have had a curious history. The Egyptians, a culture no less vain than it was early, slavishly polished stones until they reflected their maker's image. The ingenious Greeks, as practical as they were sublime, found in the mirror a weapon of startling power. It is said that a battalion of soldiers, each equipped with a 5-foot polished bronze mirror, was once assembled in the hills surrounding the threatened harbor of Syracuse. By simultaneously concentrating the sun's reflected rays on one ship after another, they burned the entire invading fleet at anchor in the harbor. Archimedes is credited with having had the idea, as well as having presided at the event—an early instance of genius in the service of the military. . . . I once attempted to describe one of my mirror works to a blind friend and caught myself up short. "But
what can a mirror mean to you?" I asked. "Well, I know what mirrors are supposed to be," she replied.[1]

Obviously Morris is interested in both the illusionistic and philosophical aspects of reflective surfaces: Mirrors play with the mind as well as the eye. Throughout his career specific sculptural attitudes continually resurface (like his two-decade-old infatuation with mirrors) and are incorporated into his sculpture: Early concepts about process, dichotomy, and paradoxical space show up even in his most recent work.

As might be expected Morris had an interesting, varied educational and professional background; no doubt this has conditioned and shaped his artmaking activity. He is a Midwesterner by birth and grew up in Kansas City. Morris enrolled at the University of Kansas as an engineering major but soon transferred to the Kansas City Art Institute. After two years of study there he moved to San Francisco to attend classes at the California School of Fine Arts.

Robert Morris, *Untitled* (photo by Rudolph Burckhardt; copyright Robert Morris/ VAGA, New York, 1989).

Two final years of study at Reed College in Oregon completed his undergraduate education. After Reed, Morris returned to San Francisco where he painted and worked with friends on improvisational theatre pieces and film. In 1961, he moved to New York City and entered a graduate art history program at Hunter College, earning a Masters Degree with his thesis on Brancusi, a Modernist European sculptor known for his disarmingly simple and elegant forms. Concurrent with his early sculptural activity Morris worked with the modern dancer Yvonne Rainer on a series of sculptural events at the Kunstakademie in Düsseldorf, Germany.

During the period 1963 to 1965, Morris composed five dance pieces that expressed his enlarged conception of sculptural activity, including: *Site*, in which he wore a rubber mask of his own likeness and moved various props around until the dancer Carolee Schneeman was revealed in the pose of Manet's "Olympia"; *21.3*, a piece that incorporated a reading of a classic art-history text, Panofsky's *Studies in Iconology*; and *Waterman Switch*, his most notorious performance event. Morris and Yvonne Rainer appear, nude but closely pressed together, face to face — partly concealing each other — and slowly move back and forth along a narrow wooden track. Like the story of the Japanese sword that reconciled two opposites Morris simultaneously represented both concealment and nakedness in this piece.

During the summer of 1967, while on a trip to Aspen, Colorado, Morris constructed a series of cut felt works that signaled the beginning of a major shift in artistic sensibility for this artist. Large sheets of industrial-weight felt were cut into strips which hung down and touched the floor. Soon Morris was creating large-scale pieces from felt and scraps of copper, aluminum, zinc, rubber, steel and large mirrors.

Generally the configurations of these environments seemed powerfully random in nature. Essentially, the presence of the material itself was stressed rather than a precisely ordered structure. Curators and contemporary critics labeled this activity as *Antiform* art. The term — if taken literally — is misleading and brings to mind Arakawa's humorous discourse on "gray" and "nongray." No work of art can be antiform per se; what is meant is that these works seem to go against the predetermined nature of Morris' earlier structures. But form is arrived at by a different route. These pieces are arranged according to a complex process of random distribution rather than predetermined selection and ordering systems. There are interesting parallels in science; one of them is the concept that views disorder on a grand scale as the equivalent of order on a small scale.

Untitled, 1970 (illustration), is an interesting transitional piece for Morris that bridges the gap between ultra-simple forms of the past and enormously complex random orderings of his later process works. On a large 6- by 25-foot section of industrial felt Morris made five equally spaced horizontal cuts that stopped short of cutting through the whole piece. When this sculpture is hung on the wall, looped felt "necklaces" drape down, touch the floor, and distort the rectangular shape of the felt; Morris even made use of gravity as part of his process.

At about the same time he did the felt pieces, he executed various "scatter" installations. Large rooms were filled with copious amounts of industrial waste materials. Morris would scour the light manufacturing districts of Manhattan collecting these substances for his enviromental sculpture. An overall image or "field" composition is created by this method of large-scale random distribution. In some of these process pieces Morris carefully placed perpendicular mirrors at various locations. Because the distribution of the materials is so regular, the mirrors create a strange illusion of slightly disjunctive space. As with his mirrored cubes, movement attenuates this effect.

In the early seventies — for his retrospective at the Whitney Museum — Morris built one of the largest indoor sculptural works ever seen in America (illustration). It extended for more than 1000 feet and took up the better part of an entire museum floor. Raw materials were brought to the Whitney and construction took place while the exhibit was open. Morris felt strongly at this time about sharing part of the artistic process with viewers, rather than presenting them with a finished product. In a direct way his concern with "process" at this time went back ten years to his early *Box with the Sound of Its Own Making*. The artist's original idea was to make the Whitney piece out of recyclable materials: standard lengths of steel pipe, long sections of timber, and rough-quarried stones that might be resold and incorporated into new buildings. After consulting with architects and structural engineers, he determined that the weight of the stone blocks could not be supported by the Whitney's floor. Instead, partially hollow, performed concrete sections were used.

In full view of museum spectators Morris and his crew aligned the hefty timbers into a three-railed track and placed industrial-sized steel pipes on the timbers to function as rollers. The heavy concrete blocks were then placed on the supporting pipes. When the track was full, some of the timbers and rollers were carefully moved causing the concrete forms to fall through in a half-controlled, half-random manner. This "process" produced a powerful, angular sculpture of dwarfing scale that measured 192 feet wide and 1080 feet long. Morris created a piece that successfully competed—in terms of physical scale and power—with a Boeing 747 or a small ocean-going freighter.

Along with earthwork artists Robert Smithson and Michael Heizer (discussed in Chapter 7), Morris reaches beyond the traditional art object to produce sculptural experiences that rival in scale and power those of industry and government. Usually, the size of Morris' large-scale projects casts them into the domain of encompassing environments that tangibly interact with our minds, bodies, and collective societal experiences.

Labyrinth (illustration), constructed in

Robert Morris, *Untitled* (Milwaukee Art Museum, Gift of Friends of Art; photo by P. Richard Eells).

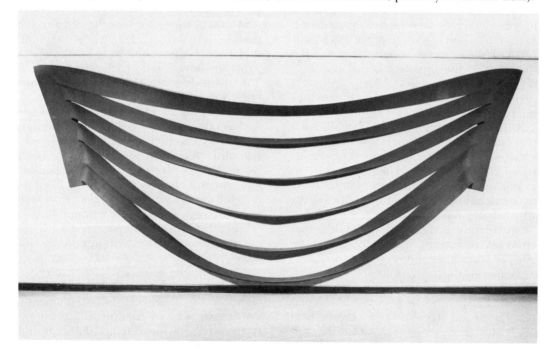

Robert Morris, large indoor sculpture (photo by Peter Moore).

Robert Morris, *Labyrinth* (copyright Robert Morris/VAGA, New York, 1989).

March, 1974, at the University of Pennsylvania's Institute of Contemporary Art, is another such work. Compressed into a circular format, 8 feet high and only 30 feet in diameter, is a 50-foot walkway. The walls are too tall to see over and the corridor—only 18 inches wide—brushes against our shoulders and creates interesting psychosocial confrontations when we meet another equally lost soul wending his or her way through the sculpture. Some would shyly try to squeeze by, others would turn around and head back to avoid the necessity of physical intimacy. It is a true maze; there are no dead ends, only circular switchbacks and false paths that would temporarily foil the novice in single-minded pursuit of the target's central core. Once in the center, the process would be reversed as you attempted to make your way out. Although it is possible to view the interior plan of *Labyrinth* from a second-floor observation area, when the participants enter it physically their visual memory is disoriented by the change of location and the immediacy of endlessly curving Morris-gray walls. The constant turns and confounding loops wipe out any attempt to memorize the "true path" from above and follow this mental map to the heart of the maze. By disrupting our perceptual and physical orientations, the artist creates new pathways of experience that employ nonpassive modes of interaction.

Mirrors continue to play an important role in Morris' work of the late seventies. Although his art has ranged from objects to performance, and process to environments, these recent mirrored works bring him back to sculptural "objects" again. But Morris has managed to express many of the concerns of these past forms in his new work. Despite the shifts in format and style there is a thread of continuity: All of his work is grounded in his profound concern for paradox, contradiction, and experiential thought. Mirrors reflect these concerns in many ways—historically, conceptually, as well as illusionistically.

Untitled, 1978 (illustration), consists of a large, complex grouping of violently curved, plexiglass mirrors and lengths of copper architectural moldings. This sculpture, despite its playful funhouse appearance, frighteningly affects our perception and psychological orientations. Baroque reflections of architectural features combine with hallucinatory images of our own bodies; fantastical visions are generated that shift and evolve as we walk within its reflective arena. From a safe distance it benignly sits patiently waiting, a complex collection of copper columns, isolated rails, lengths of molding, and odd sections of curved mirrors. As soon as you step inside its sphere of visual influence, however, everything changes. It is activated by our presence and, in turn, succeeds in modifying our actions. Like Morris' *Labyrinth*, it distorts our senses of space and vision. Ordinary movement is transformed into performance; we become dancers rather than gallery goers. Morris has thus incorporated many of his past concerns in these late mirror works: viewer participation, performance, environments, "process," and illusionistically paradoxical elements.

Untitled (1978 for R. K.) (illustration) is similar to the piece previously mentioned. This one, however, abandons its frontal pose and functions entirely in the round. Four simple, elementary concrete shapes—two cubes and two cylinders—sit on slabs of an elemental metal: lead; diagonally painted black lines connect the four inner corners of the lead base. Visions of ancient alchemical diagrams are evoked by the interaction of these visual and metallic elements. Three large convoluted mirrors surround and contain these primary forms. Once you have stepped inside the "mystic circle," no matter which way you turn, a complex hallucinatory vision confronts you. Sections of curved cylinders and corners of cubes combine with your own reflected image to assault the senses. Considering Morris' long-standing interest in reconciling the seemingly irreconcilable (the sword again), this and other pieces like it successfully combined two-dimensional illusion with three-dimensional form. As a result of this paradoxical interface, Morris creates a rich and complex psychological–visual *space* that exists in our minds as well as in the room.

Robert Morris, *Untitled* (photo by Eric Pollitzer; copyright Robert Morris/VAGA, New York, 1989).

Robert Morris, *Untitled (1978 for R.K.)* (photo by Eric Pollitzer; copyright Robert Morris/VAGA, New York, 1989).

In conclusion, all of Morris' works—sculptural process, performance, environmental—deal with opposition and transformation. Through them vision is transformed into insight, perception of form into concepts of space and the viewer's role from observer to active participant. Ultimately, Morris effectively combines traditional elements of Modernist sculpture with post-conceptual artistic concerns to produce works of a hybrid nature that establish new modes of thought and experience.

Jackie Winsor

Robert Morris' work in the early sixties exemplifies an aesthetic stance that sought to establish a new order of rationalist thought in art and effect a bold departure from artistic methodology of the past. For the most part Minimalists avoided the rough-and-tumble expressionism of the early "New York School" and crafted sleek, machine made, nonorganic forms using a variety of industrial fabrication methods and materials.

Jackie Winsor is a sculptor who responded to the primal, haunting beauty of simple geometric forms—spheres, cubes, rectangles—but transformed these shapes into highly autobiographic terms by utilizing organic materials and handmade processes. Although some aspects of Primary Structures remained, her sculpture took on a ritualistic look because of choice of materials and construction methods. In fact, Winsor's sculpture was greatly reminiscent of tribal art in feeling and mood.

Many emerging artists in the seventies reacted against the impersonal, machine-made look of formalized art and sought the interior world of psychological feelings and personal thoughts. "Basically, you make things out of the structure of who *you* are," Winsor remarked recently. This, then, is the basic difference between Morris and Winsor. Morris begins with a theory and concept from which the art is created. Winsor delves into the internal structure of who she is to produce compelling, personal works that are autobiographical in nature. Minimalist structures commented on corporate scale, technological tools, and environmental context; Winsor's forms hermetically refer inward; we relate to them on the basis of intimate, interior feelings and thoughts. One of the dangers of this kind of personal art making is that it can become too isolated and remote, so inward and removed that it speaks to no one but the artist. In this case it remains merely as therapy for the maker. But Winsor's work—because of its lucid connection between forms, materials, and process—avoids this pitfall and engages viewers in a lively dialogue with their own thoughts and perceptions.

Rather than antithetically reject all of Minimal arts' beliefs and convictions, Winsor was able to redefine and extend some of its concerns well into the late seventies. She is a mediator between Late-Modernist theory and the developing aesthetic that places the artist's personal life and autobiographical presence directly into the fabric and substance of the artwork itself. For instance, sometimes when she talks about her work, Winsor can sound like a purely Formalist sculptor from the sixties. Sculptural qualities that interest her include static forms, solitary images, seriality, and a pronounced sense of peacefulness and repose. But the sculptural works that she turns out represent definite departures from Primary Structure art produced in the mid-sixties. Two aspects of her work account for this uniqueness: Weathered, organic materials are employed along with a laborious, repetitive, handmade fabrication process. At various times Winsor has worked with sculptural materials such as tree limbs, hemp, rough cement, wooden lathing, nails, and copper wire. There is a simple elemental quality to these raw materials. Because of her time-consuming methods and thoughtful approach (sometimes she will take days to decide whether to move something a sixteenth of an inch over), Winsor completes only a few pieces each year. She generally makes no drawings, prints, or scaled-down models in preparation for a large work, preferring to start the piece slowly and make careful adjustments on it as she goes along. For this artist the meditative act of *making* is of primary importance. Sometimes halfway through she will decide the look of a work is not quite right and undo countless hours of work to go back and make major revisions. The deliberate, autobiographical unfolding of the piece is important to her. Despite the repetitious deliberation of her processes, there is no sense of mechanical execution in her work. It unfolds in a natural, organic way and is subject to revision and adjustments all along the way. When her pieces are finished, they have the look of an organic form that grew according to specific laws of art and nature. Winsor brilliantly does just that: She combines organic, personal materials and forms with geometric configurations and mechanical shapes. Winsor's preoccupation with the act of making is an essential part of her work, along with an intimate knowledge and sensitivity to her materials. She literally "lives" with each piece, in a state of continuous meditative reflection, while forming it. Winsor is in no hurry to speed the work processes up; in a sense,

like her, they are alive and must be allowed to mature and develop at their own rate.

Personal events are sometimes embedded and contained in her work, but not in an obvious way. When Jackie Winsor was nine years old, during the building of the family's house, her father gave her an enormous bag of nails, told her to nail down certain boards, and left for work. When he came home that evening, the planks were fastened all right; in fact, they were literally covered with nails. In her youthful zeal, she had used about twelve pounds of nails where one pound would have been sufficient. Winsor's father was furious about the wasted nails. She recalls that, "they made such a fuss about it that it left quite an impression on me."

Twenty years later, in 1970, that event still affected her thinking and led to the construction of her first wood sculpture *Nail Piece* (illustration). Winsor started out with nine 7-foot-long, 1- by 8-inch boards which weighed about fifty pounds. In a cathartic

Jackie Winsor, *Nail Piece* (courtesy of Mr. and Mrs. Charles H. Carpenter, Jr.; photo courtesy of Paula Cooper Gallery, New York).

way, Winsor evoked that traumatic event of childhood when she joined the first two boards with hundreds of nails and hammered the remaining seven together, one on top of the other, until all nine were fused into one piece. According to her calculations, she used fifty pounds of nails in the process. Thus the weight of the nails and the weight of the wood were brought into balance.

There are secret and hidden elements to this piece. Only the top plank hints at the enormous number of nails used and energy expended; the work becomes an artifact of an esoteric and mysterious event. In this way Winsor's work relates to tribal art; both make use of highly visible forms, the thematic meanings of which remain partially closed to the uninformed and uninitiated. Powerful, physical forms cloak hidden secrets that surge beneath the surface, unseen but potent.

Winsor was performing a private act when she worked on *Nail Piece*; old "taboos" were broken; authoritarian censure and ideas of "common sense" were flaunted by her excessive use of nails. Winsor said she was "interested in the feeling of concealed energy. I like the fact that each layer has tons of nails in it and that can't be seen."

Sexual metaphors—of a subtle nature-—reoccur throughout her work. The coupling of two dissimilar materials and interpenetration of forms is common to many of her pieces. In *Nail Piece* the penetration of nails into the yielding wood carries with it just such connotations. Although Winsor is not personally involved in feminist activism, her work, subconsciously perhaps, embodies certain attributes that the women's movement considers feminist in nature. Some of these elements include repetitious domestic acts (quilting, sewing), soft rounded forms and central imagery (suggesting the female body and vaginal connotations), and the use of materials that might be found in any household (string, wire, pins). However, it would be unfair and misleading to conveniently categorize Winsor as a witting or unwitting practitioner of "Women's Art." In its sensuality and sexual duality, her work relates equally to both male and female experiences. These binary sexual aspects are clearly illustrated in Winsor's only sculptural performance piece, titled *Up and/or Down*, presented at 112 Greene Street, June 29, 1971. For the first part, a quarter of a ton of 4-inch diameter industrial rope was laboriously hauled from one level to another through a hole in the loft floor. A "long, lean male," was positioned on top; on the bottom deck was a "soft, rounded female." During the first half of the twenty-minute performance, the woman fed the rope to the man. Later the process was reversed and the rope was lowered onto the body of the curled up female, completely covering her. The sexual implications of this piece were unmistakable.

Winsor's personal experiences in college and her beginning years in New York made her well aware of the difficulties women in the arts faced. But her family background provided strong role models to aid her. She recalls that during her childhood, acquiring an education meant studying to be a nurse, or at best a school teacher. "Women had broods," she observed, "there was nothing for women professionally." Luckily, Winsor's mother was an active, energetic individual who provided a strong example for her young daughter. The house that the Winsor family built together may have been planned by her father, but it was largely built by her mother's efforts. While her husband was at work, Jackie's mother did carpentry, plumbing, and a variety of tasks traditionally relegated to men.

When Winsor is asked why a woman would be attracted to hard labor and brute physicality usually associated with masculine endeavors, she is likely to become annoyed:

Look, what I got when I was very young was that women worked very hard. My mother is very hardworking and she is as strong as my father. The idea that men are stronger than women is a matter of what your culturation is about. As far as I'm concerned, most people's capacities are the same. I've had women assistants who came in being very limp-wristed. They came in with culturation that said women weren't physically as strong as men. They were not willing to use their hands or their bodies. But once they began working with me, they soon were as strong as I am. It has to do with a willingness to use one's abilities at full capacity. Actually, so-called masculine pursuits aren't being done very much any more — bricklaying, for example. What it's supposed to come down to is that only males use their muscles, and that's absurd, because women can use their muscles just as competently. Of course, if a woman wants to put stopping blocks in front of her, there are stopping blocks everywhere![2]

Many of Winsor's sculptures evoke the presence of the outdoors. Perhaps they reflect in subtle ways the artist's childhood in Newfoundland, Canada. Winsor was born in this wind-swept, austere, bleakly beautiful province in 1941 to a family of ship captains, constables, and farmers who had settled in this barren country over 300 years earlier. Life in Newfoundland had its own pace and its own time structure. Quite literally, Winsor observed, when her family left Canada and emigrated to the United States, they "came out of a different century."

In 1951, her father, a factory supervisor, secured a position in Boston and moved the entire family there in search of greater opportunities for himself and his three daughters. Even though Boston probably possesses more nineteenth-century charm than any other city in America, Winsor was shocked by the contrast with her native land. Until she was eighteen years old, she spent every summer in Newfoundland staying with various aunts.

By the time she entered high school, the initial awe of America subsided; but some aspects of her studies proved difficult. Art was one subject with which she had no difficulty, however. During her junior year, with the encouragement and help of two teachers, she attended Saturday morning art classes at the Massachusetts College of Art and the Boston Museum Art School. It took Winsor an extra year to graduate from high school, and during this final year she worked doubly hard in the hopes of being accepted by a decent art school. No one in her family had ever attended college before, but by saving money earned in various jobs she was able to enroll at the Massachusetts College of Art in 1961. Win-

sor thrived in this new environment; during her junior year of college she won a scholarship to the Yale-Norfolk Summer School and, "for the first time I met artists rather than art teachers. They argued a lot about art."

During this summer program she was also introduced to nature photography which sharpened her abilities to view the world in a new way. At the time Winsor believed that this process of observation seemed to be a way of "making form important."

Following her graduation from Massachusetts College of Art, Winsor chose to pursue graduate study at Douglass College, Rutgers University, largely because of its proximity to the New York art world. Surprisingly, despite many years of formal study at various art schools, she had never studied sculpture to any extent. Her paintings during graduate school were decidedly figurative; muscular, convex shapes intrigued her and were incorporated into the canvases. The story of her transition from two-dimensional work to sculpture is told in a magazine interview with the critic John Gruen:

I had paintings and I had photographs, and what finally emerged was form. I spent time discovering two-dimensional form — and internal space. Well, from these forms and the drawings I had made of them, it occured to me that I could do a shape three-dimensionally. And that was the break which led into sculpture. It was like going over a bridge and meeting another land mass. Of course, there was a lot of "I don't know" around all that. But it was very exciting. It was like opening a door — and I went right through it![3]

By the time Winsor graduated, her momentum carried her to New York City. Like so many other artists, she located a studio in the Canal Street–Bowery area with a view of the Manhattan Bridge. Despite extensive "urban renewal" in the neighborhood — which caused many small manufacturing concerns to look elsewhere for space — this area is still very much a "rope district." All kinds of natural fiber rope are manufactured here, from packaging twine to marine-grade braided line used to tie up ocean-going ships. Like many artists in the downtown area Winsor made extensive use of industrial materials found here.

Double Bound Circle, 1970 (illustration), is an early example of the way Winsor incorporated these indigenous materials into her work process. Wreathlike double coils are wrapped and formed by hand from lengths of used rope purchased nearby. There is a heavy, ominous quality to *Double Bound Circle* that arises from the impressive size and weight of the material, as well as the dark tar that stains the natural brown fibers.

Winsor disavows any thematic connection of the rope to her seafaring ancestors; instead she claims to be interested in its limp, amorphous form and "unruly" qualities. Her task as a sculptor is to form and energize this material, giving shape and weight to an ambiguous substance. She was particularly delighted when a burly delivery man came to her studio, spied *Double Bound Circle* on the studio floor, and kicked it with all his might. It did not budge. Win-sor felt that the truck driver walked away from the piece with a sense of respect for its sheer size and weight. She consistently talks a lot about this kind of "muscle" in reference to her work — not only the physical force contained in the wrapped and bound forms, but the energy it takes to produce them. Even the closed circular configuration of *Double Bound Circle* alludes to a continuously flowing field of energy. It quietly radiates a feeling of fullness and containment that is mysterious, primal, and even a bit frightening.

Another concept that very much interests Jackie Winsor is the notion of a secret core; a place out of sight and hidden, but very much alive: an inner "seed." She expresses this idea in *30 to 1 Bound Trees* (illustration), which was executed out of doors in a sparse, barren region of Nova Scotia in 1971. At the core of this site-specific sculpture was a live tree. Around this center piece, Winsor individually bound thirty stunted, white birch trees (the severity of this tundralike region accounts for their diminished size). Once the trees were all

Jackie Winsor, *Double Bound Circle* (Purchase with funds from the National Endowment for the Arts and the Member's Guild. Permanent Collection of the High Museum of Art, Atlanta, Ga. Photo courtesy of Paula Cooper Gallery, New York).

Jackie Winsor, *30 to 1 Bound Trees* (photo courtesy of Paula Cooper Gallery, New York).

plexity, detail, and slow deliberation only she can give to the work. *Four Corners*, 1972 (illustration), offers a splendid example of this process. Winsor worked four full days a week for six months constructing this 1500-pound wood and hemp sculpture. To begin with, weathered strands of rope were painstakingly unraveled to obtain thousands of feet of twine, a more linear and primary element. Then two logs 4-feet long were notched, crossed at right angles, and bound together with the twine. Winsor then wrapped and braided each of the log ends, creating four enormous, 27-inch diameter spheres that completely covered the wooden support. They look vaguely African or Oceanic: perhaps unknown ritualistic devices from an undiscovered aboriginal tribe. Placed in an anthropological museum next to blackened wood carvings, feathered constructions, and woven reed baskets, their authenticity would probably never be questioned by the general public.

In a companion piece executed the next year, *Plywood Square*, Winsor abandons the complexities of binding pieces of wood together and focuses on the compulsive act of wrapping per se. She simply and exhaustively wraps and covers a 48-inch-square piece of plywood with twine. Only the descriptive title and the semisquare shape provide clues that refer to this work's buried core.

One of the most primal and primitive pieces in Winsor's whole wrapped-wood series is *Bound Square*, 1972 (illustration). Four 72-inch-long logs are joined at the corners with twine to form a large wooden square. As in the other works, the binding is excessive and forms large balls of twine where the timbers meet. Much of the dynamics of this piece has to do with the way the rough bark and textured twine juxtaposes with a regular geometric shape, which would never occur in nature. The juxtaposition is striking and enigmatic. The twin corners provide a smooth transition from organic material to geometric form; the subtle meditative play between these two elements accounts for much of the piece's effectiveness.

After Winsor completed her "tree" se-

fixed in this manner, the group was trussed with rope to form a curious bundle. There was a primitive and ancient look to them, as if they expressed an archaic agrarian form of storing wood or fodder. Visually, they echoed the austere, wind-swept terrain of this Canadian maritime province. Although the live tree is completely hidden, Winsor feels the piece conveys the importance of this unseen element:

Otherwise how would this structure that's 20 feet high and 5 feet across stand up with the wind blowing over the top of the quarry? It would blow over. They all stabilize each other. As I was making the piece, I got more and more concerned with the fact that the live tree was being nestled inside. . . . I saw the live tree as the pivotal part of that work.[4]

The Protestant work ethic that played such a big role in the growth of America is very much alive and functioning in Jackie Winsor's art. Although she is in a financial position to farm out most of the tedious work, she prefers to do most of the fabrication herself; it is her firm belief that the sculpture's life is dependent on the com-

Jackie Winsor, *Bound Square* (Collection: The Museum of Modern Art, New York. Joseph G. and Grace Mayer Fund in honor of Alfred H. Barr, Jr., and James Thrall Soby).

ries, she became interested in plywood—more specifically, in the concept of plywood—which is a curious blend of natural wood in a synthetic configuration. To a certain extent the artist believes that various substances have a distinct presence—both apparent and hidden—and that her job as an artist is to work with these ma-terials balancing two factors, what they appear to be and what the artist envisions them *becoming*.

When I started working with ropes, I thought I would never work with plywood. I mean it was so trashy! However, after a point, and after I went through all those things about who I am and what I had done, plywood began to merge with the form I had taken, and so I began to use it. When the personality of the plywood became clear to me — how it could come together with me — then I began to sit on the floor and began laminating it and as I did that I could see all those tiny eight-inch layers going up. The point is, my opening up in its direction made me understand the personality of plywood. And that's true of all the materials I use. It's a question of really identifying with them, responding to their realness — to what they bring to you and what you can bring to them. But that involves a lot of time and a lot of work.[5]

One of the first sculptures that Winsor built entirely out of plywood was made by glueing seven 48-inch-square, ¾-inch-thick sheets of this material together and crudely gouging and hacking a shallow trough in it. Another similar plywood block was scored on the surface to produce a rough gridlike pattern. These were transitional pieces — intriguing but not completely successful — that eventually led to the construction of a series of remarkable, complex, and fully three-dimensional sculptures.

Fifty-Fifty (illustration), built in 1975, seems to indicate a departure from Winsor's earlier rough-hewn, overtly organic, "process"-oriented sculpture. But this piece — despite looking substantially different — still embodies many of the characteristics found in her older work: solid forms with corelike interior spaces, simple shapes with complex details, and a penchant for meticulous processes. *Fifty-Fifty* (half of the enclosed cube is air space) expresses with great clarity, economy, and dazzling detail many of Winsor's long-standing concerns.

The work is constructed entirely out of ¾-inch-square pine rods that measure 40 inches in length. The modular sections are stacked, log-cabin style, one on top of the other and precisely nailed together to form an intricate, latticelike cube. From the proper angle one can look completely through to the other side. Moving around it produces flickering effects as various spaces appear to open and close.

All of Winsor's work — from the earliest to the latest — seems to specifically comment on the duality inherent to human-

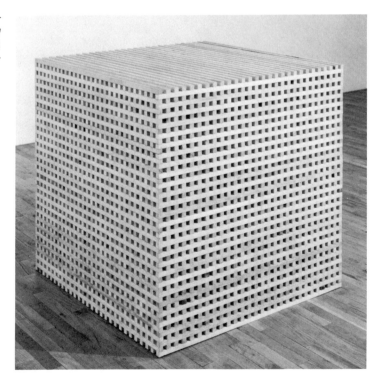

Jackie Winsor, *Fifty-Fifty* (photo by Geoffrey Clements; courtesy of Paula Cooper Gallery, New York).

kind: We are both an organic *part* of nature and makers of forms and structures that stand in opposition to the randomness of the natural world. Critic Robert Pincus-Witten, in reference to this aspect of Winsor's work, quoted from Plato's dialogues: "What I mean . . . is something straight, or round, and the surfaces and solids which a lathe or a carpenter's rule and square, produces from the straight and round . . . things like that, I maintain, are beautiful, not like most things, in a relative sense; they are always beautiful in their very nature."

Sheet Rock Piece, 1976 (illustration), further illustrates the artist's concern with transforming ordinary—sometimes ugly—materials into works of art. Sheet rock is similar to plywood in that it possesses no natural, picturesque, aesthetic qualities. It is common, inexpensive (mass-produced in enormous quantities), and designed to be eventually covered with paint and plaster. Each side of the 3-foot cube has a small,

centrally placed, square opening; through the openings sheet rock is visible twenty-layers deep. By aligning the apertures, it is possible to see through to the other sides; at the center of the cube is a small, empty white space. Winsor presents us with an interesting situation: An object of great opacity and mass is visually penetrable and has at its core an empty "seed" of light. It tantalizes us with an unreachably pure space at the center of its "being": a space that physically keeps us out; a space that can only be reached in our minds.

Ideas, forms, and materials that were developed in the past keep surfacing in Jackie Winsor's recent work. Stylistic continuity is not based upon strict linear development. Old concerns are mated to new materials, producing different visual effects and psychological feelings. Just such a felicitous recombination occurred in *#2 Copper*, 1976 (colorplate). Using vertical rods of 1- by 1-inch clear pine (similar to those used in *Fifty-Fifty*), Winsor wrapped each

Jackie Winsor, *Sheet Rock Piece* (photo courtesy of Paula Cooper Gallery, New York).

rod, starting midsection, with heavy copper wire to form two large balls of gleaming metal. The hazy red-orange color of the copper contrasts beautifully with the natural, pale yellow color of the pine. Although each rod and sphere appear to be independent, they are physically interlaced and bound together. Winsor expresses an underlying unity by this subtle joining; the parts exist as one thing, a "whole" despite the strong contrast of their materials.

Copper, because of its strong association with conduction of electricity, symbolizes man-made, technologically sophisticated power. The duality of these two materials is striking, not only visually but conceptually: Wood is a good electrical insulator, copper an excellent conductor; copper is refined and made by humans, trees are a natural organic product; wood is tough and brittle, copper is soft and malleable. A magnetic energy of great potency seems to be firmly holding these spheres together. The redundant circuit of wires suggests that forms of energy — visual and electromagnetic — are contained and bound within these simple, elemental materials and forms.

In 1978 Winsor introduced a new catalyst to the forms, structures, and materials with which she had been working for over a decade: fire. *Burnt Piece* (illustration) was to a certain extent a sculptural work envisioned as a performance piece. To begin the process Winsor first built a cube made of alternating bands of rough wood lathing and sections of concrete. Winsor is always thinking about the nature, capabilities, and limitations of the materials she uses. What would happen to a half-wood, half-concrete cube — like *Fifty-Fifty* — if it were set on fire? How would the action of the flames affect it visually? Winsor was intrigued with the idea of fire — an elemental phenomenon. But many difficulties needed to be overcome before *Burnt Piece* could be realized.

Jackie Winsor, *Burnt Piece* (private collection; photo courtesy of Paula Cooper Gallery, New York).

Winsor wrote about these problems:

. . . the main unresolved area was what kind of wooden structure would support itself as well as the weight of the cement (1400 lbs.) and also provide for the cement to be one continuous piece after the fire had burned the wood structure out. Somehow the problem didn't lend itself to a very easy solution, and I spent half a year imagining how to do it. I asked a lot of people about mixing fire and cement together, and the main advice I got was: Don't do it! I wanted to, so the image I kept in the back of my mind to guide me was that of a house burning. One thing that interested me about a building was the thickness of walls in relation to the possible size of a fire. I figured out a wood frame that fulfilled all the requirements and was strengthened by the five layers of different-grade mesh that were used to reinforce the six-inch-thick mesh concrete and keep it structurally together during the firing. A second big concern was with stressing the cement with fire. Cement is not like ceramic clay that cures with firing. If there was any flaw in the construction, the nature of the cement would cause the piece to explode during burning. Sand, which is usually mixed with cement to make concrete, could have elements in it that might be unstable during firing, so I replaced it with grog, a prefired clay that looks just like sand but can withstand the stress of a lot of heat. Trapped water could also cause stress, so I cured the concrete for a long time, let it dry for three months, and burned the piece on a dry summer day. I had no idea how easily it would burn, since a lot of wood was inside the concrete structure without air to aid its burning. As I expected, little pieces of concrete popped off. The popping was probably caused by bubbles of air trapped in the concrete. The air expanded with the heat and finally, when pressure built up, shot the fragments fifteen feet from the pieces.[6]

But on the whole the structure maintained its integrity and survived the firey transformation intact. What remained, after the ashes were brushed away and the concrete cooled, was a sculptural piece that combined many aspects of Winsor's past work: the blackened look of the tar-encrusted *Double Circle*, the complex interior connections of *Fifty-Fifty*, and even the performance element of *Up and/or Down*. *Burnt Piece* incorporated many of the concerns of her past work with heat and fire added as new ingredients.

As an artist, Winsor thrives on discovery, challenge, and hard physical labor. Perhaps the most significant aspect of her art is the remarkable transformation simple, elemental materials undergo in her work. Winsor understands, in a deeply felt way, the intrinsic nature of each substance she uses; but her sculpture does much more than merely showcase "interesting" materials. Subtle interactions take place between what the materials seem to be in their raw state and what Winsor envisions them becoming (cooked). The result is works of art that mediate between natural appearances and geometric configurations, wildness and order; all of them possess a quiet austere beauty not unlike the landscape of her native Newfoundland.

Reflecting on the artmaking process and personal background that formed her aric sensibility Winsor recently wrote:

Making art is one of the most pioneering things one can do. When I looked at my past, I see a girl who by the age of 13 had moved ten times, and my origins and family were in 19th-century England, in Canada. . . . I was being pulled from one century into another — from a rural to an urban situation, and the thing I learned most was the notion of change and adaptability.[7]

Lucas Samaras

There are many sculptors today who work with unusual materials and processes, even many individuals who view art as the result of a process. Lucas Samaras uniquely considers *himself* as an on-going artistic process, material, and work of art. "What are you?" he quizzed himself in an autobiographical interview. Mysteriously, he answers: "Inwardly I am an erotic sadness, outwardly I am a homemade process for unravelling meanings."

Samaras is an obstinately original, highly inventive individual who appropriates many nontraditional materials like five-and-dime plastic toys, pins, and photographs, and transforms them into complex constructions that defy classification. He is a lone movement and a singular school of art. Many of his pieces have such forceful and fantastic imagery that they repel (in a curiously seductive way) as much as they attract and fascinate.

Samaras is also a cunning master of camouflage. Nothing in his art is obvious or transparent (he deals in convoluted Byzantine machinations of form and content to explore the hidden psychological secrets of the world). To a certain extent nature is his greatest guide and mentor—not the visible aspects of sky, trees, ocean, but the inner landscape of psyche and self. All of his work expresses the natural opulence and awesome secrets of a vast and compelling interior world.

Thus Samaras' art proceeds from inner revelations and solitary meditaiton; he chooses to work in the open, free spaces between conventional aesthetic camps. Long before the terms *Postmodern, Eclectic,* or *Personal art* were popular art-world catch phrases, Samaras was enthusiastically mixing ancient themes with modern forms and was working in a variety of unusual "styles." In the sixties he obsessively investigated the meaning of the "self," almost to the point of pathological narcissism, when it seemed like the whole art world was fixated on the anonymous art object. His use of body imagery, personal mythmaking, and photographic manipulation predated current interest in body art, videotapes, and biographical narrative.

Samaras remains faithful to his ideas but expresses them in a wide variety of media. He is the maker of primitive pastel portraits, strange drawings, and exotic, patterned tapestries, sewn together from garment industry remnants, but his sculptural constructions and multimedia assemblages are perhaps his most successful work. Despite the profusion of materials and formats, all of his artistic themes can be reduced to a few essential concerns: transformations, the repetitive manipulation of simple everyday objects, and an investigation of inner feelings and thoughts.

Many of these outwardly "eccentric" concerns have had a broad impact on art making in the eighties. Perhaps a large measure of Samaras' isolative individuality stems from the fact that he was born in a small Greek town in Macedonia but immigrated, at the age of eleven, to the New York City area. Samaras knew little English at the time, so artmaking in primary school was the easiest and most rewarding subject for him. Because of the language problem, his "differentness" was further exaggerated by being placed back a few grades to the third grade. Samaras finished high school with an excellent command of English, but his early years in America were spent largely at home working on art and craft projects.

Samaras' family life was not very pleasant. Fights with his father were commonplace; pressure was put upon him, as he recalls, "not to paint, not to draw, or anything . . . not to play with clay, because, I guess, my father was afraid that it would lead me into art, and of course he knew that being an artist was very tough. . . . So consequently, I guess, this pressure against it helped me go into it. Like doing something you shouldn't be doing." The allure of the "forbidden" is quite important and attractive to Samaras and he manages to subtly weave this quality into most of his constructions and artworks.

Because of his ethnic background, Samaras has been immersed in and surrounded by Greek "things" and rituals all his life. Locked in his personal past are childhood memories of Greek Orthodox churches, Byzantine icons, mosaics, and liturgical implements used in mysterious religious rites. His home village of Kastoria, an old settlement overlooking a large inland lake, figures greatly in his personal history. He remembers how different the transplanted garish world of New York was from the town in which he spent his boyhood. During the summer of 1967 Samaras returned to Greece and traveled to Kastoria. Everything seemed strangely foreign yet familiar. The barren road from Salonica was occasionally relieved by crumbling mansions and green meadows sprinkled with red poppies; on the mountainous road overlooking the town a summer sun glistened on Lake Kastoria. Freshly whitewashed houses paralleled narrow winding roads. The marketplace jogged his memory with pungent smells of dried fish, olives, cheese, spices; there was a visual profusion of colors and shapes. Samaras wrote about his trip and what it meant to him in an article for *Artforum* magazine called "A Reconstituted Diary: Greece 1967":

Hundreds of remembered images, feelings, hungers, pleasures and fears of my childhood there commingled into a quick, thick, intolerable identification, Kastoria. Sky and water were the same phosphorescent, pale, dull blue. The city was suspended in between; I had expected to see it by daylight, perhaps because that's how I left it, and that's how it is depicted in postcards. How generous that I saw it at night. How gentle not to get a quick, cold shower of a twenty-year change. It looked like an Antiochian tapestry and also like a Flash Gordon city. And this view I saw through cypress trees, with my family waiting in the car.[8]

In 1955 he entered Rutgers University, as the winner of a competitive art scholarship. Alan Kaprow, Chairperson of the Art Department at that time, recalled that of twenty applicants, Samaras' relatively sophisticated semi-Cubist crucifixion drawings were the most interesting of the lot; many of the other works consisted of amateurish portraits of movie stars and teenage idols. He was an intense, quiet, articulate student who had long hair, wore his sport jacket like a cape, and refused to eat in public with friends. George Segal, who was teaching at Rutgers then, recalls that for years when Samaras visited him, he would silently sit at his dinner table, not touch a bit of food, but would accept with a grand gesture one glass of plain water.

During his first year at Rutgers he became passionately interested in the sport of sharpshooting and joined the college rifle team. Years later target sheets with some of his best scores still adorned the walls of various studios. Curiously, he wrote in *Autobiographical Preserves* that as a child in Greece, "more than anything else penetration was my favorite occupation. To cleave the twigs and leaves from a long skinny branch, transform it into a spear and never stop thrusting it into the air, earth, or objects . . . was physically direct, elegant."

The image of the target could easily be described by these words also. One of his earliest paintings done at Rutgers in 1956 reveals, amid Abstract Expressionist brush strokes, a dark, mysterious targetlike center that vaguely resembles a face, perhaps Samaras' face. Jasper Johns' target paintings emerged about this time and Samaras claims that the Looney Tunes cartoon trademark he saw as a boy in New York movie houses also influenced him. The simple form of the target was iconographic and frontal in ways that directly relate to Byzantine art; its meaning was directed inward. As far as Samaras was concerned, hitting the target clearly meant reaching your inner "self." His preoccupation with biography, personal iconography, and self-revelation can be seen in the early portraits, poems, stories, and a curious construction that took the form of a closed box and mirror. When a lid was opened, the mirror revealed the viewer's face.

Despite a protest from the administration over some handmade books in his senior show that contained "obscene" words, Samaras graduated from Rutgers in 1959 and returned to live with his family in West New York. During this transitional period he entered a graduate Art History program at Columbia University and studied early medieval art with Meyer Shapiro, but he never completed his degree.

While studying painting at Rutgers, Samaras became interested in acting and was a regular performer with several theatre groups on campus. His art instructor, Allan Kaprow, produced events of a theatrical nature — eventually called *Happenings* — that involved, in a collagelike structure, discarded materials, people, sounds, and repetitive movement. These early presentations were prototypes for the development and flowering of performance art more than a decade later. They also predisposed Samaras towards an artmaking process that would involve and exploit the human body. His photo setups, executed ten years later, can be considered as direct outgrowths of his early, experimental theatre activities. Samaras views the human body itself as an artistic material, to be used by others in Happenings and by himself in his own artworks.

Samaras remembers always being interested in things that were useless and discarded; after college he became intrigued with the opportunity to transform and "rescue" these castaways and give them a new life. Soon a whole series of boxlike constructions took form as "vessels" for stray feathers, bits of cloth, personal snapshots, and brightly colored lengths of yarn.

Box 8 (illustration) features four identical 8- by 10-inch photographs of his face, two displayed propped up inside the box and two on the inside of the lids. About this time pins and tacks obsessively began to appear in Samaras' constructions. The portraits are repeatedly pierced with strategically placed pins; hanging down from the pins are threaded lengths of wire that drape down and partially cover his face. An intricate, starlike, cats-cradle arrangement of yarn is stretched over the photograph on the far left. On the face to the extreme right a twisting braid of string, held in place by several pins, crosses his image like a snake. The interior of the box holds a variety of objects including a stuffed bird, beak pointing to a corner, and a dime-store plastic box. In marked counterpoint to the psychologically painful "voodoo" effect of sharp pins and wire is the base of the box, which is covered with a rainbowlike arrangement of multicolored yarn. Lush, geometric beauty and obsessive, symbolic violence link up to create a prickly icon of self-revelation and wonder.

"[D]o pins mean anything to you?" Samaras rhetorically asked himself in one of his self-interviews. "Let me find out," he answers.

(1) When I use them with the flat paintings they create a net pattern which breaks up the flat picture and creates a strange illusion. (2) Pins are marks, lines and dots. (3) They are relatives of nails. My father spent some time as a shoemaker. I was raised by a very religious family. The nailing on the cross. As a child I often played with pins at my Aunt's dress shop. Nailing pieces of cloth. My father spent many years in the fur business stretching and nailing furs. The pin is to an extent a part of the family. One of my earliest and strongest memories deals with seeing an Indian fakir on a bed of nails in a World Book Encyclopedia.[9]

Samaras compulsively covers the surface of these early constructions, bombarding our tactile and visual senses alike. Strong

Lucas Samaras, *Box 8* (photo courtesy of the Pace Gallery, New York).

but vague psychological undercurrents emanate from these elements and forms. They are contemporary examples of "horror vacui"—a label coined by art historians to describe the highly embellished art of certain cultures in which every square inch of a work of art is covered with decoration. Soon Samaras was using books—which after all are containers for words and ideas—as surfaces to completely cover with sharp pins. *Book 6 (Treasures of the Metropolitan)* (illustration) is simply an art book, propped open and menacingly encrusted with thousands upon thousands of enmeshed straight pins. They could just as easily be viewed, according to Samaras, as "silver straws" or "silver grass" or even "frozen rain" or "steel." These books have an aggressive air of untouchability that might be related to certain religious objects in the Greek Orthodox Church that, according to custom, must not be handled by anyone other than the priest. *Book 4 (Dante's Inferno)* (illustration) represents this "painful" quality even more forcefully than *Book 6* by incorporating a knife, scissors, and the ultimate in sharpness and danger: a razor blade.

In 1968, at the suggestion of his dealers, Samaras consented to produce a limited edition of prints. This suite took the form of a book and was appropriately titled *Book* (illustration). The genesis of this project goes back to Samaras' early involvement with writing and the pin-covered books. Unlike those fearful constructions, this printed artwork seems to invite opening, handling, and perusal. It consists of ten pages of thick cardboard, each ten inches square, that are punctured by die-cut shapes and covered with silk-screened dots and cartoonlike imagery. As you leaf through the book starting at page one, the dot grows in size but its value diminishes. By the last page it is the size of a nickel but beautifully pale and transparent. There is a deceptively cheerful, bright, decorative feeling to this particular work. In fact Samaras wanted it to look like a children's book and chose the heavy cardboard pages to reinforce this idea. But there are all sorts of warnings posted along the trail that indicate that this is no innocent toy for children: Within these pages the scatalogically outrageous, erotic adventures of *Dickman* and *Shitman* are printed amid exotic patterns and decorative shapes. Thematically, the stories can be compared to the writings of Sade, Genet, and William Burroughs in their use of complex, repulsive imagery. The picture, on the second page, of a hand holding a lighted match issues a subtle but distinct warning: Proceed with caution, things might not be what they seem; that is, you could unexpectedly get burned. Various stories appear throughout the book, cinematically fading in and out. Tipped to one of the oversized cardboard pages is a miniature book about 1 inch square with unbelievably small type.

Throughout the book small sections appear with images that refer to Samaras' body and past work: X-ray outlines of his hands, a fork with something speared on it (hints of cannibalism), a paintbrush, even his thumb. *Book* is a summation of many previously used motifs and images. Metaphysically, it is about disappearances (ultimately his own through death) and appearances (psychologically perhaps his father's).

When Samaras was three years old, his father left the family to find work in America. It was a traumatic time for him. Young Lucas did not see him again until he was ten years old. "Yes, I think I am still waiting for him to come in the form of a letter, a package, a party, an explosion, an icon, a natural event. When I make art, I am making a father," he poignantly wrote in *Autobiographical Preserves*. For Samaras the special act of art making becomes a magical means of recapturing the past and a relevant process for perceiving the present.

Although this book is loaded with obscure references, arcane symbols, obscene stories, and storybook images, its ultimate meaning is inaccessible, mysterious, and remote. Crudely and touchingly scrawled on the back cover is the word *goodby*, along with a small accordion pullout showing a faded photograph of the artist, an outstretched hand blocking his face.

Primed by his early involvement with Happenings and a brief stint at a Manhattan acting school, after graduation from

Rutgers, Samaras began building "sets" in his studio and performing in front of the camera. He refers to the Polaroid photos that result as "auto polaroids" and has been making them since the early seventies. The artist places the camera on a tripod, sets the self-timer, and visually interacts with the special scene or set. He uses a variety of photographic "devices," such as double and triple exposure, blurred movement, and multisource lighting. Many of these Polaroids show Samaras emerging from a wildly patterned background, embracing himself, crouching in a fetal position, writhing on the floor, or covered with white bed sheets. Samaras totally transforms the passion of amateur snapshots to capture images of pets, children, and relatives to fit his own psychological and artistic need for self-revelation and discovery. He alters and affects

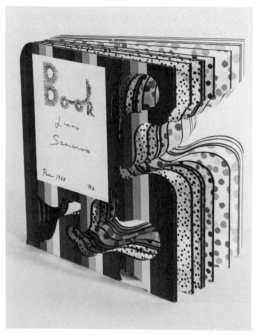

Lucas Samaras, *Book* (Collection, The Museum of Modern Art, New York. Gift of the artist).

Polaroid SX-70 prints even before they are fully "born." To achieve the desired results he draws, scratches, and paints on the photoemulsion while it is still undergoing chemical development. Even after the print has been transformed in this way, he still might paint on it, or draw ink lines, or color areas in with magic markers.

Phototransformation 10/28/73 (illustration), is an altered print that reveals the close connection between Samaras' photoworks and his pastel paintings. In Samaras' eyes any material or process can be used in a personal way to, as he puts it, "unravel meanings." As in the chalk drawings, a spatially ambiguous "set" is created to contain and play against the blurred bodies. In this phototransformation Samaras creates a visually lush and psychologically frightening "Wolfman" effect by emphasizing his bared teeth and natural beard. Through the means of a carefully composed double exposure the artist appears caught midway in a Jekyll and Hyde transformation — from human to animal. Multicolored elliptical shapes ra-

diate outward from the centrally located snarling teeth. Through the means of "instant" photography, set design, performance, and painting, Samaras records personal fantasies through obliquely self-referential imagery.

Few artists working today can match the sheer energy and visual resourcefulness of Lucas Samaras. Uncannily, he has the ability to keep reinventing his art while never waivering from the path of personal expression he has explored for over twenty years. Often Samaras will return to a theme or medium he worked with in the past and find a new way to restate old motifs. Such was the case with his one-person exhibit at the Pace Gallery in 1987: This exhibit featured a series of transformed chairs, reminiscent of sculptural assemblages done in the early seventies yet quite different in many ways.

All of the pieces in this show were fashioned out of heavy wire coathanger material twisted and bent into the shape of nonfunctional chairs. But, rather than emphasizing the sculptural identity of the chair, these wire armatures were interlaced with and wrapped around the wildest collection of cast-off materials one could imagine. It was as if a cyclone carrying chairs had swept through a flea market. Attached to the chairs were artificial flowers, pencils, knives, plastic figurines, imitation pearls, household items like eggbeaters and bent rulers, to name only a few of the materials. One chair was so encrusted with gaudy plastic flowers (colorplate) that the supporting wire framework was barely visible. This work contained so many of these imitation blossoms that it appeared to have undergone a profound sea-change. Flowers clustered like barnacles and other underwater growths that thrive on the skeletal remains of sunken ships. As if to reinforce this thematic possibility, Samaras attached a smooth-bodied plastic fish to the structure which stands out in striking counterpoint to the textural opulance of the flower encrustation. Despite the forced gaity of the colorful flowers there is a subtle undercurrent of danger and hint of death in this sculpture — flowers are a common accompaniment to fu-

nerals. After viewing this apparently whimsical piece for a while, the humor fades and feelings of a darker nature surface and inform our thought. The chair itself has been strangled and obliterated by these decadently cheerless imitations of life mocking the fragile and genuine beauty of real flowers.

Although the physical size and format of this sculptural series remain constant, each chair assumes a different visual and psychological character. *Wire Hanger Chair (Open Shoe)*, 1986 (illustration), clearly reveals the armature and structural features of the chair. Objects appear to be organically attached and to have formed symbiotic, biological relationships with the chair. But it is a strange and fearful symbiosis that breaths life into this sculptural form. The growthlike appendages are comprised of a variety of bizarre and dangerous-looking objects: A bright-green shoe is stuck with nails, green- and pink-colored upper and lower false teeth are entwined by the wire, and a sharp carving fork is impaled in one

of the chair's feet. Samaras is a master at controlling surfaces and linking forms. After a while we forget about the disparate items we are viewing and see the sculptures as unique and enigmatic organisms—attracting and repelling us with equal force.

Whatever materials and techniques this controversial artist chooses to work with in the future, we can be sure it will in some way employ symbolic, self-referential objects. Samaras has always viewed the "self" as an inescapable source of art. During the sixties he worked as an isolated individual against the formidable opposition of purely abstract Formalist Art. Twenty years ago his work uncannily predicted the importance of private and personal expressiveness in art. Time has not diminished the visual power and magic allure of those early works which incorporate and transform pins, photographs, yarn, and discarded objects of all types. Through these common materials and personal effects, transformed by the artist's hand and mind, a new world is presented to us—psycholog-

Lucas Samaras, *Wire Hanger Chair (Open Shoe)*, 1986 (photo by the Bill Jacobson Studio; courtesy of the Pace Gallery, New York).

ically arcane, visually dazzling, and unfathomably profound. Samaras never lets us forget that beneath these personal effects and homemade processes lurks one artist's omnipresent, "erotically sad," autobiographical self.

Jonathan Borofsky

Jonathan Borofsky's relationship to the process of artmaking is highly personal and even whimsical — in many ways he is a one-person movement who runs counter to prevailing fashions. Although temporal, site-specific art is no longer as popular as it was in the early seventies, Borofsky has made use of many aspects of installation art in his recent work. Walking into one of his installations is like entering a Postmodern fun house put together by a group of artists who maintain their individuality while sharing certain affinities. Borofsky does not adhere to one consistent style but rather uses styles the way other artists use elements such as line and color. His sculptural setups are energized by a unique, kaleidoscopic vision: images taken from dreams, painted self-portraits, found objects, and various sculptures are scattered on the floor, hung from the ceiling, and painted directly on the wall. One can hardly imagine a more diffused or encompassing "total" environment. The whirlwind look of such an exhibit makes us think that Boroksky makes use of every idea, dream, media, and concept that enters his mind. In a sense his outpouring of images and techniques is the equivalent of a literary stream-of-consciousness style. To walk through one of Borofsky's exhibits is like walking through a landscape of his inner thoughts. His disjunctive style of artmaking resembles the rambling monologues of our own inner thoughts as we consider everything from mundane concerns to philosophical thoughts.

Despite the fact that Borofsky's art is primarily image-based and narrative in structure, his artistic origins are rooted in a dematerialized form of conceptualism. In 1968 Borofsky's work was based solely on writing down his thoughts and addressing notes to himself. Seeking to break out of this circuitous mental activity, in 1969 he began to fill 8½- by 11-inch sheets of paper with sequential numbers that started from 1 (1, 2, 3, 4, 5, etc.). By now he has counted to over 3 million. This activity has created *Counting* (illustration), a sculptural piece that stands waist-high and is encased in a plexiglass box like a religious shrine. Borofsky uses it to provide continuity between his original ideas and his latest work; often it shows up in his sculptural environments as but one of many conceptual references and features.

Borofsky's mature work appears to bear little resemblance to this early piece, but *Counting* played a seminal role in his development and firmly established the conceptual underpinnings of his work. Borofsky also claims it taught him that "thoughts are objects." But after a few years of writing as

his primary art activity, Borofsky conceded that "concept art was reaching nobody. At least mine wasn't. Nobody wanted to go up and read those texts closeup and try to understand them. All these diagrams, they didn't matter much. Yeah, I thought I wanted to reach a larger audience without prostituting myself, but I wanted to reach people in a common language—words and images."

During his two years of counting—often for five hours a day—Borofsky found himself more and more frequently making gestural drawings on the margins of the page. In 1972 he went to an art store—and as a rank beginner might—bought a boxed set of paints and painted a crude scene of a tree in the rain with a head attached to it; in the background was a vaguely defined figure. To make a connection between his past counting activity and this painting, he drew the number he had reached in the corner of the canvas. Almost everything Borofsky does today incorporates, somewhere on its surface, the total reached in his counting activity.

Jonathan Borofsky, stacked papers *(Counting)* (photo by Geoffrey Clements; courtesy of Paula Cooper Gallery, New York).

For an exhibit at the Paula Cooper Gallery in 1980, Borofsky transformed the space into a remarkably diverse installation piece (illustration). Dream images, archetypal symbols distilled from popular books and magazines, words painted on banners, and objects rescued from the limbo of ordinary life were used to create a compelling tableau. No matter where one looked in this multifaceted space, Borofsky's fertile imagination asserts itself. Instead of striving for visual unity and thematic resolution, he creates a milieu similar to what we might find in his studio — works were shown leaning against the wall, sketches were informally pinned to the wall, and scattered papers littered the floor. Borofsky even incorporated a ping-pong table with a sign nearby announcing to the gallery-goers: "Feel free to play."

One important aspect of Borofsky's work is its apparent renunciation of professional art standards. It is overtly amateurish in many ways — casual arrangements and crude executions — yet decidedly professional in terms of the critical attention and professional settings it seeks. Furthermore his work refuses to take any one aesthetic or political stance — it thrives on contradiction. Some of it is part rational, part absurd. Eastern philosophy and pragmatic Western attitudes are juxtaposed with abandon. Even the emphasis on dream imagery is more tongue-in-cheek than serious. Borofsky trivializes the concept of dreams as a manifestation of preconscious truth: Witness his childlike drawing *I dreamed I was taller than Picasso* (illustration). Perhaps in response to the widespread use of surrealistic imagery in advertising, Borofsky, with mock-naivete, presents his dreams as parodies of other popular American dreams — "I dreamed I was rich," "I dreamed I was powerful," etc. Lurking beneath the surface of his facetious art is a broad questioning of contemporary values, beliefs, and achievements. By contradicting accepted standards of art, Borofsky makes us think about just what those standards imply and how they embody our beliefs. In essence his art is subversive; yet because of its amateur elements, we allow ourselves to take it in without erecting defensive barriers.

Jonathan Borofsky, installation piece, 1980 (photo by Geoffrey Clements; courtesy of Paula Cooper Gallery, New York).

Jonathan Borofsky, *I dreamed I was taller than Picasso* (drawing) (Collection: Martin Sklar, New York. Photo by Pollitzer, Strong & Meyer; courtesy of Paula Cooper Gallery).

In many ways Borofsky's art must be considered for what it is *not*: It is not self-consciously serious; it is not consistent, abstract, illusionistic, or filled with the pre-fabricated emotion seen in so much work labeled "expressionistic." In essence, this is an art based on contradiction and mutability. Rather than interpreting these attributes as an expression of underlying cynicism and negativity, it might be more accurate to read them as aesthetic mirrors of our modern society. Because of the breakdown of traditional institutions and belief structures we live in a world that abounds with contradictions and conflicting signals.

In recent years Borofsky's work has taken on overtly political overtones. This new stance can be traced to a dream Borofsky had while living in California and which he used in a piece he did at the Venice Biennale in 1980. The artist recalled, "I dreamed a Hitler-type person was not allowing people to roller skate in public spaces, so I decided to assassinate him but my friend told me Hitler was dead a long time and if I wanted to change anything I should go into politics. This seemed like a good idea since I was tired of making art and wondering what to do with the rest of my life."

This dream allows some interesting interpretations in terms of Borofsky's art. Hitler might symbolize the Formalist art world, dictating what was or was not accepted in painting or sculpture. As pointed out in the dream, Hitler had been dead for a long time — certainly the stranglehold of the New York School had greatly diminished by the eighties. Borofsky may have been telling himself in this dream that it was no longer necessary to wage his private war against professional art standards, and that he could now go on to another stage of his life and art — the expression of ideas that embodied a new political consciousness.

Borofsky found the perfect setting in which to create a work of art with political overtones when he was invited to the seventh Documenta exhibition at Kassel, Germany, in 1982. Here, in a space shared by nineteenth-century marble statues, he placed five enormous, black, motorized steel figures with perpetually moving arms. *Five Hammering Men* (illustration) was a mysterious piece that could not help being read in terms of a political statement. Exactly what the statement was, however, was open to a variety of opinions. Some viewed the 16-foot tall figures as symbolically threatening to break up the marble statues in the room. Because of the white serial numbers on each figure, others were reminded of Nazi extermination camps from World War II. Still others interpreted *Five Hammering Men* as an embodiment of American "cultural imperialism." Many interpretations were possible, but most viewers agreed that these large hammering figures were potent contemporary symbols.

Wisely Borofsky refuses to make work that might be seen as propaganda — art that represents only one point of view. It is important to realize that Borofsky defines "political" rather broadly to mean the social reality he experiences in everyday life. All of his symbolic images are mutable and change to accommodate new ideas or the requirements of a site. The history of his

Jonathan Borofsky, *Five Hammering Men*, 1982 (photo by Tjeerd Frederikse; courtesy of Paula Cooper Gallery, New York).

use of the hammering-man image is an interesting case in point. Borofsky traces the genesis of this archetypal symbol to "a picture in my old *Book of Knowledge*—a guy in a striped shirt hammering at a wooden shoe in Tunisia or something." In 1981, a year before the Kassel installation, Borofsky became interested in the political strikes going on in Poland and lettered the word "STRAJK" (Polish for strike) on one of his hammeringman figures. Borofsky gave his work a multifaceted meaning with this added word: Strike could refer to laborers who work by hammering an object, or it could refer to a political protest.

Borofsky's most recent work has changed considerably from his earlier environments. Gone—at least for the time being—are the random elements he used to arrange with such chaotic energy and abandon. Fewer components appear in each installation, and they seem to be carefully crafted and well chosen. Even his painted canvases, which once were executed in the style of a precocious teenager, are now quite technically accomplished, even sophisti-

cated. *2,845,317 Green Tilted El Salvador Painting with Chattering Man* (illustration) is a miniature environment of sorts. Borofsky places an oversized wooden figure with mechanized jaws in front of a painted enlargement of a green Salvadoran stamp (from Borofsky's boyhood stamp collection) that is hung crookedly. Synchronized with the rapidly moving mouth, an accompanying audio tape endlessly drones the words "chatter . . . chatter . . . chatter . . . chatter." The image on the enlarged stamp shows President Roosevelt congratulating an anonymous Salvadoran official. As with *Five Hammering Men*, the political overtones surrounding these charged images are implicit but suitably ambiguous. The chattering man facing the political functionaries can be seen as an interested observer or as a dissident who is leveling charges. In spite of the heated debate surrounding Central America, Borofsky's installation takes no clear stand in favor of either liberal or conservative positions but makes us think about the circumstances and history of American involvement in these countries. Borofsky's

172

Jonathan Borofsky, *2,845,317 Green Tilted El Salvador Painting with Chattering Man* (photo by Geoffrey Clements; courtesy of Paula Cooper Gallery, New York).

artworks present us with situations loaded with political overtones but devoid of specific political rhetoric.

"Chattering" is Borofsky's term for the constant inner dialogue of thought we all obsessively engage in when we mentally recount our beliefs, fears, fantasies, and concerns. *El Salvador Painting with Chattering Man* is essentially a self-portrait, ". . . but it's everyone I've talked to," says Borofsky, "I mean, I know I'm not alone. We're all chattering." It is this quality — his ability to see the universal archetype in the mundane — that distinguishes Borofsky's work and gives it special relevance to our complicated, shifting, and often contradictory world.

New British Sculpture — Tony Cragg

Until the emergence of new sculptural sensibilities in the seventies, the history of contemporary sculpture read like a modern fable or prime-time soap opera. During the prosperous fifties sculpture fell in love with the power and visual panache of technology, deserting traditional themes and ma-terials for the realm of industrial processes and slick finishes. But, this marriage of art and technology was doomed from the start. Guilt descended with the realization that the art world's partnership with industry was one-sided. The industrial complex shared nothing of the artist's dream for the integration of art and life; glossy sculpture only succeeded in diverting people's attention from the disastrous environmental effects of industrialization and the rampant consumerism that supported it. After the divorce, sculpture seemed to seek forgiveness for its indiscretions and purge itself of its polyurethane sins. Seeking a healing solitude, it left the corrupting influence of the Gallery–Collector–Museum circuit for the purity of unspoiled realms of desert, forest, and ocean. Earth, water, and air were the elemental materials sculptors used to restore balance and meaning to their life and art. And with body art in the seventies, artists had reached inward to discover irreducible expressions of their perceivable universe — the self. Aesthetic limits were tested as perhaps they had never been tested before, and the boundaries of art extended farther than thought possible — or, in the minds of some critics, desirable.

After this period of disenfranchisment from the brave new world of the "artificial" and the modern, sculpture in the eighties (like painting) sought to rediscover its origins in the myths, materials, and themes of older, traditionally oriented artforms. Art had advanced so far it found itself returning to its beginnings. What artists in the late fifties rejected (the expressive handcrafted object), artists of the eighties embraced as the salvation of their profession. Let some critics call this reactionary — art as a personal belief system was fighting to reassert its primacy in a world increasingly dominated by computer terminals and larger and more controlling corporations.

But some artists have questioned the relevancy of going back to preindustrial or early industrial aesthetic models in a postindustrial world. A young generation of English sculptors in particular have grappled with this problem, and in the past ten years they have created a body of work that deals with the problems of the urban environment and the evolution of the industrial age. In a sense England is the perfect location for such work because of that country's economic and social problems. Unemployment, social unrest, and the physical decay of England's factories have formed the background and set for these artists.

In the midseventies, English sculptor Tony Cragg turned his attention to the urban and social environment of his native country in order to produce complex and meditative works of art out of bits of discarded plastic gathered from gutters and vacant lots. Using these bits of rejected and "consumed" pieces of plastic like individual pieces of a mosaic, Cragg fashioned images such as human figures, European flags, and maplike outlines of various countries. What characterizes this sculpture is the questions it raises about what is "natural" today and how our environment has shaped the way we view the world.

Cragg's work had undergone significant evolution during the seventies. At that time he worked with natural materials such as stones, shells, and driftwood, arranging them into geometric configurations (similar to the English sculptor Richard Long). Later he investigated the effects of gravity: Bricks were piled high until they fell; boxes were pushed against a wall until they tumbled down. In 1976 Cragg created seemingly chaotic environmental sculptures by smashing china dinnerware and spreading the fragments in ever-widening circles until isolated pieces seemed to stand alone. These concepts of fragmentation, aggregation, and the isolation of parts are issues which still inform his work today.

Red Skin, 1980 (illustration), was the first fragmentation piece that was arranged to form a macroimage. Cragg found a toy plastic figure of an American Indian with raised battleaxe in hand and used hundreds of discarded plastic articles to form its enlarged outline on the floor. On the far wall of the gallery in which it was installed, Cragg hung the small toy "Red Indian" and the title of the piece.

Walking around this 22-foot-long piece invariably makes one feel like an urban anthropologist from the future. The most diverse collection of red plastic flotsam and jetsom imaginable confronts us: plastic cups, baskets, basins, ice-cream spoons, imitation flowers, combs, whistles, and even a nipple from a baby's bottle mockingly stare back at us. *Red Skin* makes us realize we are surrounded from birth by this common material — plastic — but because of its ubiquitous presence we rarely notice it. Cragg views plastic as a primary shaper of our environment and psyche. This metaphor takes on even more profound meaning when we consider that one of the dictionary meanings of plastic reads: "having power to form or create." Cragg's environmental sculptures also remind us of how much our society is created and defined by the ceaseless production (and consumption) of goods — so many of them ill conceived, temporal, and poorly made.

Many conceptual aspects of *Red Skin* touch upon the innate nature of our industrialized world — the world of fast-food and discardable products. Some of the individual articles of this floor sculpture still have bits of earth clinging to them — remnants of their previous burial before Cragg

Tony Cragg, *Red Skin* (courtesy Marian Goodman Gallery).

resurrected them from the earth. In comparison to their short-lived utilitarian life (being easily broken and discardable), these plastic artifacts defy chemical decomposition. Their half-life can be measured in centuries, not years. Burial in fact helps preserve many forms of plastic from the deteriorative effects of sunlight. Cragg's use of plastic is not intended to be a smuggly superior put-down of this "lowly" material. Rather, Cragg views this material as the natural expression of the industrial age. Viewed in light of the decline of England's industrial empire, *Red Skin* offers us ways in which we might begin to make sense out of the wealth of sheer objects spawned by the age of the machine. There is a conceptual quality to Cragg's work that is appealing and even optimistic: Our first thoughts upon viewing *Red Skin* might be horrific — we are beholding what we have become. Are we like this unearthed polyethylene — disposable, cheap, fragmented, inconsequential? After spending some time with this comprehensive collection, however, our initial feelings may be changed into those of fascination and wonder at the richness of forms produced by our collective efforts. Cragg's transformative vision may even allow us to leave the gallery with a view of the postindustrial world as an embodiment of potential richness and mutability rather

than a smoldering garbage dump of alienation and ugliness.

Part of *Red Skin*'s visual allure was its elusive image. The small pieces of plastic which formed the large image of an Indian were like pieces of a puzzle. Considering the size of the image formed, and its position (the floor), viewers had to work to assemble the image in their minds. Much of Cragg's aesthetic interest is directed toward encouraging the viewer's active participation.

Self-Portrait with Sack, 1980 (illustration), however, provides us with an easily read image — that of the artist engaged in the act of collecting the plastic material for his sculptures. Cragg contrasts his rather primitive collection process with an awareness of the sophisticated manufacturing processes that produced the goods he gathers. Interesting oppositional themes are set up in this sculpture. The cloth sack placed next to the technologically sophisticated materials used in the plastic-bit figure seems archaic by comparison. Similar hand-woven bags were used by food-gathering people before recorded history. Cragg's modern journey — gathering discarded plastic for his work — becomes curiously connected to ancient processes of subsistence and survival. "Modern Man," represented by the figure made up of bits and pieces of red, blue, yellow, and green plastic, gazes in a fixed stare at the bulging olive-drab sack as if in preconscious understanding of this connection. Cragg fuses these two worlds —gleaners of the ancient past and the present—and makes us consider, from a new vantage point, the realities of the present-day urban environment.

24-Hour Cycle, 1984 (illustration), moves away from his earlier fragmented floor and wall assemblages and becomes fully three-dimensional. Found objects are presented in this piece whole rather than in fragments. In *24-Hour Cycle* Cragg animates the inanimate objects that litter our industrial world by linking a metal case, cardboard suitcase, and a small well-worn wooden table with plastic pipes that seem to surround and embrace these abandoned artifacts the way a tropical jungle would soon envelop any abandoned objects with a dense growth of vines and vegetation, and slowly digest them. The twisted and convoluted orange plastic tubing evokes the image of human intestines whose job is to extract nourishment, and energy, and life from the injested food. Perhaps *24-Hour Cycle* alludes to this life-sustaining function symbolically. Certainly Cragg's use of what most people would consider nonartistic and "consumed" materials suggests that with the transformative vision of art, it is possible to resurrect these discarded objects and give them new life.

The structure of *Citta*, 1986 (illustration), also makes allusions to a complex biologic function: Cragg has fashioned a shoulder-high, figure-eight helix out of small wooden house-shaped blocks that mimic the molecular structure of DNA—the genetic blueprint for life. This structure stands next to a painted gym locker and kitchen table. Cragg wants us to look beneath the visual appearance of objects and recognize that on a molecular level strong connections exist between seemingly disparate forms of

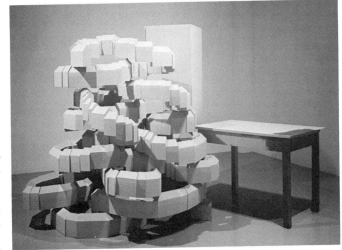

ABOVE: Tony Cragg, *24-Hour Cycle* (courtesy Marian Goodman Gallery).

RIGHT: Tony Cragg, *Citta*, 1986 (courtesy Marian Goodman Gallery).

matter. Since Cragg was a biochemist before turning to art, his understanding of the basic chemical processes at work in the world helps him make connections between the way chemical processes form life and the way scientists manipulate chemicals to create plastic compounds.

All of Cragg's recent work challenges the romantic notion of an unbridgeable gulf between the natural and the created environment. Cragg clearly visualizes, because of his scientific background, the place of plastic and other manufactured materials in the natural order of the world. Through his work the artist presents us with a vision of aesthetic and conceptual unity: bits of brightly colored plastic, as well as large objects like kitchen tables and gym lockers, are derived ultimately from the same chemical hopper as sea, sand, mineral ores, and rainwater. And the "natural" products of an urban, industrialized society are viewed as potentially full of as much meaning as those of field and forest.

All of the sculptors in this chapter attest to the diversity, inventiveness, and broad philosophical stance three-dimensional work has assumed over the past two decades. The possibilities open to sculpture now are broader than they have ever been. Attitudes about purity of form and the necessity for sculpture to be viewed in the round no longer limit today's artists. They are as likely to use color and environmental space in their work as marble and bronze. Future developments will no doubt create an even richer mixture of sensibilities, processes, and means of reaching the public.

5 Photography

Landscape to Narrative

In 1977 a large photographic survey exhibit, *Mirrors and Windows*, was presented by the Museum of Modern Art and traveled to major cities throughout the country. John Szarkowski, curator for the show, observed that during the past twenty-five years the preoccupations of photography have shifted from public concerns (social documentation) to the pursuit of private interests (personal expression). One could argue that the great American photographers of the thirties and forties — Walker Evans, Dorothea Lange, Paul Strand — were always following individual visions as well as social goals. Nevertheless, a great deal of recent work has clearly moved away from the documentary realism of the past towards more expressive styles that we usually associate with painting and sculpture.

Ironically, this new, personal direction has been propelled in large measure by the acceptance and aesthetic success of socially relevant photographers like Evans, Lange, and Strand. All of these individuals believed it was their obligation as privileged witnesses to accurately record as a public document the social dynamics of their era. The domination of reportage photography during the forties and fifties can also be traced to the large number of news magazines published during that period that provided employment and outlets for their work. By 1970, however, electronic journalism, with its advantage of instantaneous distribution, drove many of these publications out of business.

Another important factor in the evolution of fine-arts photography was painters' and draftsmen's adoption of photographic tools and processes. Notably, Andy Warhol and Robert Rauschenberg were instrumental in broadening the field by their use of photographically generated images in their multimedia paintings. Although purists decried these endeavors as bastardized forms of photography, no doubt they helped pave the way for the acceptance of "straight" photography within the art world.

During the late sixties and early seventies conceptual artists like John Baldessari, William Wegman, and Jan Dibbets made widespread use of the camera in their idea-oriented work, thus helping to remove the stigma of "craft" that had been attached to photography for so long. What essentially interested these artists about this medium was its "dematerializing" presence: The photograph was not an object like a painting, drawing, or sculpture, but a means of notation and conceptualization. Not surprisingly these artists were labeled "anti-photographic" by the old-guard members of the documentary school. Nevertheless, conceptual work undeniably helped to establish our contemporary recognition of photography as a viable, mature artform.

In terms of technical process photography is an optical-mechanical-chemical means of accurately recording light. Depending on where one points the camera and on how one interacts with this modern process (actually it is older than most people think, having been invented more than 150 years ago), it can faithfully record a scene or document a setup that the artist creates. John Szarkowski's survey show of 1977 spoke to both possibilities; its title, *Mirrors and Windows,* refers to the documentary and expressive sides of the medium.

Szarkowski stated in his exhibit catalogue:

The distinction may be expressed in terms of alternative views of the artistic function of the exterior world. The romantic view is that the meanings of the world are dependent on our own understandings. . . . It is the realist view that the world exists independent of human attention, that it contains discoverable patterns of intrinsic meaning, and that by discerning these patterns, and forming models or symbols of them with the materials of his art, the artist is joined to a larger intelligence.[1]

Szarkowski's metaphors of mirrors and windows can help us to understand the many aesthetic options available to photographic artists today. One must be careful, however, not to view this distinction as a rigid formula, thereby avoiding the issue of *individual* creativity and response within the broad world of contemporary photography. Some artists, perhaps the most interesting ones, seem to work in both camps simultaneously creating works of art that defy any categorization. Many are able to make subtle social observations *and* personally revealing aesthetic statements.

Photography as a form of artistic expression received a powerful boost when the Museum of Modern Art staged its *Family of Man* exhibition in the midfifties. For the first time, a museum dedicated to showing the most important developments within Modern art staged an enormous show of recent, nonhistorical photography. The public's response was phenomenal. By any standard the show was a blockbuster, generating favorable publicity and widespread interest in the aesthetic potential of a medium long neglected by the art world.

Not everyone was pleased, however, with the thematic concept of humankind as one big family: that underneath our varying costumes and social habits, we are all really the same. Some critics found it too sugary and simplistic—overloaded with platitudinous captions such as "eat bread and salt and speak the truth." Regardless of its aesthetic limitations, it did plant the seed for the emergence and acceptance of photography within the art establishment.

Eventually, the combined pressure of increasing numbers of photoartists, collectors, and viewers won for this emerging artform the understanding and recognition denied it for so long. During the oil embargo and consequent recession of 1973 to 1974, this medium received another boost when the art market turned to photographs as good, lower-priced investments. By 1977 leading art galleries like Janis, Castelli, Sonnabend, and even the prestigious Marlborough Gallery had climbed aboard the bandwagon with major shows by important photographic artists.

During this period of rapidly increasing awareness, historical figures from the past, as well as previously overlooked living artists, were "discovered" and given large exhibitions by these trend-setting galleries.

Along with an interest in classic black and white photography, the midseventies saw the emergence of color photography as an accepted medium. Until this time, color photographs were viewed with suspicion and distrust—many photoprofessionals regarded the effects of color as a distracting element. But Joel Meyerowitz, a photographer who worked almost entirely in color, has revealed that in the hands of the right person this medium can yield surprising results. For one thing, he works with a 40-year-old view camera that uses 8-by 10-inch sheets of film. Instead of enlarging from a negative, Meyerowitz contact prints from this film; the images this process yields are no larger than 8 by 10 inches but are unsurpassed in sharpness and tonal gradation.

Meyerowitz is fascinated with the landscape because this theme allows him the opportunity to explore the constantly changing beauty of natural light as it reveals and transforms the countryside. Through his ability to organize photographic form and control color, Meyerowitz presents these places to us in a special way—through the unifying vision of his color photography.

Diane Arbus' work represents a unique and highly original contribution to the history of contemporary photography. Her work reveals the existential underbelly of the world conspicuously absent in the *Family of Man*, exhibition. Perhaps the Museum of Modern Art recognized this previous omission when they gave her and two other photographers a show in 1967 titled *The New Document*. Unlike the *Family of Man*, in which everyone and everything was beautiful, Arbus revealed a different, more personal version of photographic truth: She presented us with a variety of disturbing portraits that documented the everyday reality of urban ugliness, psychological despair, and human bravery. By photographing the forbidden, Diane Arbus brought us much closer to a new awareness of our condition which, however unpleasant, might provide us with more genuine optimism than scores of pretty and ultimately deceiving pictures.

Duane Michals views photography in a different way: He stages pictorial dramas of contemporary life that encompass and yet go beyond the limits of everyday occurrences. Using a serial-image technique, he creates photographic fictions that intermesh seamlessly with experiences we may have dreamed of or taken part in. All of his works — some poignant, some comical — share a heightened sense of wonder with the way our daydreams and fantasies interact with mundane elements of life.

Ralph Gibson's work manages to synthesize elements of "street-photography," a documentary art form, and visual aspects of Formalism normally associated with recent schools of painting. He is able to combine stunning design qualities, such as strong geometric forms and interactive plays of black-and-white shapes, with the notion of photographic "realism." The result is a body of work that draws simultaneously upon the influences of Surrealistic painters like Magritte and documentary photographers like Walker Evans. In this sense Gibson represents the attitude of many emerging artists who are reluctant to fit into rigid categories of the past.

Now that photography has come of age in terms of acceptance as an art medium, it can more freely choose the aesthetic arena it wishes to work in. Also, with this acknowledgment has come the development of a reflexive stance — photography is now used as a means of introspectively looking at itself. The work of Cindy Sherman expresses this position. Sherman's large-scale color self-portraits probe the psyche of popular cultural myths seen in the movies — the "seductress," the "girl next-door," the "poor girl," "the rich girl," etc. Through these familiar stereotypes Sherman explores the emotional topography of mainstream media-dominated life.

Because of the vigorous activity within the photographic marketplace and the breakdown of constraining aesthetic rules, a wide variety of personal interests and sensibilities can be found in the work of today's photographers. While some artists are choosing to pursue further the possibilities inherent in documentary work, others are following directions suggested by developments in painting, conceptual art, and environmental work. Above all, most contemporary photographers, like artists in other fields, realize that individual perception and personal statement are the central issues of the day.

Joel Meyerowitz

Over the past decade Joel Meyerowitz has firmly established himself as a photographer who works almost exclusively in color. While this may not seem to be a particularly radical position today, until the midseventies an overwhelming majority of serious fine-art photographers had rejected the color process as unsuitable for important aesthetic statements; it reeked of a crass commercialism.

Historical precedent also had something to do with this decision; practical color emulsions were not invented until relatively late in the game. It was difficult enough to develop films that responded to the variable intensity of light: Early black-and-white emulsions were extremely insensitive and required inordinately long exposures.

The problems of recording the full-color spectrum of light were even more difficult and complex. But the dream of a film process that could accurately render color extends all the way back to the very beginnings of the medium. Niepce, the Frenchman who is credited with inventing photography, performed pioneering experiments with color processes in the 1820s.

It was not until 1935, however, more than 100 years later—that color photography emerged from the laboratory and experimentation stage and took its place as a commonplace process. At that time the Eastman Kodak Company introduced Kodachrome, a special, multilayered, dye-coupled emulsion that soon won great acceptance among commercial and amateur photographers alike. Today, it is a rare commercial film or national magazine advertisement that does not appear in glowing color. Almost of equal rarity is the black-and-white television screen, home movie, and wallet snapshot. In fact, it is almost impossible to purchase amateur black-and-white film emulsions today. Full-color film processes have entered our lives to such an extent that the memory of an earlier entirely monochromatic world has been all but erased.

But if one were to look back at most of the photographic exhibits in major museums and commercial galleries over the past two decades, one might be surprised to see that black-and-white images still reigned supreme in this area. This aversion to color is in part a response to the wide use of color photography to satisfy the consumer demands of a mass audience. Many serious artists in this medium purposefully stuck with monochromatic emulsions to reaffirm the nonmercantile aspects of their work. Also, many artists concerned with the photograph as a document of Social Realism—such as Walker Evans and Paul Strand—quite rightly viewed the element of color as "prettifying" and essentially trivial.

Recently, a new generation of photographers, with a different set of aesthetic concerns, have rediscovered within the color process a world of striking luminosity, beauty, and significance. Joel Meyerowitz is just such an artist; his meticulously crafted and elegantly conceived color prints of various landscapes — both rural and urban — have established him as a leader in this area.

Meyerowitz was born in New York City in 1938 and grew up in a neighborhood where knowing how to fight was an extremely useful skill. His father, a former vaudeville actor and boxer taught him to defend himself. From his father, Meyerowitz learned the value of timing and careful observation, watching for one's opponent to telegraph moves before he makes them. Meyerowitz credits his father and the tough neighborhood in which he grew up for instilling in him a special awareness of the world.

After studying painting and medical illustration at Ohio State University (he liked to paint loose and draw tightly, he says), Meyerowitz returned to New York City and found a job as a designer and art director with an advertising agency. One day his office sent him to supervise the work of a photographer who was taking pictures of some female models in an apartment setting. The photographer was Robert Frank, a Swiss-born artist, who had recently arrived in America. Meyerowitz watched with rapt attention as this man performed a veritable ballet with his camera. Frank was in constant motion, talking to the models continuously while he photographed them. Occasionally he would swing the camera in front of his face, releasing the shutter as it passed in front of him. Meyerowitz had never seen anything like this; he was used to photographers who carefully set everything up, and requested that their models hold still while they carefully snapped the shutter. Frank's quiet intensity and unusual methodology captivated the young art director. When Meyerowitz went back to the agency later that afternoon, he told them he was taking a leave of absence as of Friday; although he did not know it at the time, he would never return to the commercial art world. Within days he had purchased a 35 mm camera, and a new career had begun. Soon Meyerowitz roamed the streets and public squares of New York trying to capture the energy and poetic drama of the city.

Fifth Avenue was a favorite haunt of the artist; there was always plenty of activity and a wide segment of humanity; smartly dressed businesspeople, pathetic "bag-ladies," beautiful magazine models, and ordinary citizens made their way up and down this famous thoroughfare at all times of the day. Before long Meyerowitz noticed other photographers like himself working the street (notably Diane Arbus and Garry Winogrand), regardless of high winds, sleet, cold, rain, or snow. Winogrand and Meyerowitz soon became friends and sometimes walked along Fifth Avenue together, observing life and comparing notes on photography.

One of the high points of Meyerowitz's early career occurred in 1963, when he met the famous French photographer, Cartier-Bresson, at New York's St. Patrick's Day parade. Meyerowitz spotted a well-dressed, older man with a camera behaving in a manner reminiscent of Frank's, jumping around, quickly thrusting his camera close for a picture, then moving on. When he cautiously approached the man and asked him if he was Cartier-Bresson, the photographer replied, "No I'm not, are you the police?" Meyerowitz apologized and said he was only trying to take pictures of the parade. When he heard that, the man changed his tune, grabbed Meyerowitz's jacket, and said, "I'm Cartier-Bresson, meet me here after the parade and we'll have coffee." Meyerowitz and his photographer friends stood in amazement as he bounded off. Minutes later they witnessed the legendary photographer in action, as a drunken spectator came after him from the sidelines and tried to grab his camera. In one easy motion Cartier-Bresson flung his Leica at the man's face, causing him to fall back into the crowd. Like a yo-yo, the camera miraculously bounded back into his hands. After years of this kind of public photography, Cartier-Bresson had learned the value of attaching his camera to a cord that was firmly tied to his wrist. After the parade Meyerowitz and Cartier-Bresson went

to a coffeehouse and talked about the methods and techniques the French photographer had developed during his career.

Within five or six years of picture taking, Meyerowitz had established himself as a young photographer to be reckoned with: In 1963 he was included in a group show at the Museum of Modern Art, *The Photographer's Eye*; he visited Europe for a year and a half; and he had a one-person exhibit called *My European Trip: Photographs from a Moving Car* at the Museum of Modern Art.

In 1973 Meyerowitz's attitude towards photography began to change. Color took on an important role in the artist's work. Even his street photography became less casual and took greater advantage of light, space, and architectural elements rather than concentrating only on the fleeting event or so-called "decisive moment." New considerations began to form in his work: an architectural precision in framing, compositional stability, and a sense of monumentality and timelessness. Meyerowitz felt he had developed the "street-photography" aesthetic to its limits. It was time to establish his own "voice" or personal signature with the camera. New tools and techniques were needed for the task. In 1976 the New York State Council on the Arts awarded him a grant to buy an 8-inch by 10-inch camera and to spend a summer on Cape Cod, Massachusetts, photographing the landscape.

Meyerowitz turned to a large-format Deardorff view camera (the 8-inch by 10-inch size refers to the dimensions of the negative) in order to increase the quality and quantity of information that the camera could record. Because the negative is so big, enlargements are unnecessary; to make a print the film is placed in direct contact with the printing paper, exposed, and developed. No sharper image is possible for that size format. Enlarging any negative causes a certain loss of detail and sharpness, regardless of how good the optical printing system is. Through the 8-inch by 10-inch negative, Meyerowitz discovered the technical means to realize his newfound vision: magical and subtle evocations of natural light, infinite space, and a dazzling fidelity to detail.

Another advantage of the large-format color image is the broad tonal range it offers. With no enlargements to degrade the print, the full spectrum of textures and hues become apparent looking at these photographs.

Bay/Sky, Provincetown (colorplate) is a dramatic study of shifting clouds, light, and sky. A tiny figure at the end of a sandbar and a beached sailboat that glows in contrast to the dark sand are important visual counterpoints to the suffusive light and brooding sky. It is a familiar scene, much photographed by amateurs. Meyerowitz lifts it out of the ordinary landscape genre through meticulous attention to color and light control. Although no artificial lighting is added (the artist shoots all of his photographs with existing light), the tonal effect is arresting and seems almost artificial in nature. Part of the explanation lies in the type of film he uses, a color negative emulsion called Vericolor II Type L (formulated for long exposure), which seems to possess the remarkable ability of accurately rendering both natural and artificial light simultaneously.

To maintain control over contrast and color Meyerowitz often prints the negatives himself rather than having an assistant or commercial color lab take care of this detail; most prints take hours of demanding darkroom time to realize the subtle details and luminosities that the oversize negative is capable of achieving. *Bay/Sky, Provincetown* registers both delicate and saturated color, along with high-and low-contrast tonality.

Scale is another element at which Meyerowitz is a master; the small but sharply detailed elements of figure and boat punctuate the composition and add to the feeling of immense, open space that distinguishes this particular print. They become human reference points that help situate the viewer in this saga of windswept space, epic scale, and dramatic color.

Red Interior, Provincetown (colorplate) was taken during the summer of 1977. This print, perhaps more effectively than any others, demonstrates the film's ability to register a variety of light sources. The human eye — unlike the camera — is able to automatically compensate for the varying intensity and color of light. Because of this we rarely notice the dramatic light changes we encounter everyday of our lives. The warm glow of an ordinary tungsten bulb is very different in color from daylight but for the most part this passes unnoticed. Color film, however, notes with unforgiving fidelity every shift in light volume and color.

Red Interior, Provincetown records at least four separate light sources with four different color temperatures: the moonlight which casts a pale blue over the whole scene; the greenish street light that paints the side of a cottage green; the reddish glow of the car's interior light, and a yellow beam that emerges on the bottom from an unseen source. All of these widely differing lights are harmoniously combined in this photograph, creating a brilliant, other-wordly mood.

Provincetown (colorplate), another dynamic color statement, combines aspects of Meyerowitz's early street photography with his more recent large-camera color visions. Most of his later photographs record simple, everyday scenes: the exterior corner of a Cape Cod shingle house, looking from inside the house through an open door to a green garden, and a simple picket fence outlined against clouds and sky. The focus of his mature work has shifted towards the act of *seeing* and away from social commentary and "important" subject matter. No doubt the technical nature of the 8-inch by 10-inch view camera has played a significant role in this shift of consciousness.

Everything Meyerowitz sees in the large ground-glass viewfinder is inverted. To take a picture he covers his head and shoulders with a black, light-proof cloth and composes the picture viewing everything upside down and backwards. Because of this Meyerowitz views the most common of subjects in an "abstract" way. As he once stated in an interview:

C U R R E N T S

Joel Meyerowitz, *Bay/Sky Provincetown*.

Joel Meyerowitz, *Red Interior, Provincetown*.

Joel Meyerowitz, *Provincetown*.

Joel Meyerowitz, *The Arch*.

Cindy Sherman, *Untitled 1981* (photo courtesy of Metro Pictures).

Christo, *Running Fence*, Sonoma and Marin counties, California, 1972–76 (copyright: Christo/CVJ Corp. 1976).

This whole inversion was exciting. I don't care if it was a gas station or if this is a rubber raft or if this is a crappy little house. That's not my subject! This gas station isn't my subject. It's an excuse for a place to make a photograph. It's a place to stop and to be dazzled by. It's the quantity of information that's been revealed by the placement of these things together, by my happening to pass at that given moment when the sky turned orange and this thing turned green. It gives me a theater *to act in for a few moments, to have perceptions in. Why is it that the best poetry comes out of the most ordinary circumstances? You don't have to have extreme beauty to write beautifully. You don't have to have grand subject matter. I don't need the Parthenon. This little dinky bungalow is my Parthenon. It has scale; it has color; it has presence; it is real: I'm not trying to work with grandeur. I'm trying to work with ordinariness. I'm trying to find what spirits me away. Ordinary things.*[2]

Provincetown is distinguished by two elements: implied motion and vibrant color. Within the rigid geometric framework of the white deck railing and the side of the house, wildly flapping multi-colored bed sheets are blurred by the relatively long time exposure. The wispy high-altitude clouds overhead underscore the feeling of motion suggested by the out-of-focus sheets. As in all of the Cape Cod series, color performs a transcendental act, lifting these scenes out of mundane genres into worlds of heightened acuity. The artist works with both strong color contrasts and close-hued harmonies; bright orange sheets are juxtaposed with a cobalt blue sky, while the blue and white striped material bridges the gap between foreground and background. *Provincetown* is a complex but tightly ordered fantasy of dancing sheets, wind-swept sky, and the simple geometry of Cape Cod's indigenous architecture.

In the late seventies, Meyerowitz was commissioned by the St. Louis Art Museum to document the contemporary life, past history, and architectural features of St. Louis. After four visits, during which the artist took over 400 8-inch by 10-inch color negatives, 107 photographs were printed and exhibited in the St. Louis Art Museum under the title of *St. Louis and the Arch.* Meyerowitz used Eero Saarinen's 635-foot arch alongside the Mississippi River as

a unifying visual theme. This arch symbolizes America's westward immigration and identifies St. Louis' role as a gateway to the West. Varying perspectives on the arch are interlaced throughout the series. Meyerowitz has photographed it from every possible angle; it appears in closeups, far in the distance, peering over a verdant ballpark, and rising above tall office buildings. It is never out of sight for long. "I found it deeply moving," said the artist, "profound. There were days when, standing beneath it, I felt I knew the power of the pyramids." *The Arch* (colorplate) seems to capture this exact feeling. Through a time exposure taken at night, Meyerowitz captures the reflections of the city's lights which ripple on the gleaming metal surface of the arch like waves on the surface of a pond. City and arch are visually united and symbolically perceived as one entity. Only the base of this enormous monument is visible, but much of its overall power and size is communicated by its monolithic metal trunk. There is something both mysterious and matter of fact about this photograph. The arch's daytime solidity and physical presence are transformed in this night view into ephemeral elements of reflected color and light. As in many of Meyerowitz's photographs, contrasting color elements are juxtaposed and reconciled through harmonious composition. Intense reds and oranges are strikingly set off by the black background and work well with the pale green tones generated by the artificial light. A tiny streak in the upper-right-hand sky, registered by the celestial movement of a star during the long exposure time, lends a subtle compositional· element and comments on the time it took to gather the light for this lyrical document.

Whether recording the shifting light over Cape Cod Bay or documenting the inner-city life of a large urban center, Joel Meyerowitz continues to refresh our visual perceptions without resorting to arcane views or bizarre angles. He applies the means of large-scale film format and color photography to expand our visual and psychological awareness of the natural world—

lending, as he once said, ". . . a sense of the eternal to the commonplace."

Diane Arbus

Diane Arbus is best known for her remarkable photographs of the unfashionably homely, the bizarre, and the economically dispossessed, yet she herself was born into a comfortable, upper-middle-class Jewish world of culture and refinement. Her parents owned prosperous retail stores on Fifth Avenue, and she was raised with a great consciousness of style and beauty. Nothing in her childhood hints at her adult obsession with the underbelly of a culture inundated by the glamorous fictions of the media. Yet this may be precisely why Arbus sought to portray aspects of life shunned by the public images found in advertising. The truthfulness of Arbus' dark underworld was inevitably more meaningful to her than the superficial vanities of fashion photography.

Some of the characters in Arbus' photographic dramas assume frightening, mythical roles: Giants, midgets, prostitutes, pimps, and the mentally retarded stand before us. What saves all of these subjects from cheap sensationalism and lurid exploitation is the implied presence and vulnerability of the photographer herself. Diane Arbus enters into an intimate, sympathetic dialogue with all of her subjects. She views them as individuals strangely ennobled by the burden of humanity they carry and the isolation they feel. Although she avoided a condescending attitude of pity towards her subjects she did see elements of herself in many of their lives. They shared a tragedy but not the *same* tragedy. At the same time, she was in love with the *idea* of who and what they were and how her world and background were totally different from theirs. These misshapen individuals and social misfits allowed Arbus to feel something she felt she had been cheated out of as a child: misfortune. "One of the things I felt I suffered from as a kid was I never

felt adversity," she once remarked. "I was confirmed in a sense of unreality which I could only feel as unreality."

In part, Arbus photographed in order to feel. It was not a question of experiencing joy or pleasure, but of confirming the existence of a broader, and perhaps more interesting, world beyond the confines of her childhood milieu. Arbus' images are travel photographs of a sort, mementoes she has gathered on a trip, reminders of where she has been and whom she has visited. One of her favorite things, she was quoted as saying, is to go where she has never been before.

In terms of craft, one might almost say that Arbus was uninterested in technical means — at least the aspects of the medium that center around equipment, film emulsions, and darkroom techniques. She is anything but a photographic Formalist; clever composition, meticulous craft, and printing tricks were never what interested her most. Her photographs, however, are stunning and masterful examples of straightforward and effective visual and thematic organization. Rather than approaching her art from a background of design and formalized aesthetics — where everything is artfully posed and camera angles are carefully chosen — Arbus approaches the subject directly and unerringly.

She was first and foremost a "subject" photographer; the content of the picture was always much more important than visual and compositional elements. For her the camera was a legitimate means of reaching what really interested her, people. In this sense she is a documentarist like Eugene Atget and August Sander, the great German cataloguer of human archetypes. At the same time, she is a different kind of notetaker, very much interested in recording the inner psychological terrains of particular individuals as well as their outer appearances.

In the *Aperture* monograph devoted to her work, Arbus talked openly about her experiences photographing the kind of people at whom our parents always taught us not to stare:

Freaks was a thing I photographed a lot. It was one of the first things I photographed and it had a terrific kind of excitement for me. I just used to adore them. I still do adore some of them. I don't quite mean they're my best friends but they made me feel a mixture of shame and awe. There's a quality of legend about freaks. Like a person in a fairy tale who stops you and demands that you answer a riddle. Most people go through life dreading they'll have a traumatic experience. Freaks were born with their trauma. They've already passed their test in life. They're aristocrats.[3]

Despite the great notoriety and fame of her "freak" photographs, it is important to note that these subjects represent only a fraction of Arbus' total output. In fact, one reviewer—speaking of the Arbus monography—wrongly states that, "the majority of the photographs [in the book] are of midgets, giants, a hermaphrodite and transvestites." Exactly nine such individuals appear in a book that features almost eighty portraits. A simple majority are of physically ordinary people; the rest of the photographs feature topless dancers, circus performers, or devotees of nudism. This gross misperception can only be explained by the mind's attempt to preserve a threatened image of itself. Arbus hits dangerously close to home when she witnesses and records the freakish overtones of so-called "normal individuals."

Witness the solid majority of ordinary people who turn up in her photographs revealed and transformed by the uncompromisingly harsh light of her flash-gun: an outlandish woman in a strange, feathered bird mask; the blue-jeaned couple sitting on a park bench in Washington Square Park; two Brooklyn girls in oddly matching bathing suits; a heavily made-up blond with iridescent lipstick; a pudgy couple happily dancing.

The truly deformed souls who appear in Arbus' photographs function as "plants," placed there to lull us into complacency and give us a false sense of security — thank God we do not look like them. But after viewing a retrospective selection of Arbus photographs the line between us and them begins to break down. By the end of the exhibit we feel as if we have witnessed scores

of "freaky" and deformed people, whereas we may have only seen a handful. This then is the basis for the profoundly disquieting nature of her work, the recognition that when anything is looked at long and hard enough it becomes strange and foreboding. Arbus depicts both the deformed and the "normal" as unwitting victims of life. In *Woman on a Park Bench on a Sunny Day, New York City* (illustration), a woman gazes at us with a disturbing look that becomes more frightening the longer we stare at the print. Through the unerring directness and selectivity of her vision Arbus reveals an unseen world. "I mean if you scrutinize reality closely enough," she said, "if in some way you really, really get into it, it becomes fantastic. You know it really is totally fantastic that we look like this and you sometimes see that very clearly in a photograph."

One of Arbus' best-known photographs in the category of the physically odd is her portrayal of Eddie Carmel. The photograph is titled *A Jewish Giant at Home with His Parents in the Bronx, N.Y. 1970* (illustration). There is an almost scientific and anthropological tone to the title that belies the amazing vision Arbus presents to us in the picture: a mythical apparition, amusing and tragic at the same time. The verbal description pretends to present us with the necessary biographical data one would need to decode the image. Nothing makes it easy to interpret, explain, or even "see" this incongruous scene. The picture refuses to fit any previous category of photograph we may have encountered — family snapshot, fashion illustration, or newsphoto. There is no "fine-arts" look to this piece, no tasteful composition, no beautiful subject matter, no lush natural light — just the brilliant glare of a flash-gun that illuminates the trio. The harshness of the artificial light reveals with minute detail every aspect of this ordinary–extraordinary scene: the shrouded sofa and armchair, the cellophane-covered lampshades, his mother's delicate printed dress, scraps of paper on the couch, framed reproductions of classical paintings, his father's formal attire (no doubt donned for the photograph), and, of course, the enormous height and little-boy look of Carmel himself. His mother looks up at him in self-conscious puzzlement and wonder; his father gazes abstractly into the distance past him.

The photograph looks suspiciously out of kilter; it appears to have been taken in a specially constructed room, the kind one finds at science expositions that change and distort perspective. The longer we look at this documentary photograph, the stranger things become. In fact, with no great mental gymnastics it is quite easy to turn the picture around and envision a normal-sized

Diane Arbus, *Woman on a Park Bench on a Sunny Day, New York City* (Collection, The Museum of Modern Art, New York. Mrs. Armand Bartos Fund. © 1972 The Estate of Diane Arbus).

Diane Arbus, *A Jewish Giant at Home with His Parents in the Bronx, New York 1970* (Collection, The Museum of Modern Art, New York. Mrs. Douglas Auchincloss Fund. © 1971 The Estate of Diane Arbus).

Eddie Carmel juxtaposed with Lilliputian furniture and tiny parents.

One is struck with the remarkable consistency and continuity in all of Arbus' pictures; even though she photographs the broadest range of individuals, she seems to be showing us the same thing. Perhaps this uncanny ability to reveal the hidden and frightening nature of humanity is the reason she has gained the reputation of mainly documenting "freaks." The irony is that quite often in Arbus' work these deformed people appear to have an acceptance of themselves and life that often surpasses the attitude of her "straight" subjects. Witness the confident assurance of a carnival hermaphrodite at home in his trailer or the perfectly cozy scene of the Russian midgets in their apartment on 100th Street in New York City.

One of the most haunting photographs of Arbus' career is the widely reproduced *Identical Twins, Roselle, N.J. 1967* (illustration). There is nothing overtly grotesque about these two handsome girls. They stand at attention before us like dual apparitions —immovable, inexplicable. Most viewers are captivated by this striking image yet cannot determine just what affects them. Perhaps part of their fascination lies in the mind's inherent love of comparison. Their similarity compels us to study them and note the slightest nuance of dissimilarity. For instance, the girl on the right subtly smiles, the girl on the left does not. Despite their genetic sameness there are significant differences that reveal themselves the more we study this odd duo.

Inevitably, this photograph functions on a level beyond conscious perception, entering us swiftly and incomprehensively. If one looks at it deeply enough, the innocent girls take on a frightening appearance— they become twin messengers from a mythical *Village of the Damned* about to tell us something we would rather not hear.

Despite their Siamese-twin similarity— Arbus even visually fuses their dark corduroy dresses together—they seem worlds apart psychologically. Arbus' own words hauntingly seem to comment on this study of sameness. "What I'm trying to describe is that it's impossible to get out of your skin into somebody else's and that's what all this is a little bit about." Perhaps our uneasiness in the presence of this photograph stems from Arbus' depiction of the ultimate psychological isolation: to be so physically close to an identical copy of yourself yet mentally miles apart.

The underlying theme of aloneness— especially in photographs that feature couples—is present throughout Arbus' entire body of work. She shows us, with compassion and truth, the tremendous odds we face trying to find comfort and solace in a less-than-perfect world. Nowhere is this attitude expressed more touchingly than in *Teenage Couple on Hudson Street, N.Y.C. 1963* (illustration); in this photograph she captures something at once brave and sad. The diminutive body size, adolescent features, and awkward posture of the posing couple is juxtaposed with their adult demeanor and attire. The boy is dressed in a sports jacket, tie, overcoat, and dress shoes with his hair rakishly combed in the style of the day; his eyes shyly avoid confrontation with the camera. She, half repressing a nervous smile, looks at us directly, proud to be photographed with her man. Her clothes are also "grown-up"—camel hair coat, print dress, hose, and stylish shoes. Arbus captures this couple out on a date. Everything about them, their look, attitude, and stance, suggests a valiant effort to mimic the adult world they want so desperately to enter. Hand gestures play an important role in this teenage narrative: His right hand clutches her shoulder, holding her near; his other hand is insecurely thrust into the pocket of his overcoat. She has her hidden arm around his waist; the other is awkwardly held to her side.

The environmental context in which they are immersed adds to the picture of brave optimism in the face of sociological and economic adversity. The brick wall with its cryptic graffiti messages from individuals with no social standing, the edge of a dark doorway that obliquely juts into the picture, and, above all, the street debris—cigarette butt, soda-straw wrapper, and crumpled paper—are all important details

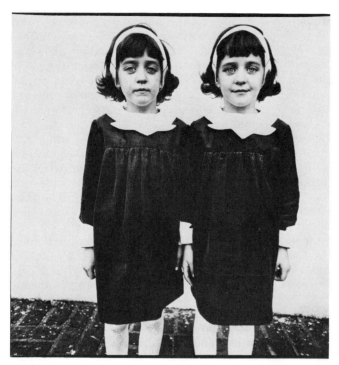

Diane Arbus, *Identical Twins, Roselle, New Jersey 1967* (Collection, The Museum of Modern Art, New York. Richard Avedon Fund. © The Estate of Diane Arbus).

Diane Arbus, *Teenage Couple on Hudson Street, New York City 1963* (Collection, The Museum of Modern Art, New York. Purchase. © 1972 The Estate of Diane Arbus).

in this dark prognostication of the couple's socioeconomic fate. Arbus seems to ask, "What lies before them?" Their attitude, half-aware and half-oblivious, makes the scene all the more touching. Arbus is no social documentarian, however, preferring to invent her own vision of reality rather than capture it. "Lately," she said in her monograph, "I've been struck with how I really love what you can't see in a photograph. An actual physical darkness. And it's very thrilling for me to see darkness again."

Reading Arbus' words, one is struck with the degree to which her aesthetic sensibility is reminiscent of the character of Holden Caulfield, the teenaged hero of J. D. Salinger's *Catcher in the Rye*. The piercing sensitivity and alternately amused-horrified outlook on life show up regularly in all of Arbus' photographs. Like the adolescent, Holden Caulfield, Arbus saw things simultaneously from two conflicting points of view: the idealistic, wide-eyed wonder of a child and the oppressive, "adult" vision of diminishing possibilities and rapidly fading hopes.

Every stranger who might be encountered on the street had for Arbus an unfathomable potential for both innocence and evil. Arbus captured both of these qualities in one of her most provocative and disturbing images, *Child with a Toy Hand Grenade in Central Park, N.Y.C. 1962* (illustration). Despite the benign context of the piece the young child playing in the park appears almost supernatural and frightening. The terror-stricken eyes and distorted mouth transform this no-doubt perfectly normal child into a disturbing image. As in most of Arbus' photographs, the subject is simply placed in the center of the composition and is allowed to incongruously contrast with the background — in this case, a sun-dappled and tree-lined walkway. Somehow he becomes all the more evil in appearance because of this innocent environment. His mother or nanny peeks out fuzzily from behind his head and patiently waits in the background; a toddler and a baby in a carriage approach, completely unaware of the drama taking place in front of them. Only

Arbus sees the truth of this particular moment; as she once stated, a good photograph, ". . . is a secret about a secret."

Again the details of the print hammer home Arbus' message of hidden despair and reveal the tawdry underpinnings of human existence: his cutesy print shirt and little-boy shorts with the loose shoulder strap; one hand holding a plastic hand grenade, the other tensely clutching, claw-like, an imaginary one; his no-doubt normally angelic face distorted and transformed beyond normal bounds; and, again, a litter-strewn ground of scrap papers, wooden ice-cream spoon, and straw wrapper. Innocence and horror commingle in ways that seem to go far beyond rational explanation. Because facial expressions are so mobile and fleeting we often miss them. But using the instantaneous recording ability of the camera is a way of fixing these moments of short-lived reality that pass before us, usually unrecognized. Arbus uses this instrument as a means of breaking through to another world, a world physically close but conceptually distant from our ordinary perceptive. "I really believe," she said, "there are things which nobody would see unless I photographed them."

In the summer of 1971, at the age of 48, Diane Arbus took her own life. Much has been made of this tragic event, in light of the remarkably probing images that established her as one of the most original and powerful photographic artists of the last two decades. Since her death her work has been shown at the U.S. pavilion of the Venice Biennial (she was the first American to be so honored) and in a large and enormously popular retrospective at the New York Museum of Modern Art. Arbus changed the parameters and territory of photographic art and her influence in this area is indisputable.

Although some critics and commentators have observed that this tragic fate was inevitable in view of her work — she went too far and looked too deeply — this attitude seems too patronizing and glib. In a significant way her photography stands apart from her final self-destructive act. All of her pictures connect her, and us, to a revelatory perception of humanity that optimistically links us to life.

Those who knew her remember her as a shy, perceptive, trim woman of extraordinary grace and charm who had the courage to look closely at the things most of us avoid. Diane Arbus will be remembered as an adventurous artist who held up her camera in wonder and amazement to the broadest range of humanity possible — courageous giants, crying babies, and bewildered adults.

Diane Arbus, *Child with Toy Hand Grenade, New York 1962* (Collection, The Museum of Modern Art, New York. Purchase. © 1962 The Estate of Diane Arbus).

192

Duane Michals

Like Meyerowitz, Duane Michals became introduced to photography by chance circumstance. During a trip to Russia someone lent him the use of a camera, and he occupied himself by taking pictures of buildings, people, and scenery along the way. What particularly fascinated Michals was the willingness of the Russian people to be photographed by a visiting westerner. Old peasant women, factory workers, and cherubic children lined up in front of his camera to pose for him. Michals discovered, quite early in the game, the power a camera exerted over people.

While the artist was establishing himself in the field of fine-art photography, fashion and magazine work formed the bulk of his educational experience in this field. Until fairly recently few degree-granting institutions offered programs in the study of photography. During the last decade, however, this situation has changed radically; it is now a rare art school or university that does not have a well-staffed and adequately equipped photography department.

Michals started out as a reportage photographer, taking essentially documentary images. He steadily developed an approach to picture taking that centered on the setting up or creation of events and pictorial situations. Unlike photographers of the rapid-fire school, Michals does not attempt to capture the fleeting moment but to re-stage it. Through this method he hopes to reveal an otherwise hidden structure of the world generally unperceived in everyday life. Michals carefully conceives and sets up scenes, usually of a narrative nature, that often tell a strange, mystical story about personal relationships. What he produces is in a sense the opposite of documentary "street" photography: Whereas that form of picturemaking seeks to reveal the underlying Surrealistic structure of random events, Michals painstakingly stages fantasy events that take on a disturbing air of reality. In a book he produced, titled *Real Dreams*, the artist comments on the difference between "straight" photography and his own creations: "Some photographers literally shoot everything that moves, hoping somehow, in all that confusion to discover a photograph. The difference between the artist and amateur is a sense of control. There is a great power in knowing exactly what you are doing, even when you don't know."

Michals starts with a preconceived idea or story plan, basing the structure of the photographs on the needs of the concept. Quite often language is used—the artist writes commentary or titles on the photographs—which functions as a verbal conditioner for the images we view. His captions shape our experiences of the pictures with startling effectiveness. Generally, Michals uses more than a single photograph to sustain the narrative element and provide a sense of flow to the story. Although much work in recent times has tended towards the reportage or documentary approach, it must be remembered that the story-telling or illustrational possibilities of photography were perceived from the beginning. The anecdotal quality to Michals' work—although new in appearance—actually extends back to the early days of the medium. Two nineteenth-century photographers, Julia Margaret Cameron and Henry Peach Robinson, staged elaborate allegorical setups for the camera. During the late 1800s they created dreamy pre-Raphaelite photographs that seemed to hover in spirit and feeling between this world and the next. Cameron's *The Whisper of the Muse,* for instance, shows a heavily bearded man with half-shut eyes, holding a stringed instrument, in the company of two ethereal young girls—Nymphs. One is drawn close to him about to whisper privileged information in his ear; the other coyly sits at the bottom left corner of the photograph, directly confronting our gaze.

Robinson's most famous composition, *Fading Away,* is no less otherworldly in spirit. Like many of Michals' photographs, it thematically encompasses powerful emotions: anguish, a sense of loss, emotional trauma, and a feeling of helplessness in the face of life's uncertainties. This photograph, taken in 1858, shows a beautiful young woman lying prone on her death bed. It is twilight;

the father stands pensively with his back to the camera, staring out the window into the fading light. The sick girl's mother is near her, offering what comfort she can. Both Robinson and Michals invent the reality of their photographic vision by employing multiple images and techniques like double exposure, printing one image on top of another.

The Human Condition, 1969 (illustration), is an early narrative piece by Michals which surprisingly resembles Robinson's and Cameron's aesthetic in feeling, mood, and methodology. The first image of this sequence simply shows a young man with sports jacket and bushy hair, standing on the subway platform of the 14th Street station and looking directly at the camera. All around him people seem intent on boarding the nearby subway car that has just pulled in. In the next frame a mysterious light seems to flood the central figure from above, bleaching out his features and figure almost to the point of nonexistence. The third frame introduces snowlike reticulation in the dark areas to the left and right; the image of the man is fainter. By the fourth and fifth frames, the background setting of the photograph has gradually shifted from the subway station to the far reaches of the galaxy, conceptually sweeping the figure into a blurred and gaseous nebula. The sixth and final image shows a pristine, sharply focused Milky Way floating serenely in space. The subway patron has disappeared, absorbed and transformed by this stellar event.

In this piece Michals comments poignantly on the theoretical origins of our existence as a species (that the earth—and therefore humanity—was formed from gaseous galactic material) through provocative and poetic juxtaposition. Scale differential plays a vital role in this piece; Michals takes advantage of photography's capacity to make large objects small and to blow up the tiniest of forms to great proportions. Thus, a six-foot-tall figure and a celestial body, millions of light years across, can both fit into an 8-inch by 10-inch photographic print.

Another sequential "photo-story" (Mich-

als' term) is the mysterious *Chance Meeting* (illustration), photographed in 1970. This piece, carefully staged on a New York City street, may make wry reference to documentary photographers who work the same territory in hopes of capturing such a fleeting event. In a certain sense Michals sees himself as an "antiphotographic" photographer. Whereas many artists rely on purely visual givens and documentary methods, Michals makes use of literary concepts and symbolic elements; and if most photographers shoot single, hermetic images, this artist produces sequential pictures that tell a story and are meaningless if seen alone.

In the first frame of *Chance Meeting*, two men, similarly dressed, approach each other along a littered city sidewalk. The man on the right, walking away from us, almost fills the frame; the other man is in the center of the composition, walking toward us. Sequence two shows them nearing each other, about to pass; the next frame documents the moment of passing, with the figure on the right staring directly ahead, and the man on the left turning at this instant to continue looking at the other fellow. Frames four and five show the gentleman walking towards us turning around and looking contemplatively back. In the final frame the man whose back has always been turned to us has now stopped and is looking back quizzically.

Michals' narratives are fascinating because, despite the clear structure and straightforward photographic illusionism, they are marvelously open in meaning and content. More questions are raised than are answered by this puzzling sequence of "street photographs." Did these men go to high school together many years ago, once live on the same block, or work for the same company at one time? Or could this be the beginning of a furtive homosexual encounter? Only enough information is given to set our own imaginations in motion. Michals constantly infuses his work with themes of recognition, loss, and memory; the act of photography becomes a means of creating thematically elusive dramas and recalling forgotten dreams.

One of the best-known and widely ad-

The Human Condition

2

3

Duane Michals, *The Human Condition* (©
Duane Michals).

mired narrative artworks of Michals' career is the curious piece called *Things Are Queer*, (illustration) completed in 1973. Like *The Human Condition* it envelopes us in a confounding set of spatial illusions. This particular work also confronts the issue of photographic reality; Michals suggests that perhaps we should not accept the truthfulness of a photograph as readily as we do. This series expresses an *Alice in Wonderland* sensibility: What you see may not accurately represent what exists. Photo-

graphic vision, the photographer reminds us in this sequence, depends on what we think we know about the subject in question.

Things are Queer opens with what appears to be a photograph of an ordinary bathroom: white porcelain sink with a small framed print over it. But the next image immediately causes us to doubt the veracity of the preceding image. An enormous foot stands next to the now miniscule plumbing fixtures. In the next sequence the camera

Duane Michals, *Chance Meeting* (© Duane Michals).

"pulls back" even further to reveal a bearded man standing in a full-sized room with a large mirror to the right, which clearly reflects the toilet and sink but only a faint image of the man. The fourth image transforms the scale and "reality" of the series even further: We now see the preceding photograph as an illustration in a book with a large thumb covering part of the text and image. In the fifth and sixth sequences the camera appears to move back still further to show a man holding a small book in front of him. Sequences seven and eight show the photograph framed and hanging over a sink. The last photograph is exactly the same as the first; we have come full circle in a remarkable journey through photographic space and time which, as Michals shows us, is entirely mutable.

One of the most emotive artworks in Michals' repertoire is, surprisingly, a single-image photograph, titled *This Photograph Is My Proof* (illustration) with a short text written underneath. This work un-

derscores the point that despite the popular saying "Photographs never lie," quite often they do. The image in this photograph shows a smiling couple, apparently in love, sitting together on a bed. The woman endearingly rests her head on the man's shoulder and wraps her arms around him. In many respects this is quite an ordinary photograph. But Michals' written language cuts through the radiant image like a knife. The title itself, *This Photograph Is My Proof* sets up a defensive posture that contradicts the apparent meaning of the image. Language interacts with the picture in such a way as to make us doubt its veracity. The statement below the image rings of loss, loneliness, and disillusionment: "This photograph is my proof. There was that afternoon when things were still good between us, and she embraced me and we were so happy. It did happen, She did love me. Look see for yourself." Because of the insistently pleading nature of the language, we begin to doubt the reality of the image.

Things are Queer

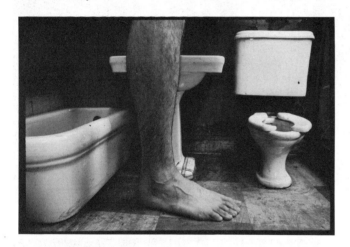

3

Duane Michals, *Things Are Queer* (© Duane Michals).

CURRENTS

Things are Queer

The writer and the subjects in the photograph remain anonymous; perhaps they are unrelated; perhaps the writer faked the photograph and invented the melodramatic story from thin air. Ultimately, the power of this narrative lies in the strong possibility that verbal fiction is compounding a visual lie — that *nothing* about this clear image and easily read text is true.

Confronting a Duane Michals' piece is somewhat like peering into another person's daydreams and fantasies. In the final analysis one of the most remarkable aspects of Michals' work is the way he completely transforms photography into a medium that undermines the foundation upon which it claims to be built — the objective recording of visual appearance and reality. His strange personal journals project us into an hallucinatory domain where things are not what they appear to be. Philosophically, Michals believes that the visible world which most of us view with certainty is only an apparition, a figment of our collective imagination. In support of this view, he once wrote:

"I am a reflection photographing other reflections within reflections. It is a melancholy truth I must always fail. To photograph reality is to photograph nothing."

Ralph Gibson

Ralph Gibson, like many of his peers, sharpened his photographic skills and earned a living for many years doing freelance work for advertising agencies and large corporations. During the sixties he worked on assignment for Eli Lilly Pharmaceuticals, Bantam Books, and *Look* and *New York* magazines. The commercial world of photographic illustration formed his training ground and apprenticeship.

Gibson got his start in picturemaking during a tour of duty in the Navy; he spent three years as a photographer's mate taking a variety of documentary photos of personnel and equipment. After the Navy he attended the San Francisco Art Institute, one of the first art schools in the country to offer a degree program in photographic

Duane Michals, *This Photograph Is My Proof* (© Duane Michals).

studies. Believing that apprenticeship is one of the best ways to learn the craft and business skills necessary to survive in a highly competitive field like fine-art photography, Gibson secured a position as darkroom assistant to Dorothea Lange, a well-known documentary photographer of the Depression years. Gibson spent a year and a half learning how to print and run a free-lance photography business.

Like Meyerowitz, he was also impressed with Robert Frank and his influential book, *The Americans*. One of the reasons this artist-produced publication had such an impact on the field of "art" photography — aside from the outstanding quality of the pictures — was that it offered new distribution opportunities for the serious photographer. As the big picture-news magazines folded one by one, there were fewer places in which to publish serious photographic work. Editors of the surviving journals were less and less interested in the aesthetics of the picture story — unlike the golden era of the late forties and fifties — and more concerned with getting the news photograph quickly, easily, and cheaply.

Frank's book showed artists, whose best efforts were thwarted by insensitive editors and cost-conscious managers, an exciting alternative for publishing their work. The book format, carefully controlled in content and scope by the artist, seemed to offer the greatest rewards. Above all it was the *concept* of *The Americans* that distinguished it from other photographic publications. It seemed less a collection and more like a well-thought-out visual essay, stating a thesis and developing it step by step, photograph by photograph. The book derived its power from its unified statement and completeness of vision that most photographic collections lacked.

Gibson well understood this aspect of Frank's work and adopted it as an important model in the development of his own style and work. Because photography was still in its infancy — as far as public recognition and acceptance were concerned — it was necessary to seek other means of reaching a greater audience. The artist-produced book, in which the photographer retains complete control over contents, design, and production, seemed to be one of the answers to the distribution problem.

Showing remarkable enterprise and initiative, Gibson and some associates founded their own fine-arts photographic publishing house, called Lustrum Press, and in 1970 produced a highly popular and critically acclaimed book of Gibson's work, *The Somnambulist*. The book resembles Frank's *The Americans* in structure and style; but, in terms of thematic content, it very much represents Gibson's own personal style and aesthetic predilections. The subtle vision that pervades the book can only be described as a kind of documentary Surrealism. Gibson takes the deadpan, high-contrast realism of Frank's reportage style and transforms it into an enigmatic, pensive view of the world by isolating parts or details of things (a detached pair of hands, the corner of a building). There are no distorted, dripping watches, à la Salvador Dali, or displaced scale juxtapositions like Rene Magrite; nevertheless the unmistakable tone of the book is one of a disquieting unreality. Gibson uses the structure of the book form itself, with its sequential linearity and double-page layout that consistently places two images side by side, to reinforce this effect. No titles appear to destroy the feeling of wholeness; what we see are two photographs that Gibson wants us to see and consider in relationship to one another. One double page, for instance, features a picture of a nude woman sitting cross-legged on a low hassock with the dark outline of a man about to take off his hat in the foreground; on the next page is a strange photograph of a man's face distorted in a weird grimace on the bottom of the print; to the right appears a shadowy silhouette of a baby with outstretched hand. The next page contains an equally striking juxtaposition. The photograph on the left shows a high-contrast image of a beautiful woman standing in a park with her outstretched hand interlocking with another hand (perhaps the photographer's) that enters the frame from the bottom left; the image on the right documents a burning store, the Snip and Clip beauty salon. Throughout the book

Gibson compares and juxtaposes a variety of striking images: people enveloped in clouds of smoke, a sunflower towering over a house, a man sitting in a parked car, and carefully composed sections of bodies severely cropped and starkly lighted. Looking through the book one gets the feeling that this could be the storyboard outline for a film — not so gratuitous a perception, considering that Gibson occasionally worked on 16-mm films as a cameraman (including the CBS documentary called *Conversations in Vermont* that Robert Frank made).

Everything about *The Somnambulist* is well thought out and subtly designed to reinforce the theme of duality that runs throughout the book — reality and dream. On the first page a photograph of a sand dune and sky appears with a superimposed image of a man's hand holding a pen, as if it is drawing and inventing the reality of the camera's image. A verbal statement on the right reads in part, "While sleeping a dreamer reappears elsewhere on the planet, becoming at least two men. His sleeping dreams provide the substance of that reality while his waking dreams become what he thought was his life."

One of the most striking photographs in *The Somnambulist*, which occurs in the very beginning and sets the mood for the entire work, is an image that shows a hand mysteriously reaching through a doorway, casting a shadow on the opposing wall (illustration). Within the geometric framework of the hall, open door, and empty room, the presence and power of the lone hand is compelling.

Gibson creates a unique experience by isolating the hand in this way. By doing so he invokes a powerful human presence through economical means; this part of the body is one of the oldest and most powerful images in the history of art. Its unique dex-

Ralph Gibson, from *The Somnambulist* (photo courtesy of Castelli Graphics, New York. © Ralph Gibson).

terity and ability to perform complex functions also symbolize the essence of our humanity. Gibson invokes a feeling of timelessness in this print; he reminds us of Neolithic cave works in which early artists placed their outstretched palms on the damp cave walls and covered body and stone with colored pigments. When the hands were removed they left behind permanent negative images. Through their painting ritual, they created a permanent graphic record of their presence that immortalized their fleeting presence. Gibson's deserted building assumes the role of a contemporary cave. No traces of occupancy are visible—no furniture, clothes, or other artifacts—just the dominating presence of a human being symbolized by the reaching hand.

The underlying theme of a human body in a precarious balance with the environment is repeated throughout the book. A foot or a section of the face or hand appears with regularity; we never really know to whom these parts belong or what the people might be doing. Gibson obscures details that might ground the photograph in clear but mundane references to everyday reality. He never supplies answers to questions about location or the subject's activity, preferring to provoke and titillate aesthetically rather than explain. Gibson achieves these effects without means of trick photography by carefully selecting camera angles, suppressing details, and juxtaposing objects that might occur naturally in everyday life. Despite the studied, almost formal organization of Gibson's photographs, they look like a heightened form of fiction that comes disturbingly close to reality.

Ralph Gibson, from *The Somnambulist* (photo courtesy of Castelli Graphics, New York. © Ralph Gibson, 1982).

Ralph Gibson, from *Deja-Vu* (photo courtesy of Castelli Graphics, New York. © Ralph Gibson, 1982).

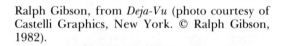

Another photograph that describes the multidimensional world in which Gibson revels shows a man walking through a swirling cloud of vapor (illustration). In sections of Lower Manhattan, live steam—sold to office and commercial buildings by Con Edison, the local utility company—escapes from leaky pipes and surfaces. Gibson makes apt use of this ready-made Surrealistic environment to create a mysterious and diabolical urban underworld. By simply holding his camera askew, he creates the illusion of a body floating through a cloudy sky. Simple, bold elements—the man darkly outlined against the white ground, the diagonal lines of the sidewalk, and the atmospheric illusion—create an engaging composition of power and mystery.

Gibson obviously holds on to many elements of documentary photography gleaned from individuals like Dorothea Lange and Robert Frank. But his work is clearly distinguished from theirs by its strong, almost abstract elements and conceptual overtones. One photograph that appears in Gibson's second book, *Deja-Vu*, clearly illustrates the way he has combined aspects of documentary street photography with influences of Minimalist art. Like the previous photograph of the steam, this image also includes a fragmentary figure that partially emerges into the picture frame from the left (illustration). Gibson consistently makes use of high-contrast printing techniques and the inherent grain structure of fast films to create powerful graphic images. The relationship between the long rod the man is carrying and the painted street line is considerably heightened by these processes. Through strict control of technical means—the use of film developers and high-contrast printing papers—

Ralph Gibson, from *Days at Sea* (photo courtesy of Castelli Graphics, New York. © Ralph Gibson, 1982).

Gibson achieves a synthesis of Realism and Formalism.

Often figure–ground relationships are set up that transform commonplace objects into semiabstracted shapes and patterns. In this respect Gibson has been profoundly influenced by the work of many different painters and artists. His early training at the San Francisco Art Institute stressed the interrelationship between all the visual arts rather than the separate identity of photography as a craft. Recently the work and writings of Kasmir Malevich, the Russian Constructivist, have particularly inspired him. In an interview with Allen Porter, Gibson stated:

. . . his (Malevich's) writings in "The Non-Objective World" have fuelled my mind and given me courage . . . he says, "We are bound to acknowledge that the subconscious mind is more infallible than the conscious." An interesting and certainly true statement . . . I always come to understand my work after I study it for long periods of time, but I rarely understand it immediately. . . . It takes my conscious mind, or intellect, time to catch up with my subconscious.[4]

Gibson's recent book project, *Days at Sea*, extends his concerns for contrasting light

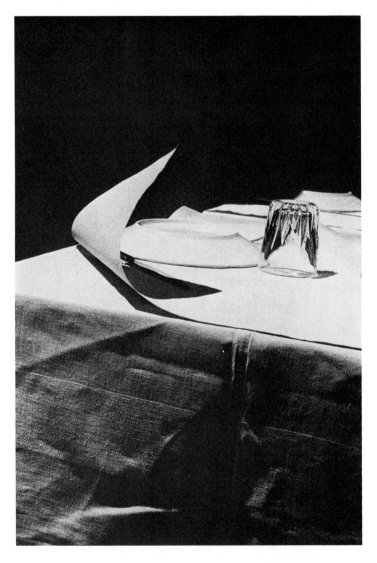

Ralph Gibson, from *Days at Sea* (photo courtesy of Castelli Graphics, New York. © Ralph Gibson, 1982).

and surface textures into an even more refined mode. The essay opens with a photograph that shows a man sitting on a dock beside the ocean (illustration). Only one leg is visible as it extends into the picture frame clothed in an elegant pin-striped material. Although the subject is wearing a shoe, he is not wearing any socks and his white skin stands out in marked contrast to the dark water. Gibson achieves many of his visual effects by carefully controlling the light in his photographs. From the appearance of the shadows raking across the bottom of the print, it seems that this shot was taken either early in the morning or late in the day. The ocean, bare skin, and weathered texture of the wharf evoke feelings of the sea and voyages to exotic lands. Appearing throughout the book are sensuous photographs of female nudes, curvilinear architectural details, and isolated sections of the body like thighs, hands, and arms.

Gibson's vision in *Days at Sea* is broad yet unified; we float effortlessly in a subtly erotic assortment of human and architectural forms. The artist's ability to compose subject matter, control lighting, and arrange sequential order achieves thematic unity.

Gibson confers a certain status to everything he photographs. All of his subjects assume equal importance; he is not particularly interested in the social significance of what he turns his camera on. He is, however, concerned with evoking a unified vision of the world that is dependent on his disciplined visual acuity. Witness his arresting photograph of a paper-covered dining table with overturned plates and a glass (illustration). There is something subliminally sexy and elegant about this particular piece. Again, it is the exclusion of extraneous details, the oblique lighting, and the simple organization of the photograph that causes it to become at once an ordinary table scene and a sensuous, metaphysical landscape. Gibson often walks a fine line, rarely straying into the mundane territory of the banal or the rarified atmosphere of purely formalistic design.

The result of this synthesis is a form of Neosurrealism. Unlike the historical movement of the thirties, however, the artist is not overtly concerned with obvious symbolism. His objects and people float in a realm of open-ended meaning and exploration. In this way Ralph Gibson's photography departs from the limiting sphere of documentary realism and becomes an engaging exploration of the physical vastness of the visual world.

Cindy Sherman

Cindy Sherman uses photography to expose the psychological manipulations that lie beneath the surface of slick media presentations. Her large-color photographs of herself are set in poses reminiscent of Hollywood films of the fifties, in which the female lead is invariably cast as a helpless and hapless individual unable to assume control over her life. But Sherman's images are not didactic—they lure us into their patently make-believe world subtly and without overt feminist criticism or sermons about the evils of the media. Ambiguity is very much a part of their aesthetic stance—are we meant to view these images humorously as sentimental camp, or as serious manifestations of her psyche, or perhaps both? Sherman's work allows us to consider both scenarios equally. Her finely balanced images work for the most part in this meaningfully contradictory state. Sherman is certainly not alone in using ambivalence as an ingredient in her art—David Salle's images purposefully contradict each other thematically, visually, and psychologically. This aesthetic position denies the existence of any single truth. Perhaps this polymorphous stance is in response to the enormous changes that have occurred in the art world during the last several decades. Artistic styles and beliefs have emerged and receded with alarming rapidity. Artists like Sherman and Salle have positioned themselves equidistant between a tenuous loyalty to late Modern traditions and haunting doubt that many of the tenets of Modernism are outmoded and irrelevant. We must remember that Sherman and Salle's generation were raised entirely on television, film, and countless magazine ads—images and voices introduced them to manufactured wonders and urged them to participate in a collective technicolor dream. Not surprisingly—given the formal education of most artists—they have learned much about the idealistic goals of "high" art while immersed in a sea of seductive "low" art. Sherman's photographs demonstrate the tension that exists between these two camps and enable us to think critically about the role of the media in the visual arts today.

Sherman was born in 1954 in northern New Jersey but moved with her family to Huntington Beach on Long Island when she was three. Growing up, Sherman recalls spending a lot of time sitting in front of the television set and drawing. Encouraged by her teachers, she continued to draw and paint through high school and entered the State University of New York (SUNY) at Buffalo in 1972. Sherman's mother suggested she major in art education so she could teach when she graduated. At this stage of her career she claims she never really gave much thought to being a full-time artist.

Near the SUNY campus was a building —orginally an old ice house—owned by the Ashford Hollow Foundation, containing studio spaces that could be rented by State University art students. It was here that Sherman and her friend and fellow student Robert Longo founded Hallwalls, a wildly successful alternative art space. After operating only one year they received an $8500 grant from the New York State Council on the Arts and used the money to bring a variety of contemporary artists to show at Hallwalls.

Sherman was secretary of this alternative space and within a short time met a host of New York conceptual and performance artists. It was about this time that Sherman, believing painting was dead, decided to get into photography.

Although the technical end of photography did not particularly interest her, she became quite involved with the conceptual issues that were emerging in the midseventies. One course assignment was to use a series of photographs that represented the passage of time. Sherman was excited by the project but could not decide what to do. Longo, by now a good friend of Sherman, knew about her fondness for making herself up and dressing in outrageous outfits to amuse herself and her friends. Longo suggested doing a series of photographs of putting on makeup. "So I did this transitional series," Sherman recalls, "—from no makeup at all to me looking like a completely different person. The piece got all this feedback. It dawned on me that I'd hit on something." At this point Sherman realized that photography was the ideal medium in which to explore her ideas.

After graduation Sherman and Longo moved to New York City, sublet a loft, and took jobs to support their artmaking. Late in 1977 Sherman began a series of black and white photographs called "Untitled Film Stills." These images were inspired by some soft-core pornographic photos she saw that an artist friend who worked for a Manhattan publisher brought home. Sherman was intrigued by the thematic formats of the photographs. They seemed to be from fifties movies and showed women in a variety of situations. "What was interesting to me," Sherman noted, "was that you couldn't tell whether each photograph was just its own isolated shot, or whether it was in a series that included other shots that I wasn't seeing. Maybe there were others that continued some kind of story. It was really ambiguous."

To make those "film stills" Sherman relied on her friends to trip the shutter when she felt the pose was correct. But one of Sherman's most successful early black and white images, *Untitled Film Still #48*, 1979 (illustration), was taken by her father. While visiting her parents in Arizona (they had retired there when she was at SUNY), Sherman set out to make a whimsical version of *Heidi*. Dressed in a plaid skirt and blond wig she wanted to jump a fence into a cow pasture and have her father take a few pictures, but the cows proved very uncooperative. While driving home she asked her father, just for fun, to take a few pictures of her standing with her suitcase (full of wigs and costumes) alongside the road gazing into the distance. The photograph that

Cindy Sherman, *Untitled Film Still #48,* 1979 (photo courtesy of Metro Pictures).

resulted from this session has been called "The Hitch-hiker." Probably the most successful of her early images, there is a wistful, melancholic quality about it. Sherman had succeeded in combining many romantic but ambiguous pictorial elements into one compelling picture. Compositionally, the white stripe of the highway leads our eye close to and around the luminous image of the hitch-hiker forlornly standing by the road side. The background is as idyllic as any framed reproduction one might find in a Western motel. Pine trees, rock outcroppings, and a sunset-streaked sky are perfect foils to the poignant image of Sherman standing next to her suitcase.

In 1981 Sherman began to photograph indoors in order to gain more control of elements such as lighting and background. At this time she also switched to color printing and exhibited large photographs that measured 2 by 4 feet. The focus of this new work had shifted. These images were cropped tighter and greatly emphasized her facial expressions. *Untitled 1981* (colorplate) reveals Sherman cast into the archetypal role Hollywood regularly assigns to women: that of the frightened and helpless maiden. This posture, according to cul-

tural myth, is designed to evoke erotic feelings in the perceptions of male viewers.

The women Sherman portrays in this series all fit neatly into stereotypic categories; they are passive, defenseless, and dependent on exterior forces. All of her characters have withdrawn into dream states and are seemingly unable or unwilling to do anything about it. Sherman portrays most of them as players in a drama in which they have no control — they are waiting for outside events to intervene and change their fates.

Even when one of these women is seething with anger it remains for the most part repressed beneath the surface. *Untitled 1983* (illustration) shows Sherman dressed in a classic black suit, standing rigidly, with her hands clenched in anger. The small portion of her face not covered by the blond wig is made up with red cosmetics to give the impression of someone angry enough to burst. This is the closest Sherman's women get to expressing anger and actively trying to control their lives.

Regardless of how we interpret Sherman's images, there is a degree of recognizability to them that accounts, in large part, for their popular and critical success.

Cindy Sherman, *Untitled 1983* (photo courtesy of Metro Pictures).

Cindy Sherman, *Ad for Diane B* (photo courtesy of Metro Pictures).

Sherman's invented characters evoke the haunting feeling that we know these people, yet we cannot quite place them with any certainty—it is like seeing people on the street who look like famous television actors but remaining unsure of their real identity. Unlike much art of a decade ago, we do not have to know much about recent art history to appreciate Sherman's photographs; they are inspired by and subtly participate in the business of advertising and media manipulation.

Considering the close relationship that exists between Sherman's work and advertising art, it may come as no surprise to learn of a new series of photographs that blur the lines between art and commerce even more.

In the early eighties Diane Benson, the avant-garde clothing designer and owner of the Diane B Company, asked Sherman if she would create a series of ads using her fashion line. Since there were no restrictions from her client other than using Diane B clothes, Sherman accepted the offer with the understanding that she would retain the right to exhibit and sell any photographs she did for the ad campaign. "They let me do whatever I wanted to do," Sherman said, "and even liked some of the funniest and ugliest photographs the best." Sherman was an ideal choice for Diane B since the store projected a culturally sophisticated and aesthetically progressive image. Since Sherman was not obligated to feature the client's product, what resulted was a series remarkably similar to her earlier work. In one image from the Diane B series (illustration) Sherman is seen enveloped by the oversized Diane B clothing that she clutches in both of her hands. Sherman's persona and pose are quite similar to those projected in past work. Her gaze is directed inward and she is seemingly lost in a reverie.

Inevitably, Sherman's images encourage us to freshly perceive and critique many aspects of our media-dominated environment. By assuming some of the manipulative stances of advertising and transposing these affectations into the realm of "fine art," she exposes the media for what it is —a finely tuned machine that turns out seductive fantasy images designed to mesmerize us and sell us products. Sherman's images simulate these commercial dream images and lure us farther and farther into convoluted conceptual labyrinths. In essence her photographic work represents the psychological control that seems to be at the very core of our complex and highly structured society.

Photography, for many years insecure about its acceptance into the pantheon of the "Fine Arts," can now rest assured in terms of its contribution to Modernist thought. Arbus, Weston, and Cartier-Bresson are now considered exemplary masters of contemporary art. Furthermore, in addition to classic modern photographic work, recent photography produced by artists such as Cindy Sherman, Richard Prince, and Sherrie Levine has concentrated on the critical appraisal of Modernist theory and practice. In many ways photography (perhaps in retaliation for the hazing it once received from the formalist art establishment) is now busily formulating its own independent critical vocabulary and aesthetic values. What started as a means of recording everyday visual information has now been transformed into a keen instrument for the exploration and redefinition of our cultural milieu.

6 Performance/Video

The Parameters of Time and Space

Both performance and video art were hailed in the early seventies as the harbingers of a new era—one in which these forms would play an important social and aesthetic role. By using experimental media, proponents believed visual art would shake its centuries-old dependence on the materiality of the object and communicate more directly to an audience. Through the heightened reality of a live event or the all-encompassing presence of television—an almost inescapable phenomenon today—larger groups of people could be reached and a greater impact could be made on the general public.

This particular stance was highly desirable to a group of artists who had rejected the materialism of a society they believed to be madly bent on blind consumption. In those days there was a messianic zeal to the vision of performance/video artists—and some adventurous critics—who believed that these were *the* artforms of the future. Leo Castelli, the dean of avant-garde art dealers, set up an ambitious distribution system for video tapes in partnership with a well-known European gallery. Castelli counted heavily on people's buying tapes just as they would drawings, prints, and special edition books. But, except for a brief flurry of rentals from museums and university art departments, the average collector shied away from such purchases. People may have been interested in viewing them, but owning a tape was another matter.

Part of the problem regarding audience acceptance had to do with the nature of video in the early seventies. Many of the established aesthetic sensibilities of the sixties—such as Minimalism, lack of interest in nonart audiences, and a calculating strategy of Modernist dogma—were extended wholesale into these new forms. Often the result was a predictable reaction from public viewers that was short and to the point: "Tedious and boring," they said. But artists felt differently. They believed they were collectively trying to expand the consciousness of a population dulled by innocuous ads, inane television programming, and humdrum publications.

The ultimate sin for commercial art— boredom—became an identifying feature of Modernist forms in the late sixties and early seventies: One "art" film repeatedly showed a closeup of a hand blindly trying to catch objects falling within its reach. Although the general public found this obtuse, Formalist critics viewed it as profound; if only a small segment of the population liked it this was proof of its value—here was an alternative to the commercially successful, and "sinful," world of kitsch and prime-time television.

Some artists in these new media, however, began to question some of the highly formalized mannerisms and sterile gestures that they viewed. So much of the content appeared to be an extension, into the mediums of live action and video, of Minimalist and conceptual concerns. Many of these pieces seemed conceived for and performed to other artists rather than to a broader, more general audience.

During the early eighties—with its fiscal constraints and longing for security seen in nostalgic revivals of traditional art—old-time video and performance work seemed a thing of the Formalist past. "Decorative" art (long a dirty word in establishment art circles) returned in the late seventies with a vengeance. The salable art object—now that the stringent realities of the marketplace loomed ominously—reestablished itself as a preeminent fixture of the gallery world and art scene. The boom days of unlimited economic and aesthetic growth were over and with it the utopian visions of not a few early pioneers.

This does not mean, however, that challenging art forms like performance and video are no longer viable. It is probable that these modes of expression will change and respond to the contingencies and dynamics of contemporary life in a new and positive way. Already there are signs of this happening in the work of a new generation of intermedia artists.

Performance art as it is practiced in the mideighties has leaned more toward the world of experimental theater rather than the pristine visual environments of galleries and museums. The work of many young performers, such as Eric Bogosian, may best be described as satiric, one-person theater. In 1986, Bogosian's Off-Broadway evening of monologues, called "Drinking in America," ran for four months and drew rave reviews. While it is difficult to consider this form of theater visual art, nevertheless, the world of ideas that informs the art world no doubt affected Bogosian's work.

One of the rallying cries that united performance/video in the early days was its rejection of the material object and celebration of temporal—theaterlike—visual experiences. Doug Huebler, a prominent conceptual artist, made a statement to the effect that the world was already too full of objects and he did not want to add any more through his work. This became the credo for a host of artists working in performance and video modes. This remark, it should be noted, was made during the late sixties, a time of economic growth, when NEA funding and gallery sales supported, encouraged, and allowed this kind of speculative thinking. By 1978, a short decade later, the world had radically changed for everyone, artists and public alike.

As the economy muddled along under various constraints, people faced the challenge of making less do more. Audiences began to doubt the meaning, effectiveness, and value of artworks that seemed to function in a self-reflective vacuum. High art fell from its pedestal. What used to be an unfashionable attribute—popular, broad-based appeal—became sought after by a new breed of artist.

This transformed perspective might be one of the keys to explain the art of the eighties, which seems to differ from the seventies most dramatically in its attitude towards the public. Indeed, a younger group of artists now perceives of the audience as a partner in the completion of the artwork, rather than as an adversary.

The historical origins of twentieth-century performance art extend back to the Italian Futurists of the early 1900s. Filippo Marinetti, a member of this group, first performed an evening of special events in Trieste, Italy, in 1910. Six years later the Dadaists put on a much-heralded series of performances in the Cabaret Voltaire in Zurich. In a certain sense the structural organization of their events was a theatrical response to Cubism; they aggressively collaged and juxtaposed people, objects, and movements into one wild melee. Despite their comic overtones these performances constituted a frantic, despair-ridden response to the ineffable horrors of war, economic ruin, and political cynicism. Most of the actions were nihilistic in character: the poet, Tristan Tzara, danced like a wonton harlot; Hullsenbeck ceaselessly beat a huge bass drum; and participants wore African-like masks, which caricatured their own faces. The horror of the recent war in Europe psychologically precluded their doing art that was concerned with classical beauty and aesthetic refinement. In the grip of something far greater than the traditional concerns of art, they were obsessed with the reality of their time.

Many of these same concerns and actions reappeared in the late fifties under the heading of live-action, painters' events called Happenings. Well into the twentieth century, artists like Allen Kaprow, Red Grooms, Claes Oldenburg, and Jim Dine were instrumental in developing and enlarging the scope of performance-based art. Quite a few of these midcentury Happenings were multimedia events, and made use of recorded and live sound, projected images, and body movement. They fused aspects and elements of theatre, literature, dance, and music. Before long the primal, amateurish power of the first Happenings gave way to polished circuslike events that often mixed everything together indiscriminately. Wall-to-wall pop music, flashing lights, and beautifully projected slides became the rage; commercial interests adapted the experiments of the Happenings and produced sound and light shows that were quite popular.

The midsixties, however, saw the emergence of Joseph Beuys, a highly original and powerful German performance artist who produced some of the most compelling live-action events of the past two decades. Until his death in 1985, Beuys (pronounced "boys") was profoundly influenced by the same social conditions that had spawned the original Dadaists. Because of the traumatic and frightening experiences he had undergone in the German air force during World War II, his life and his attitudes about art were dramatically altered. After the war, making traditional sculpture and paintings had lost most of its meaning for Beuys; his performances (or "actions'," as he calls them) sought to create special situations through which people could question the meaning of events in their lives through the manipulation of symbols. Some of his actions restaged the personal experiences he had growing up in the strange little southern German town of Cleves; others dealt with coming to grips with life—and death—through a creative interplay of psychodrama and unusual environments. All of Beuys' performances addressed themselves to the increasingly important role imagination and creative growth could play in modern life.

At the same time—the midsixties—Nam June Paik, a musical composer and philosophy student by training, was consciously tackling the problem of how to squeeze art out of a seemingly artless medium most people of refined sensibilities had abandoned long ago—television. Previous to making this decision, Paik had been working in electronic music and viewed the complex flow of electrons in sophisticated devices as an interesting counterpart to the workings of our brains. Video was a whole new territory for Paik, and it represented a synthesis between live performance (the visual aspects) and the technological workings of electronic equipment. Due to his training as a musician he approached the medium more in terms of time, sequence, and event rather than on a strictly visual basis.

Quite a few performance and video artists today, besides Paik, make use of sound and music in their work—notably Laurie Anderson, who is profiled in this chapter. This synthesis of sight and sound can, in part, be attributed to the thought and practices of John Cage. Through Cage's philosophical writings and popular courses at the New School for Social Research in the midfifties, many artists were influenced by his enthusiastic support for the interweaving of many artforms into one. Cage proposed that dance, visual art, language, theatre, and particularly sound be recombined in new and exciting ways. Kaprow, considered by many the father of contemporary Happenings, was a devoted student of Cage's and a proponent of live-action art. Paik also followed this disciple, putting his electronic music background to good use through the venue of performance and video art.

More recently, a generation of younger artists—having been raised since birth with telecommunications — recognized the enormous emotional power and popular appeal of music and set about to incorporate it into their work. Laurie Anderson remains one of the most effective musical performance artists of this younger generation.

Anderson's work represents a distinct departure from most of the work of the past; she and other young performance artists have restructured their aesthetic priorities and have questioned the validity of traditional avant-garde polemics. Anderson is extremely interested in creating a new dialogue with American mainstream culture; she is concerned with effecting changes as a participant rather than as an outside agitator. To this end, her recent work assumes the format of a strange metaphysical rock concert, utilizing—and taming in the process—a vast array of complex electronic equipment. Although her performances—unlike the cool, Formalistic works of the recent past—are humorous and entertaining (they make use of songs, stories, multimedia visuals, and electronic music), there is a chilling undercurrent of impending doom to her performance-concerts. Anderson synthesizes a variety of contemporary experiences into one subtle, free-flowing meditation and narrative about life in the United States. By the end of the performance we are not quite sure what we have witnessed: an avant-garde rock concert or a diabolically clever indictment of our culture's values and social priorities.

In essence, all of these performance/video artists work with the nontraditional but commonplace materials of space and time, transforming the basic elements of their lives (their bodies, actions, and dreams) into the substance of their art. Through twentieth-century tools and sensibilities, they invite us to participate in an on-going adventure: closing the narrowing gap between art and life.

Joseph Beuys

One look at the intense, incredibly gaunt, hollow-eyed face of Joseph Beuys (illustration) confirms one thing: His features seem to express the essence of Germanic history and art. In him one can see traces of beatific suffering found in Medieval religious paintings, transcendent expressions seen in Dürer's engravings, and the tormented angst characteristic of German Expressionism. Beuys is the contemporary European standard bearer of all of these influences and traditions. His art, however, extends beyond these historic sources to create a contemporary social and aesthetic dynamic too compelling to ignore, too disturbing to forget.

During the early seventies Beuys was a popular art instructor at the Düsseldorf Academy of Art. Wearing an old Homburg—one of his trademarks—blue jeans, and a multipocketed fisherman's vest with a patch of animal fur attached, he held sway over hundreds of students from around the world: a contemporary Pied Piper of Hamlin. Beuys insisted that his most important artwork was his teaching. Even though this charismatic individual was the most widely exhibited and hottest-selling European artist of the past decade, he steadfastly maintained that the actual artwork is only the "proof" of the more important theorem. His *art* existed not in the product, but in the expanded awareness of people who witnessed his performances and sculptural environments and were forced to deal with a battery of complex notions about their lives and the contemporary world. In essence, his strange actions and unusual artifacts were directed towards the expanded awareness and heightened emotions of his audience. This broad, didactic vision of art runs counter to the prevailing attitude of many teaching artists, who, when they achieve a measure of success and financial security, leave the academic world in order to concentrate more completely on painting, drawing, or sculpture—cultural products of a sort. But Beuys viewed things differently; he saw artmaking as a broadly educative endeavor of the highest order. Through his enlarged notion of creative activity and artmaking, he attempted to reach and affect all sections of life—including the political arena.

By using a wide variety of processes and materials—sculpture made out of animal fat, chalkboard drawings about mythical stags, and ordinary objects like sleighs that are altered and transformed into mysterious and compelling artifacts—Beuys functioned as a poetic social critic and teacher of humanity. Although he worked in a variety of media, making sculpture, multiples, and complex drawings, no doubt his performance pieces, which he prefered to call "actions," have had the greatest impact and have come closest to his concept of art as "teaching."

Innately Beuys was a tireless activist. In his mind passivity was equated with ignorance. He believed that only through purposeful creative activity can humankind evolve and realize the unrealized potential of the individual and society. "I am no optimist, and no pessimist," he said, "I endeavor to see reality. This is the line of evolution, the avant garde." Through his striking public performances and odd sculpture, Beuys cast himself in the role of metaeducator—fervently involved in speculations about how creativity and the arts can bring about a change in the development of contemporary society.

Beuys possessed the archaic soul of a romantic visionary and the mind, heart, and predisposition of a contemporary, radical reformer. This combination proved so threatening to many of his fellow teachers and the administrators at the Düsseldorf Academy that eventually he was dismissed.

At the very heart of the disputed issues was the question of what role the artist should play in today's society. Beuys, unlike many of his colleagues who were content to pursue traditional careers in art, believed the process of making art should be actively directed towards expanding the potential of our lives—not merely towards making pleasing drawings or beautiful sculpture. Because art is inspired by life, it should be made for all people, Beuys reasoned. In this broad sense it is inherently political and needs to be firmly committed to nothing less than the questioning of societal values, structures, and beliefs. In the artist's opinion art activity and life experiences are not unrelated entities; they are one and the same. All creative work, including sculpture, drawing, painting, environmental work, and performance pieces, should be directed towards a fusion of experience that could effect a powerful transformation of human nature.

Transformation was the key term for Beuys; this concept ran like a powerful undercurrent through the entire body of

Joseph Beuys, from *Without a Rose, It Doesn't Matter* (photo by Eeva-Inkeri; courtesy of Ronald Feldman Fine Arts, Inc., New York).

this controversial artist's work. Indeed, Beuys transformed in his work the widest variety of materials imaginable—Volkswagen buses, skulls of animals, clothing, shepherds' crooks, even grand pianos—but the most important and mutable material, around which all others take their place, was the artist himself. At the heart of all of his performances and sculptural works was the central issue of *autobiography*. In a sense, through his "actions" and events, his life was presented to us in the form of codified gestures and symbolic forms. Beuys literally transposed his life into his art, offering it to us as an example of human potential and creative endeavor.

In light of his autobiographical bent, to fully understand Beuys we must go back to his birthplace and trace the life experiences that have formed him. The peculiar geography and the political history of his home region in southern Germany appear to have had a lasting effect on the artist's life.

Beuys was born in 1921 in the small town of Cleves, only a few miles from the Dutch border. This region of the lower Rhine no doubt greatly influenced the young artist's imagination and psyche. Caroline Tisdall's monograph on Beuys describes with great detail this particular region of Europe:

There can be few places in Northern Europe stranger than Cleves and the country that surrounds it. Outsiders call it the "terror landscape," referring partly to the superstition of its inhabitants and partly to the atmosphere that prevails over dune and marsh as the Rhine and Maas flow towards the sea. This is a Celtic and Catholic enclave in a Germanic and Protestant country, a place where the border counts for little in the minds of the people; by name and by culture many are Dutch, just as the land has been at times in the past. The history of Europe has been played out over this land, by Romans, Batavians, Franks, Germans, Frenchmen and Spaniards.[1]

Cleves was a remarkably appropriate beginning for an artist who would eventually embody a strong, politically conscious, trans-European sensibility in his art.

As an only child, large portions of Beuys' youth were spent alone in solitary play. His father, Joseph Jacob, owned a local feed store and always seemed to be preoccupied with his business. The artist's childhood was not an easy time; his relationship with his parents was distant, and he was often left to himself. At this time Beuys engaged in elaborate games of childhood fantasy; visions came to him, appearing mainly as visual images rather than concepts or words. Beuys, in a biographical statement called *Life Course/Work Course*, hints of hallucinatory experiences at the age of five and six that he has referred to and titled: "stagleader," "radiation," "the grave of Genghis Khan," "heather with healing herbs," and "the difference between loamy sand and sandy loam."

Beuys remembered that for years, as a child, he would play at being a shepherd; with a stick of wood for a crook he would lead his imaginary flock everywhere around the countryside. The swan, another important symbolic figure in Beuys' bestiary, can be traced to a prominent architectural feature in the heart of Cleves. Towering over the cityscape stood the Schwanenburg castle, which was topped with an enormous golden swan. Strange swan cults were active in the surrounding countryside, and Beuys vividly recalled the effect this had on him: ". . . when as a child I looked up at the castle I always had the swan before my eyes."

Another potently symbolic image—well steeped in archaic German legend—is the stag. Beuys depicts this animal along with the wild hare as a metaphoric nomad who traverses the boundless landmass of Eurasia, a plain that stretches for thousands of miles eastward from the northern Rhine region. In essence it is a continuous body of land that seems to ignore political borders and the cultural differences of its diverse inhabitants.

The Siberians in particular believed that the stag was an animal with magical powers that held special significance for their people. Shamans, or community religious figures, are said to be able to "die" and be reborn again in the form of magical wolves, stags, and hares who appear when the tribe is in danger.

Beuys had this to say about the symbolic meaning of the antlered beast in the folk traditions of Europe:

The stag appears in times of distress and danger. It brings a special element: the warm positive element of life. At the same time it is endowed with spiritual powers and insight and is the accompanier of the soul. When the stag appears dead or wounded it is usually as a result of violation and uncomprehension . . . (the mercurial nature of the stag is expressed in the antlers. The flow of blood through them reflects a twelve-month cycle: the mobility of blood, sap, hormones). . . .[2]

Despite the strong influence of his home region, Beuys' experiences during World War II probably had the greatest impact on his life and predetermined to a great extent the philosophical basis for his art. When he was nineteen, the artist was drafted by the German air force and was first trained as an aircraft radio operator. Later, when war casualties mounted and labor shortages arose, Beuys was retrained as a pilot and was assigned to a squadron in the Crimea. Beuys was wounded five times on various missions before his plane was hit by Russian antiaircraft fire and crash-landed on the barren plains in a raging snowstorm. Luckily, he was thrown clear of the wreckage and was eventually rescued by a band of nomads:

Had it not been for the Tartars I would not be alive today . . . it was they who discovered me in the snow after the crash, when the German search parties had given up. I was still unconscious then and only came round completely after twelve days or so, and by then I was back in a German field hospital. So the memories I have of that time are images that penetrated my consciousness. The last thing I remember was that it was too late to jump, too late for the parachutes to open. That must have been a couple of seconds before hitting the ground. Luckily I was not strapped in — I always preferred free movement to safety belts. . . . I must have shot through the windscreen as it flew back at the same speed as the plane hit the ground and that saved me, though I had bad skull and jaw injuries. Then the tail flipped over and I was completely buried in the snow. That's how the Tartars found me days later. I remember voices saying "Voda" ("Water") then the felt of their tents, and the dense pungent smell of cheese, fat and milk. They covered my body in fat to help it regenerate warmth, and wrapped it in felt as an insulator to keep the warmth in.[3]

Over the course of his professional art career, especially during the fifties and sixties, Beuys made extensive use of these elements; fat and felt were not used as Formalist materials, they were highly charged metaphors for traumatic life experiences. To Beuys they signified the healing and warmth that literally meant life to the wounded artist.

Postwar Germany was a depressed and confusing time of guilt and anxiety for millions of stunned survivors. Even the glorious golden swan on the Schwanenburg castle had fallen down and lay in ruin. The horrible deaths of many close friends haunted Beuys (his copilot in the JU-87 that crashed was pulverized on impact).

After the war he enrolled in a natural-science program at a nearby university. Beuys had always been quite interested in zoology throughout his childhood. Elaborate notebooks were filled with detailed information on his collections of insects, mice, fish, and butterflies. He even had a small amateur laboratory set up at his home for experiments. He was not at the university long, however, before he grew disillusioned with the way science was taught. After his brush with death and his "rebirth," scientific study, as it was taught then, seemed inadequate to address the central issues of life and existence that obsessed him. Beuys recalled with particular horror a lecture given at the University of Posen by a biology professor who had devoted his life to studying amoebae that phylogenically seemed to exist somewhere between plant and animal life. Here was a distinguished professor, a paradigm of rational thought and knowledge, who had devoted the better part of his life to studying this miniscule, fuzzy animal-plant, plant-animal. A feeling of dread flowed through Beuys as he listened to the droning voice of the professor and stared at the out-of-focus slide of that one-celled form. He left the program that afternoon. Throughout the rest of his life, the image haunted the artist and reconfirmed his belief that societal overspecialization has greatly contributed to meaningless work and empty lives.

In 1947, Beuys enrolled in a program of study at the Düsseldorf Academy of Art. There he discovered that science was not the only field of study to be restricted by narrow professional beliefs. Beuys was quite fond of one instructor, however—Ewald Mataré, a sculptor who was dismissed by the Nazis in the thirties and subsequently rehired in 1946. Mataré was a fascinating individual, who believed in the complete unity of art and life, materials and form. He was a poetic soul who was always dropping mystical sayings in his teaching, such as "art is a footstep in the sand." Beuys was attracted to the scope of Mataré's ideas and completed his studies under Mataré's personal direction.

After graduation from the art academy Beuys retreated from urban life and lived for ten years in a lonely region of the lower Rhine. He was still greatly disturbed by his war experiences and not all that sure of his artistic aims and how he could implement the strong emotions he felt about art and the world. In this desolate region he led a quiet, reclusive life, observing the behavior of forest animals and working on a series of strange and nontraditional sculptures.

The year 1949 was a crucial period for the artist; emotional depression led to anxiety, and this deep-seated angst led to psychological illness. Beuys' life was at a turning point; personal torment and mental pain rocked his being and tested his spirit. Later Beuys viewed this experience as a difficult but necessary voyage for him to undergo; a process of personal introspection, spiritual death, and subsequent rebirth. "The positive aspect of this is the start of a new life," the artist reflected, "the whole thing is a therapeutic process. For me it was a time when I realized the part the artist can play in indicating the traumas of a time and initiating a healing process. That relates to medicine, or what people call alchemy or shamanism, though that should not be overstressed. For me it meant the continuation of the threads in my biography that led me to scientific and biological experiments. . . ."

A performance piece Beuys did in Middelburg, Holland, 1972 recreated through audience participation many of the experiences and feeling he was undergoing a this time in his life. The artist viewed the piece as a psychoanalytical "action" that sought to actively share his life experience with others. Participants were asked to bring bicycles and were given directions which took them to a local cow barn; here they experienced the organic, musty smell of animal dung, straw, and warm milk. Then they were led to a living room which was set up in a pre-World War II style with old furniture and oil lamps. A tape-recorded woman's voice commented on the artist's life during the time of his breakdown. The voice described Beuys' life of working in the fields and barns, and how he would lie in bed in the mornings, uninterested in rising and beginning the day. It was the story of a profoundly troubled individual, in his words "a misfit."

After he pulled out of his deep depression, he was able to resume his art activity once again. Beuys worked on a large group of figurative drawings, secured several sculptural commissions, and helped to complete the bronze doors of the Cologne Cathedral, a commission given him by his former teacher Ewald Mataré.

Between 1955 and 1957, however, Beuys received few sculptural commissions and once again slipped into a depressed state, seriously doubting his own worth and ability. As part of his recovery therapy, he worked at a friend's farm through most of 1957. Slowly, he began to rebuild his health and regain his balance; slowly, the desire to begin making art surfaced again. After this crucial stage of his life, Beuys began methodically to develop the governing principles upon which his mature work is based.

In 1962 Beuys met Nam June Paik, a young Korean artist, who was one of the leaders in an international avant-garde art movement called "Fluxus" (the flowing). This group opposed the idea of art as a static commodity that was made by "career" artists and appeared to exist in an environmental vacuum. The Fluxus people were interested in merging elements of art and life and breaking down what they thought were arbitrary divisions between theatre, poetry, music, visual art, and dance. All of these modes of thinking were connected, as far as they were concerned, to the totality of human experience. They were interested in constructing a new aesthetic model which would connect everyday life with art. Fluxus derived its name and found support for its beliefs in the writings of Heraclitus, the ancient Greek philosopher who wrote: "All existence flows in the stream of creation."

Beuys felt a strong kinship to the philosophical ideas of Fluxus and welcomed the sizable audiences these artists usually attracted to their performances. These public forums seemed particularly appropriate to the kind of art Beuys was interested in producing.

Many of the original members of Fluxus had been trained as musicians, so elements of sound were emphasized in many of the group's performances. Acoustic effects, body movement, avant-garde music, and a wide variety of unusual materials were often employed in Fluxus events. Like certain classical Greek philosophers, they shared a belief in animism: the conviction that all objects possess hidden life, meaning, and vitality. This philosophical aspect of Fluxus strongly appealed to Beuys' own mystical belief in the potential life of certain materials; also, he believed that through their performance methodology everything could be illustrated. This form turned out to be a perfect vehicle for his didactic art.

Chief, first performed in 1963 in Copenhagen and later at the René Block Gallery in Berlin, established many of the directions Beuys' "actions" were to take over the next decade. This piece differed from many Fluxus performances, which sought to Dadaistically provoke the audience. *Chief* was slow, meditative, ceremonious; it was designed to elicit introspective questioning, not outrage, on the part of the observers. For nine continuous hours the René Block Gallery was transformed into a site that combined aspects of ancient primitivistic ritual with modern electronics. In the middle of a well-lighted room was an oblong roll of gray felt, measuring 2.25 meters, with Beuys hidden inside it. At both ends of the roll were two dead rabbits stretched out and guarding the entrances. On the floor to the left of Beuys was a thick copper

rod, most of its length tightly wrapped with felt. Leaning against the wall was another rod with a much smaller section of felt wrapped around its center. Wads of pale yellow margarine were placed at various intervals along the wall board. A triangle of fat covered the left corner of the exhibition space, obliterating these three planes. Lengths of coiled wire crossed the floor and led to a high-fidelity speaker propped up against the wall.

Throughout the performance, by means of a microphone placed inside the roll of felt, the artist uttered a variety of primitive sounds which were amplified and played over the speaker system. During the course of events one heard: strained breathing, gurgling noises and coughing, groaning, and whistling sounds. The most often repeated and consistently recurring sound was the deep, throaty cry of the wild stag, ööööö, which came from deep within the diaphragm. Beuys also voiced this primal, symbolic sound years later during graduation exercises at the Düsseldorf Kunstakademie. Of course, this action never endeared him to the school's administration and no doubt contributed to his controversial dismissal years later.

Beuys explained: "The sounds I make are taken consciously from animals. I see it as a way of coming into contact with other forms of existence, beyond the human one. It's a way of going beyond our restricted understanding to expand the scale of producers of energy among co-operators in other species, all of whom have different abilities. . . ."

The title *Chief* refers to Beuys' belief that he was the representative and spokesperson for the world's animals. Over the years the artist had developed an elaborate ideological scheme whereby animals were elevated to a level of importance and status far beyond that which is normally given them. In a particularly provocative comment Beuys had stated that he believed that a wild hare running from one corner of the room to the other can accomplish more for the political consciousness of the world than can a human being. He felt that the "elementary strength" of animals should be harnessed along with the logical thought that is valued so highly in modern society. Beuys in fact would like to have seen worldwide legislation that would elevate the status of animals to that of humans.

This notion represents a conscious political act. Beuys' art has always been politically aware in that it has taken up the issues of political awareness, freedom, and the ability to transform our lives through social action. Since the midsixties, Beuys had functioned directly in the political arena, confronting the status quo of the German system. In 1967 he founded his own political party, called the "Organization for Direct Democracy"; Beuys claimed all of the animals of the world are members, thus making his party the largest political organization in the world.

The basic purpose of his organization was to create a more participatory form of government for millions of disenfranchised citizens without property, status, and access to information, who entrust vital decision making to hastily elected officials. Beuys believed that to bring about a more direct and equitable form of democracy — as promised in the words of the West German constitution, "All power proceeds from the people"—a more responsive system needed to be installed. The artist reasoned that law really defines the concept and power of the state; if people are to truly have a voice, they must directly vote for basic laws, not figurehead representatives.

For years Beuys campaigned, often alone and without a great deal of financial support, in working-class sections of town. He tirelessly discussed these issues with retired factory workers, civil servants, workers, and homemakers. Beuys believed that, basically, elected officials have no confidence in the ability of their constituents; they ask for their vote and promise to represent them once they are elected to office. In opposition to this system, Beuys wanted to give every voter the ability to respond directly to specific issues. This artist viewed the serious business of politics, which affects all of our lives, as a form of art — art as an activity directly tied to our physical well-being and psychological health.

In 1976 Beuys ran for a seat in the West German Parliament as a nonparty candidate from the small town of Oberkassel. He garnered about 2 percent of the local vote. Only a few years later, in the election for Parliament, he won 3.5 percent of the vote on a nationwide basis. Some cities in southern Germany even gave him 15 percent of the vote.

Although Beuys believed that education is necessary for our physical, psychological, and spiritual survival, he tended to view the official state educational system as unresponsive and inadequate to real human needs. In terms of his own teaching he assumed somewhat of a heretical stance at least as far as other artist–teachers were concerned. Beuys maintained, for instance, that his artwork was important only as a kind of proof for his theoretical ideas. The concepts and perceptions made possible to the viewer were far more important than the physical reality of the work. The essence of his art lay in information processing, not materialism. Beuys believed that every experience, when viewed from a creative perspective, is enlightening; basically, he was opposed to the rigidity of official schools which quite often restrict learning. The artist was quick to point out that life, not school, is the real teacher of humankind; "the whole world is an academy," Beuys steadfastly maintained.

On November 25, 1965, at the Schmela Gallery in Düsseldorf, Beuys performed what may have been one of the strangest and most compelling "actions" of his entire career, *How to Explain Pictures to a Dead Hare* (illustrations). The title by itself is striking and sets up curious overtones of incongruity and contradiction. It clearly reveals the artist's awareness of language as a catalyst helping to unlock the poetic mysteries of the world. In this sense Beuys shared the Romantic ideals of the legendary German writers Goethe and Schiller. Another profound influence on Beuys' thinking was Novalis — a nineteenth-century philosopher who believed that poetry expressed the essence of the human spirit and that words were profound shapers of the world. Beuys has repeatedly stated that "language is indispensable" in that it helps us move from passive modes of thought to actions and gestures that lead to physical and spiritual liberation.

For the three-hour duration of this remarkable performance the Schmela Gallery—which housed a large retrospective exhibit of Beuys' drawings and artworks—was closed to the public. Inside the exhibition space the artist walked around and performed all of his gestures while gently cradling in his arms, as if it were a small child, a large, dead hare. No spectators were allowed inside the gallery, but viewers could watch from an open doorway or through a display window on the street. What they saw was bizarre to say the least; prior to the performance Beuys had immersed his head in honey and applied iridescent sheets of gold leaf to the sticky coating. "In putting honey on my head," Beuys explained, "I am clearly doing something that has to do with thinking. Human ability is not to produce honey, but to think, to produce ideas. In this way the deathlike character of thinking becomes lifelike again. For honey is undoubtedly a living substance."

Gold and honey were used by the artist to refer to ways of thinking and responding to the world that have nothing to do with rational modes of thought; in fact, they symbolize modes of cognition that are completely alogical. Ultimately, Beuys was questioning the belief that we really can understand the inner workings—and our lives—through normal modes of perception. Through this action the artist hoped to reach us in an unfamiliar way: creatively, imaginatively, and, above all, magically. "Explaining pictures to a dead hare" questioned our belief that art or any natural phenomenon can be satisfactorily explained in any terms. There was a curious sense of rationality to this piece, however, that called for the development of values and perceptions that extend beyond the

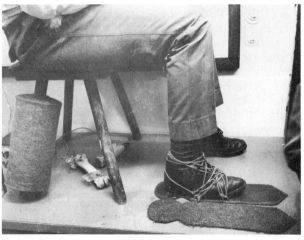

Joseph Beuys, *How to Explain Pictures to a Dead Hare* (© Ute Klophaus).

limited ken of logic. Beuys wrote: "The idea of explaining to an animal conveys a sense of secrecy of the world and of existence that appeals to the imagination . . . even a dead animal preserves more powers of intuition than some human beings with their stubborn rationality."

Many complex symbolic elements to Beuys' past appeared in this particular action: One of the legs of a stool he sat on was thickly wrapped with felt to "warm" it; a radio constructed out of modern electronic parts and animal bones was connected to an amplifier and placed under the chair; tied to his left shoe was a pointed felt sole, and an identical one of iron was attached to his right foot. As in the performance *Chief*, sound was an important aspect of this piece. Throughout the long ritual Beuys' lips were inaudibly moving, "explaining" his art to the hare. Whenever Beuys moved to show a certain drawing to the dead animal, his movement was punctuated by the clanging sound of the metal foot piece. The time span—three hours—lent a certain significance to the work. The effect of a twenty- or thirty-minute action would not have conveyed the seriousness of the artist's intention. Modern entertainments are designed around convenient time slots and limited attention spans; ritual events depend on longer periods of hypnotic repetition. Watching Beuys explain the meanings of art to a dead animal for hours on end became almost a self-fulfilling prophecy: The lengthy time period lent a certain authenticity to the event; one received the impression that the animal understood.

In May of 1969 Beuys performed what may have been one of his most visually beautiful actions, *Iphigenie/Titus Andronicus* (illustrations). The Frankfurt *Experimenta 3* festival invited Beuys and several other artists to take part in their avant-garde theatre program. Beuys was asked to come up with stage plans for two plays, *Iphigenie* by Goethe and Shakespeare's *Titus Andronicus*. After some consideration, Beuys decided to combine the two plays and feature a simultaneous reading of both.

On the evening of the first performance the curtain rose to reveal a most unusual stage plan. The left half of the stage, representing *Titus Andronicus*, was completely empty; *Iphigenie's* side featured a roped-off corral containing a pure white horse, brilliantly lighted, eating hay. Underneath the horse was a large section of metal flooring the dimensions of the enclosure; every time the horse moved, its hoofs clattered noisily on the iron plate. Beuys entered from the side, wearing a dazzling white fur piece that mimicked the natural coat of the animal. When he approached the microphone set up at the center of the stage, he took off the jacket and placed it next to a group of props.

Language and sound were important elements in the performance; readings from *Titus Andronicus* were heard through audio speakers placed on the far left of the stage and selections from *Iphigenie* were broadcast through the right channel. While the reading was going on Beuys performed various ritualized, secretive actions with the items gathered around the microphone: Cubes of sugar were fed to the remarkably quiet and docile steed; cryptic comments were occasionally spoken into the open microphone; margarine (fat) was spread on the floor; huge cymbals were clashed when the need arose (as a means of restoring order to a sometimes boisterous audience); and arcane diagrams were painstakingly drawn with chalk on the stage floor.

Out of many elements—visual, linguistic, and auditory—Beuys created a set of new experiences based on the historic texts of classical writers. All of Beuys' works express the notion of synthesis and recombination: the old with the new, the apparent with the cryptic. But it was the image of the strikingly beautiful horse itself—a symbol for humankind's interdependence with animals—that combined with Beuys' curious actions to produce the most haunting effects.

Peter Handke, a well-known, postwar playwright, was in the audience at this performance and wrote favorably about it in the German press. However, he was annoyed with the way Beuys reacted to the audience at times: When the horse occa-

Joseph Beuys, *Iphigenie/Titus Androni-cus* (© Ute Klophaus).

CURRENTS

sionally urinated and the audience applauded, the artist responded by clapping back. Handke felt Beuys should have been more aloof and hermetic on stage. The author concedes that the event has grown powerfully in his memory in the years since it was first performed. Today it functions in his mind like an unforgettable "after-image," fusing what appeared to be so many disparate images and concepts into one haunting vision, the beauty of which becomes, according to Handke, "utopian and therefore political."

In April of 1971, Beuys orchestrated another "action," titled *Celtic + ~* (illustration), in Basil, Switzerland. Deep in the supposedly bomb-proof Civil Defense chambers of the town's *Zivilschutzröume* building, the artist performed a ritualized event that combined elements of Christian symbolism with references to European paganism. Characteristically a slew of materials and implements, both common and unusual, were positioned throughout the thick-walled concrete "Kunstbunker." Large pieces of white cloth were hung from the wall as screens for two 16-mm film projectors; a grand piano — symbol of high culture — took on a powerfully sculptural presence and was set next to an ordinary watering can, a white enameled basin with a cake of soap, a large galvanized tub filled with water, an aluminum ladder, and three tape recorders.

Beuys appeared wearing his familiar hat (the German press refers to him as "the man with the hat"), blue jeans, work shirt, and his multipocketed fishing vest. The artist launched the proceedings by carefully washing the feet of about seven spectators with a scrub brush and soap in the old-fashioned water basin. Then, with about fifty spectators crowding around him, he sprawled on the floor and drew three diagrams on a chalkboard, erasing each one before proceeding to the next. Next, he abruptly got up, walked to the piano, sat down, and sat motionless for what seemed

Joseph Beuys, *Celtic + ~* (© Ute Klophaus).

to be a long time. Suddenly, two of the artist's past performance films were simultaneously projected on the cloth screens: *Eurasianstaff* and *Vacuum* ↔*Mass* After the films were shown, Beuys climbed a ladder and scraped particles of gelatin off the walls (placed there before the performance began) and piled them onto a large metal platter. When enough of the transparent, shimmering substance was collected, the artist stood in the center of the room and dumped the gelatinous mass over his head. Structurally, Beuys provided a sense of unity to the piece by ending it, just as he had begun, with a symbolic reference to Christianity. With the crowd hushed and reverent, he slowly walked to the large, water-filled metal tub, strapped two gleaming flashlights on to his thighs and stepped into the water. A musician friend of Beuys' by the name of Christiansen completed the Baptismal by ritually pouring water over his head.

Celtic + ~, like much of the artist's work, was multilayered, complex, and alogical in the way it mixed Christianity with paganism and alchemical symbolism with elements of scientific thought. Through this performance Beuys took us on a shamanistic journey of personal transformation and discovery; his actions expressed the overriding struggle he sees going on in the world: between life and death, awareness and ignorance. In participating in this process with him, our awareness of our own part in this continous drama was heightened.

Although Beuys had numerous invitations and opportunities to visit America in the late sixties and early seventies, he refused to visit the country while the Vietnam War was in progress. In 1974, after the withdrawal of American troops from Southeast Asia, Beuys finally accepted an offer to tour the United States. Advance publicity about this controversial artist had been trickling in for years, mainly through journals and word-of-mouth. Now the American art world would get a firsthand view of what the European commotion was all about. New York, in particular, was waiting with baited breath.

Inspired by the concept of America, Beuys conceived of the entire visit as a performance piece that would in various ways deal with the traumatized psychological state of the nation following its withdrawal from Vietnam.

I like America and America Likes Me (illustrations) was a three-day nonstop performance piece that began the moment the plane carrying Beuys from Germany touched down in the United States. An ambulance met the plane at New York's Kennedy Airport, where medical attendants wrapped Beuys in gray felt and drove with caution lights blazing to the René Block Gallery in Manhattan. A section of the large, well-lighted exhibition area was cordoned off with industrial chain-link fencing. Inside, a live coyote was visible cautiously pacing back and forth; a mound of hay was stacked in the corner; fifty copies of the *Wall Street Journal* (which were changed daily) were arranged in tall piles; and two felt blankets and a water dish were positioned in the center of the floor. Beuys was carried into the gallery and placed in the cage by two ambulance attendants. Here he was to live, eat, sleep, and share the space with a wild animal for a total of three days and nights.

Visitors thronged to the gallery for a glimpse of the strange cohabitation of man and beast. Beuys appeared most of the time wearing the felt blankets, as if he were a mysteriously hooded monk. A shepherd's crook stuck out of the top opening instead of his head. For the most part, the coyote cautiously ignored Beuys and spent his time walking around the parameter, eating and sleeping. Probably the most poignant visual element of the compound was the continously burning flashlight that peeked out from under the folded sheets of felt. This device seemed to function as a kind of entropic indicator, slowly running down and visually registering the consumption of energy by the gradual dimming and yellowing of the light. There was something ominous about this "clock"; in a way, it was measuring and metering our time as well. For three days and nights the flashlights burned and Beuys and the coyote engaged in, as he put it, "total communication and dialogue" with each other.

Like his ideological and symbolic references to the swan, hare, and stag, Beuys had definite views about this indigenous North-American mammal. Beuys specifically chose the coyote because he felt its history was analogous to the plight of the American Indian: Both have been persecuted and greatly misunderstood over the last 200 years.

The artist believed that native Americans were deprived of their homeland, status, and self-esteem essentially because they refused to assimilate into a European-based society and culture. One cannot fail to see a parallel between the imposition of our values on the Indians and American interference in Vietnam. Both episodes show a lack of sensitivity to the workings of the world of which the artist seeks, through his actions and symbols, to remind us.

Throughout most of the three-day performance, Beuys was engaged in a continuous dancelike action with the animal. The "shepherd" would move in a certain direction and be followed by the curious coyote. Then Beuys would accurately mimic the movements of the doglike creature, never taking his eyes off the animal. Usually, these "dances" would last an hour or two (about thirty such sequences took place over the course of three days); between sessions the

Joseph Beuys, *I Like America and America Likes Me* (photos courtesy of Caroline Tisdall and Schirmer/Mosel).

coyote would cautiously rest on the felt blankets, never completely closing its eyes or turning its back on the audience of curious spectators. Patterns of behavior quickly developed in this controlled microenvironment. Sometime during the day the animal would turn its attention to the neat pile of freshly delivered *Wall Street Journals,* tearing, mauling, chewing, and then finally urinating on them to the unrestrained delight of the crowd. Only once did the coyote exhibit any menacing behavior towards the artist; Beuys easily discouraged the snarling beast by brandishing his wooden shepherd's staff. From that moment on, they lived peacefully in mutual respect.

Occasionally, during the course of their three-day live-in Beuys would leap up, reach into his robe, pull out a triangle, and deliberately strike it three times. Just as the last crystalline note was fading, a loud, twenty-second blast of recorded instrumental turbine sounds rocked the compound. After this noisy catharsis the artist would usually retire to the hay-strewn nest in the corner for a cigarette. Often the coyote would join him there. At these moments, huddled together with the smoke curling about them, man and beast seemed to strike a note of preternatural unity and peace rarely seen in the modern world.

One of the most significant aspects of Beuys' work, and this piece in particular, was that there was no clear-cut moralist message attached to it, no propaganda about saving the coyote, or restoring lost land to the Indians, or keeping America out of Vietnam. Beuys was working with concepts and meanings that go far beyond these specific issues and perhaps address the root causes of all of these problems. In all of his performances, potent images were planted in our minds, questions about our culture were raised, and the unfathomable mysteries of the world were mirrored in his strange and compelling events. The artist was trying to reach preconscious and prerational levels of meaning impossible to arrive at through any other means.

At the heart of all of Beuys' work and teaching was the unrelenting belief that creativity is the most powerful tool for transformation of our society. He was careful to point out, however, that art is by no means the only source of creative expression. Lawyers, factory workers, housewives, clerks, and schoolteachers can perform their particular jobs with increased value to society and themselves through the liberating effects of personal creativity. Beuys believed that when the individual is stifled by inflexible government bureaucracy, unresponsive political systems, and counterproductive competitiveness, society as a whole loses and the results are highly visible: senseless violence, rampant crime, and widespread states of apathetic hopelessness.

For daring to address these very issues through his teaching, Beuys was dismissed from the state-run Düsseldorf Art Academy in 1972. For the next four years, he performed his "actions" throughout most of Europe and continued working on his political party "The Organization for Direct Democracy." He also was active in founding a "Free International University" in Ireland—a more neutral ground than mainstream Europe according to Beuys—and continued his fight against the Ministry of Education for reinstatement at the Academy. Victory was achieved when the German Supreme Court declared on April 7, 1978, that Beuys' dismissal as professor of art in 1972 was illegal.

This ruling led to a celebrative piece, performed by Beuys and his students, called *Crossing the Rhine* which was perhaps one of the most triumphant and meaningful actions of the artist's life and career. Beuys sat silently in the front seat of a long, wooden boat while five students rowed him across the Rhine to the Düsseldorf Art Academy located on the other side. There he would resume his appointment as professor of art and continue his broad-based educational program.

It is easy to understand why many of Beuys' sculptural environments and performances were so controversial and provoked heated debate in Europe and America: He questioned the modern role of art and its goals. Some saw him as a charlatan and fake; others, as the revived conscience of a forgotten time and the spiritual hope for the future. One thing is certain, it will not be easy to forget this messianic artist. Beuys' death leaves behind an important legacy of work all of which seems directed toward one goal: the transformation of life and time — our life and our time. Beuys neatly summed up the philosophy behind his work (which could also serve as a fitting epitaph) when he wrote: "The key to changing things is to unlock the creativity in every man. When each man is creative, beyond right and left political parties, he can revolutionize time."

Nam June Paik

When New York's Liberty Music Shop — a large record and high-fidelity equipment store — opened its doors for business on the morning of October 4, 1965, video artist Nam June Paik was on hand to be the first purchaser of Sony's new portable videotape recording equipment. Paik, a former student of electronic music, was no newcomer to the emerging field of video; he had been grappling with the aesthetic problems of television as an artform since 1959. At that time Paik made a conscious decision to translate some of the concerns of electronic music composition into the visual arena of video: "I knew there was something to be done in television and nobody else was doing it, so I said why not make it my job?" Since the late fifties Paik has so thoroughly pursued his interest in video that he is considered by many today to be the founder and one of the grand masters of this still-developing medium.

Through his work at the Studio for Electronic Music in Cologne, the artist acquired a knowledge of the complex circuitry of modern electronic devices; televisions in particular interested him. Paik and an electrical-engineering friend worked months rewiring thirteen old black-and-white sets they had bought used or salvaged from German junk heaps. The result of their efforts were shown at the Galerie Parnass in Wuppertol, West Germany, in what was probably the first exhibit anywhere of video art. What the bewildered art viewers saw was totally unlike any image their sets received at home. In fact, if their televisions behaved like Paik's did, they would have called the repair service or decided it was time to buy a new set. Not one operated in any normal way: Some had been ingeniously rewired to distort the broadcast signal into a wildly contorted, fun-house mirror pattern; others stretched the picture horizontally until it was indecipherable; a microphone was hooked up to another set which scrambled the picture when it picked up sounds in the gallery; and one television's circuit polarity was switched, creating negative images — white on black.

Over the doorway of the gallery — in marked contrast to the electronic art inside — was hung the freshly slaughtered head of an ox. Needless to say the controversial exhibit drew record numbers of people who pondered the severed animal head over the entrance and the curiously "malfunctioning" sets inside. No television artworks were sold but the incident sparked heated debate and discussion for months afterward; one thing Paik has never been as an artist is boring.

The violent contrast between the bloody ox head and the quietly passive electronic images is typical of this artist's whimsical sensibility and anarchistic sense of humor. Paik's mind, a kindred spirit to Arakawa's, is laced with dynamic contradictions and fluctuating patterns of thought that effortlessly moved between East and West. Typically, Paik views the modern world as a curious synthesis of these great branches of thought; he once observed that just as Istanbul could be seen as the European gateway to Asia, Tokyo could be considered the beginning of America.

Paik was born in Korea, educated in the Orient and Germany, and now lives and works in New York City; in the process of living and traveling in various parts of the world, he seems to have synthesized these various psychological territories into one electronically unified aesthetic in his mind.

Physically, Paik is a short, slender man, well dressed, well mannered, and reserved in public almost to the point of shyness. A mental image formed after meeting him, however, would never relate to his early performances: They were so wild and violent that he earned the label of "cultural terrorist" even among other radical and avant-garde artists. As a young musical student Paik was terrified by the thought of playing the piano in public; in fact, he never performed for this very reason. But after his meeting in 1958 with John Cage, an influential experimental composer and musical theorist, much of Paik's reluctance to appear publicly vanished. Paik assumed the role of performer with a vengeance and produced a series of terrifying and dangerous-seeming Happenings that surprised and perhaps even bothered many of his close associates.

Paik's meeting with Cage was clearly a powerful catalyst in the artist's career, although it would be hard to specify in what exact way. Cage himself, a man of gentle wisdom and wit, understandably takes little direct credit for influencing the early works of this one-time shy composer turned performer.

Homage à John Cage—done soon after

Paik's "enlightening" meeting with Cage—involved tipping over a precariously balanced upright piano until it crashed "musically" on the floor of the stage. Karlheinz Stockhausen, a well-known electronic composer in the audience, thought a horrible accident had taken place and rushed up to help right the valuable instrument; Paik thanked him politely but refused his help.

This was only the beginning of a madcap series of violent musical events that Paik performed in the late fifties and early sixties throughout Northern Europe. Soon critics called him "the world's most famous bad pianist," a title for which he feels great affection today.

One for Violin (illustration) was another performance of seemingly inane and cryptic destructiveness. Paik stood facing the audience in a standard preconcert pose with a violin by his side, much like any normal concert musician who was about to perform publicly. Grasping the instrument with both hands, like a baseball bat, he slowly lifted it over his head with thoughtfulness and deliberation; after a well-timed pause,

Nam June Paik, *One for Violin* (photo courtesy of Manfred Leve).

the violin was swiftly brought down on a table directly in front of him and smashed beyond recognition. The audience was stunned; many music lovers were appalled by what they perceived to be an irresponsible and wanton act. By ritually destroying a precious instrument of humanistic overtones (the violin most approximates the range and timbre of the human voice), people felt Paik was attacking the essence of European culture. Sympathetic avant-garde critics, however, viewed the formalized act as a means of clearing the air of musty, centuries-old misapprehensions as to the nature and role of art in our present day culture. They felt Paik was paving the way towards new artistic perceptions based on twentieth-century cultural realities. To be sure, Paik's events contained certain aspects of Dada—a nihilistic art movement that arose in Europe shortly after World War I. No doubt Paik's incorporation of shocking and property-destructive elements in his performances functioned as a personal catharsis enabling the artist to come to grips with his shyness and also comment on the violent nature of Western civilization. Certainly, no cynical disrespect for classical music and musical instruments of the past was intended. In fact, Paik said he was not thinking of destruction at all when he performed these pieces. What bothered him was the imbalanced reverence for the violin and piano; they had become culturally loaded symbols instead of musical instruments. The public had come to perceive them as sacred artifacts rather than as expressive musical voices—state-of-the-art technology from another time. Paik believed that *our* era demanded new aesthetic sensibilities and electronic instruments that matched the needs of the present. An element of shock was used—just as Dadaists had used it to awaken themselves and the public to the insane cruelty of modern warfare—as a means of shrugging off what the artist may have perceived as the burdensome, irrelevant, and dangerous weight of the past.

One of the most psychologically frightening and disturbing performances this composer-artist ever did bore the innocent title *Étude for Pianoforte* and opened in 1960 at the Düsseldorf studio of Mary Bauermeister, a painter friend of Paik's. A space was cleared in the loft, and a grand piano was positioned in the center; despite the slightly disconcerting presence of a gutted piano next to the new one, the "set" had all the earmarks of a traditional musical recital stage. This deceptiveness well suited Paik's anarchistic intentions. On the night of the performance the artist appeared in a suit and tie, looking for all purposes like a traditional performer of refined classical music. This effect was reinforced by the first stage of the performance as Paik played a lyrical work by Chopin. Halfway through the work, however, the mood of the performance dramatically shifted; Paik abruptly stopped playing and broke into tears. He then jumped into the smashed piano on the floor, thrashing about noisily among the strings and debris. Then, with a demonic gaze in his eyes, he produced a menacingly long and sharp pair of scissors and slowly made his way towards the audience. At this point, people began to cautiously move towards the exit. Paik unerringly made his way to the section where John Cage, Karlheinz Stockhausen, and the pianist David Tudor were sitting; he took Cage's jacket off and began carefully cutting off his shirt with the scissors. Paik later remarked that when he saw that the composer had no undershirt on he took pity on him and instead cut off his tie section by section. The group's reaction was understandably mixed; some laughed uneasily at the would-be joke; others were terrified by this physical affront to a member of the audience. Paik did not stop here; from one of his pockets he drew out a bottle of shampoo and poured the liquid detergent over Cage's and Tudor's heads. Calvin Tomkins, in an amusing article on Paik, relates that when Stockhausen saw this he cautiously started to edge away fearing the same humiliating treatment. Paik interrupted his lathering for a moment, shouted reassuringly to the composer, "Not for you!" and continued working on the musical duo. Suddenly, without a word, Paik left the soapy victims and quickly exited. The audience was stunned into docile submission by the swift succession of events. No one moved or spoke until several minutes later when the studio phone rang. It was Paik calling from a nearby phone booth; the concert was now over, he solemnly reported.

Although Paik came from a fairly well-to-do Korean family, the story of its economic ups and downs reads like a fictional television serial. No doubt this constant flux and change in his personal life helped to form his aesthetic sensibility and view of the world as essentially absurd and arbitrary.

Paik was born in Seoul in 1932, at the height of his family's fortune. His grandfather was the founder and owner of the first modern textile plant in North Korea. All of their holdings, however, were lost in the worldwide economic depression of the thirties; later, the family gradually rebuilt its wealth and acquired two steel mills by the 1940s, only to have them confiscated and turned into "people's factories" by the Communist government in 1945. The relative instability of the world made a great impression on Paik when he was quite young. When hostilities broke out between North and South Korea during the fifties, Paik's family boarded a packed refugee train and managed to escape the country, taking little with them besides their personal belongings. Eventually, they ended up in Japan where Paik studied music and philosophy at the University of Tokyo. He graduated with a degree in aesthetics and succeeded in convincing his parents to let him enter a Ph.D. program at the University of Munich. Germany, at this time, was one of the world centers for avant-garde music composition, attracting and funding composers like Cage and Stockhausen. Because of Paik's philosophical interests he was drawn to the work and writings of John Cage, who believed that no barriers should exist between new music, ideas, and the workings of the contemporary world. This composer also emphasized in his music an Oriental belief in the workings of chance and fate: elements that were deemphasized in European art.

At the same time as he was performing his terroristic Happenings, Paik was also composing conceptually impossible music scores, including one for a symphony performance that would last 100 years. Correspondence art was mailed to many people around the world; and the artist even founded a school called the University for Avant-Garde Hinduism, which essentially existed only in his imagination. Paik claims to like Hinduism because of its ideological openness. "They even like sex," he observes.

Until his involvement with the rewired television sets (which came about through his interest in electronic circuitry used in musical composition), Paik considered himself a musical composer who worked with sound. The powerful presence of the doctored televisions changed his attitude about the aesthetic potential of this commercial communication tool. TV was rapidly changing our perceptions of the world, yet it was usually ignored by the intellectual and artistic establishment during the early sixties. Paik saw great opportunity in this underrated but pervasive medium.

After his video show in Cologne (with the ox head hung over the entrance), Paik returned to Japan to work with the expanded visual possibilities of color television, not yet available on the European market. A friend of his in Tokyo introduced him to a brilliant young electronics engineer, Schuya Abe; together they began to unravel the secrets of color television's circuitry. Paik and Abe soon discovered that the workings of a color set were so complex and dependent on various optical-chemical-electronic variables that the whole system operated on statistically random factors. For instance, despite the strictest quality-control standards, color televisions rolling off the Japanese assembly lines vary considerably in picture quality. During testing the best sets are pulled off the line and sold at a premium as "monitors" for the television industry. This element of electronic indeterminateness greatly appealed to Paik; in it he saw a relationship to the human condition: We all have the same basic wiring and components but we all behave and perform at different levels.

Paik had always been intrigued with the idea of America and American culture; in 1964 he got a chance to experience it first hand. When he arrived in New York during the month of June, the city seemed unbearably hot and dirty. Although little about Manhattan appealed to him at first it soon became his home.

Through his composer friend, Karlheinz Stockhausen, Paik was put in touch with Charlotte Moorman, a young cellist, who was organizing an avant-garde music festival at Judson Hall in New York. Moorman wanted to include Stockhausen's composition *Originale* on the program, but the composer refused to allow it to be performed without the assistance of Paik (he had appeared in it many times in European concerts). Paik's appearance in New York at this time was a stroke of luck for himself and Moorman. Paik was happy to offer his assistance and replay his role in this unusual performance for the New York audience: Using a can of aerosol shaving cream, he covered his entire head with the foamy lather; then grains of rice were sprinkled on top before he plunged his head into a bucket of cold water; while foam dribbled down his face, he began to play the piano in a gentle manner as if nothing had happened.

This event marked the start of a long and successful working relationship between Nam June Paik and Charlotte Moorman; the artist was to write many unusual musical performances for this dynamic promoter of advanced art, including one that inadvertently got both of them in trouble with the law.

For a long time Paik had been interested in the idea of incorporating sexual themes into contemporary music. He reasoned that erotic overtones had been important elements in literature and the visual arts for centuries but had never directly entered the domain of music. During his days in Cologne he had envisioned producing a theatre piece that involved a nude female pianist playing Beethoven's *Moonlight Sonata*. He had to give the idea up, however; no woman piano player could be found to perform the piece.

Paik felt that Moorman would be an ideal performer with whom to work in the production of events such as this. Despite her apprehensions about nudity, she was soon convinced by the artist's sincerity and the effectiveness of his avant-garde ideas. Thus, over the next few years she appeared, both clothed and nude, in many of Paik's unusual performances on American and European tours.

Opera Sextronique, produced in February of 1969, was one such performance that did not pass without official notice. The New York City police, in one of their periodic vice crackdowns, heard about the nude event, raided the performance, and arrested Paik and Moorman. The piece was designed to be played in four acts. In the first "movement" Moorman plays the cello dressed in a bikini made of small battery powered lightbulbs; then she changes to a topless evening gown for the second act; in the third act she wears a football shirt and helmet and is nude from the waist down; throughout the finale she appears totally nude with a large bomb instead of a cello.

Although Paik and Moorman had performed this piece many times in Europe, Moorman never got to the final "aria" in the New York performance. During the topless second act, the vice squad halted the proceedings and carted Paik and Moorman off to a downtown jail. This was a disturbing turn of events for the unlucky duo; Moorman was worried about what her family would think and how the arrest would affect her professional standing as a cellist. Paik was depressed about the incident and seriously thought of returning to Germany, where he fantasized he would be welcomed as a cultural hero martyred by the conservative American establishment. During the trial their lawyer, Ernst Rosenberger, who had defended Lenny Bruce during his run-in with the authorities, was able to get Paik acquitted but could only win a suspended sentence for Charlotte Moorman. This disagreeable incident put somewhat of a damper on the artist's plans to introduce sex into musical performances.

About this time Paik became interested

in the vast untapped audience potential of broadcast television. In the late sixties, through funding from the Ford Foundation, he was able to work in the well-equipped facilities of WGBH, a public television station in Boston. He and a group of experimental film and video artists were invited to produce broadcast-quality videotapes to be aired throughout the NET network under the program title of *The Medium is the Medium*, a reference to literary critic, Marshal McLuhan's popular theories about media.

Paik's offering, *Electronic Opera No. 2* (illustration), opens with a standard shot of the Boston Symphony playing Beethoven's fourth piano concerto. After a while, however, strange bursts of abstractly patterned color invade the screen and drown out the musicians. The general effect is one of visual distortion and recombination of images from many sources. Someone at the master control seemed to be wildly switching the channels, causing images to collide and juxtapose in a myriad of ways. In a sense *Electronic Opera No. 2* is patterned after our own

experiences of channel switching in search of the most suitable and entertaining program; in Paik's hands this video event for broadcast television becomes a post-Cubist montage of wild and diverse patterns, images, and sounds.

Intercut with shots of the symphony and a plastic bust of Beethoven were flashing images of a nude go-go dancer gyrating to the syncopated beat of music other than the piano concerto. All of these cultural signs and symbols were linked by brilliantly colored sine-wave patterns that were generated by electronic devices. As a coda to this complex broadcast extravaganza, the artist set fire to the toy piano just as the music built to a climax. Tiny flames and billowing smoke, enlarged by the closeup lens of the television camera, are superimposed on the placid features of Ludwig Van Beethoven. Just as the stirring final bar of the concerto sounds, the tiny legs of the enflamed piano collapse and fold.

Paik was inspired by the complex visual effects he was able to generate with the relatively simple signal-processing equip-

Nam June Paik, *Electronic Opera No. 2* (photo courtesy of Electronic Arts Intermix).

ment available at the Boston studio of WGBH. But if he could design and build a more sophisticated device, video programs could be composed without the enormous cost of a large studio and expensive technicians. Fred Barzyk, a producer at the station, was solidly behind the project and helped Paik obtain a $10,000 research grant for the project. The artist immediately sent for his engineering friend, Schuya Abe, and together they worked for a year in nearby Cambridge, Massachusetts, developing a sophisticated video synthesizer that allowed the operator to dramatically alter the scale, form, and color of any image fed into it. By adjusting a vast array of dials and switches, the artist would be able to completely manipulate the source material. For instance, an ordinary video image of a cat can be made to perform amazing visual feats: With the press of a button the natural color of the fur can be rapidly transformed to cobalt blue, then to reddish orange, and finally to a purple-blue; with the twist of another dial the iridescent creature can be made to slowly dissolve into an amorphous puddle of throbbing color which slowly transforms into a glowing sunset. By hooking up the synthesizer to a computer memory bank, scores of video images can be recombined and metamorphosized; with existing technology, well over a million permutations can easily be created. Not surprisingly, one of the aspects of this complex machinery that most pleases Paik is the random factor built into the electronic design. By setting the controls on a sort of "automatic pilot" mode, the synthesizer can operate independently and produce results unattainable through conscious manipulation. The artist, whose inclinations tend towards a belief in the wisdom and "tao" of chance operations, quite often follows this path in his compositions.

In 1972, Paik was invited to participate in a newly formed experimental workshop, called the *Television Laboratory*, at New York's WNET. Using the station's studio facilities he completed his *The Selling of New York* videotape (illustration), which was broadcast to the metropolitan population later that year. It is a humorous, tongue-in-cheek homage to the difficulties and delights of life in the Big Apple. Although horrified at first by the magnitude, noise, dirt, and violence of the city, over the years Paik has developed a strong affection for its raw energy and innate optimism.

The Selling of New York opens with Charlotte Moorman playing the cello in her Paik-designed "TV Bra" (two miniaturized Japanese televisions hooked up in tandem and worn over her breasts); soon an image of a bureaucrat, played by Russell Connor (former member of the New York State Council on the Arts), comes on. Connor drones on and on in dry official language about the problems of crime in the city. When the television camera zooms back, we quickly realize that we have been viewing the set in somebody's apartment. A young woman taking a bath is watching the city official on her own television. Occasionally, she responds to his remarks about crime with insipid comments such as "Oh, really?" and "How true!" In a Paikian manner the program abruptly switches from a couple simultaneously making love and trying to turn the television off to excerpts from Japanese television programs and then to a man who peeks through the keyhole of a pornography shop and sees the official continuing his monologue. In the final shot of the video tape, we watch a burglar jimmy open a window, enter the apartment — empty except for the television with the prattling official reassuring us that New York's crime rate is not so bad — steal the TV, and leave the same way he came in, through the open window.

In February of 1976, the René Block and Bonino Galleries, along with television station WNET, coordinated their efforts in a tribute to Nam June Paik as an early pioneer in video art. Many of Paik's older videotapes were shown at the galleries, along with recorded examples of his National Educational Television broadcasts. His newer work, however, made extensive use of closed-circuit and live video situations: Paik referred to it as interactive television. By utilizing the live image of this invention rather than a canned image, the artist made us question in a new way the content, context, and perceptual effect of an experience that was unfolding in front of us. Various visual gags—some of them corny, all of them incorporating TVs—were situated throughout the René Block Gallery: a small portable television was placed inside an old wooden radio receiver, and another set appeared at the bottom of a fish bowl with goldfish swimming around it.

The most successful live-circuit installation of the show was Paik's *TV Buddha* (illustration). An eighteenth-century Japanese Buddha, from the artist's personal collection, sits on a white, cloth-covered table, facing a small, spherical Futurist television. A closed-circuit video camera directly in back of the monitor televises an image of the Buddha onto the screen. The statue sits there contemplating itself via a twentieth-century electronic icon—TV. Seen together they express a synthesis between Eastern and Western ideology and help bridge a 200-year-old gap between the religious thought of the past and the technological reality of the present. Another interesting installation piece in the show

that sets up similar connections between the old and the new is Paik's humorous piece titled *Moon Is the Oldest TV Set* (illustration) (the artist habitually drops articles from his English, making it hard for people to understand him at times; Paik speaks five languages—all poorly, he claims). Each of the twelve black-and-white monitors displays a different, unchanging image of a certain phase of the moon. From the title and setup of this piece, Paik implies that before electricity and television, the only prime-time event "broadcast" during the evening and seen throughout the world by countless people was the appearance of the moon.

At the same time as his retrospective at the Block Gallery, Paik installed a large multimonitor video piece, titled *Fish Flies on the Sky* (illustration) at the Bonino Gallery. Twenty color television sets are mounted in an irregular pattern on the ceiling of the gallery; to enable the viewer to comfortably view the celestial monitors, soft pads are placed on the floor underneath the televisions. Paik refers to the installation as an "antigravity" piece. Basically,

Nam June Paik, *The Selling of New York* (photo courtesy of Electronic Arts Intermix).

Nam June Paik, *TV Buddha* (photo by Peter Moore).

Nam June Paik, *Moon Is the Oldest TV Set* (photo © Peter Moore, 1982).

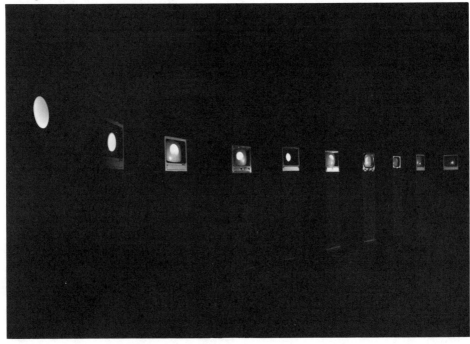

two channels of video information are simultaneously broadcast over the monitors: One half of the sets show multicolored tropical fish lazily swimming around; the other half feature military jet aircraft screaming overhead. The juxtaposition of the peaceful beauty of the fish and the frightening power of the planes is striking. Also, the whole context of viewing television changes. Paik extends the idea of looking at the natural world into the artificial world of watching television. In *Fish Flies on the Sky* the artist presents us with a multidimensional Surrealistic event rather than a videotaped television show.

Besides being one of the earliest pioneers in the field of video art, Paik has succeeded in transcending the audience limitations of the gallery-museum context in which most artists show their work. Through his participation with National Educational Television, Paik's work has reached millions of viewers, many of whom have never visited a professional exhibition space. Not only is Paik committed to finding a wider audience, but he believes the medium has yet to be explored by artists. Television, Paik speculates, holds great untapped potential. "We are now at stage of ancient Egypt with hieroglyphics," he told Calvin Tomkins in an interview. "Until recently, TV equipment is so expensive that only the priests can use it. And there is *constant* effort made by networks and by TV unions to keep production costs high. That is classical way of monopoly capital —you know? I want to find ways to cut costs so it can be opened up to others— many others. . . . Problem is not really socialism or capitalism but technology, you know—how we manage that."

Paik has confidence that much of what we view in the future as art will be an information gathering, coding, and transmitting process. The artist will be viewed as social communicator and a humanizer of technological tools. Paik is quick to point out the value of the exciting visual presentations that artists have been doing for centuries. Important information never reaches the public, Paik maintains, simply because it is too boring and unwieldy in its original form (he cites Rand Corporation reports that even the *New York Times* would not reprint because of their density and complex language). Paik feels that artists functioning as communicators are needed to take that information and make it accessible to a much broader segment of the public. The material contained in the research might affect us all so it is imperative to have clear access to it.

Paik's thoughts about art are obviously not limited to formal aesthetics. In a video-

Nam June Paik, *Fish Flies on the Sky* (photo © Peter Moore, 1976).

taped interview conducted in April of 1980, he had this to say about future art activity and the present economic crisis in the postindustrial world:

Number one, information has to be recognized officially as an alternate energy source. Information changes our life style. Dancers and yogis achieve ecstasy with 200 calories. However, racing drivers, if they want to achieve ecstasy, have to burn 200,000 calories. I found this new economical law . . . certain activities which have a higher spiritual quality use less energy. For instance, when rich man can buy a Jasper Johns for $100,000, that takes very little energy. However, if rich man buys Cessna airplane and flies around it takes a lot of energy. But contribution to national economy would be the same. . . . $100,000 spent. Same thing when you buy a videotape recorder for $500. Or if you rent a car for $500 — ten days — national economy benefits the same. But this videotape recorder consumes much less energy. So when we shift our economic priorities of consuming to more spiritual sphere then we can solve all economic problems with much less energy.[4]

Paik even whimsically relates sex — one of his on-going aesthetic concerns — to our current obsession with energy shortages and restrictions. Paik wryly notes that making love requires only an expenditure of about 800 calories to attain ecstasy, but traveling for pleasure by car or plane consumes millions of calories of energy per hour. He believes that by avoiding high-caloric activities and focusing instead on spiritually charged but energy-conservative art activity, society as a whole would benefit. "Everything counts on calories," he says, "We need worldwide dieting, lose weight."

Beneath Paik's playful analogies is a great deal of rational, sober thought; many academic experts in the energy-research field wholeheartedly agree with his basic premise: move information instead of mass. Howard Klein of the Rockefeller Foundation finds Paik without fail to be one of their most productive and perceptive consultants; despite his humor, Klein reports, Paik's mind unerringly goes straight to the heart of the matter.

In the last few years, Paik's concept of the artist as a pioneering thinker, making new concepts and technology accessible to the general public, has been vindicated. Although he has been experimenting with interactive television for years, only recently has society begun to realize the economic benefits of telecommunications. Advances in computer technology have combined with dwindling energy resources to make it desirable and perhaps necessary for businesses to cut operating expenses, while at the same time increasing information exchange and processing. By mastering the concepts and design of visual intercommunication, many people in industry could effectively complete their work without wasteful commuting and expensive out-of-town business trips.

Paik is a man with a definite vision; his work and philosophy express an optimistic belief that the role of the artist is to clarify our thinking about a broad range of social concerns, including the role technology and telecommunications could play in our lives. He believes that to survive we need to be able to contemplate and integrate timeless concerns with present-day realities — just as the artist's ancient Buddha views and defines itself through a contemporary instrument of visual communication: television.

Laurie Anderson

When Laurie Anderson walks on stage in one of her well-attended performances, she cuts a strange and foreboding figure; surrounded by enormous amounts of esoteric electronic equipment, she appears dressed all in black with a short cropped "punk" haircut and dark, sunken eyes. Her effect on the audience, however, is anything but obtuse or hermetic; Anderson attracts a broad following of thousands of people from New Wave enthusiasts to linguistic scholars. Most performance art of the recent past drew, as she once remarked, "the same three hundred people to the same artspace every year." Her interest in playing big commercial theatres to large, admission-paying audiences stems from a desire to enter and contribute to the ongoing cultural life of America.

Most of Anderson's recent performances have taken the physical and structural format of rock concerts and make extensive use of electronically enhanced songs and a smooth, professional, theatrical approach. This direction runs counter to the prevailing aesthetic climate that was operant when she first became interested in performance and musical artforms.

Anderson represents a decidedly different, Postmodern sensibility in her work. She is an indication of a new drift in art — away from a hermetic Formalism and towards a body of work that mediates between easily understood, popular forms and meaningful statements that question the very nature of contemporary life. Anderson feels she can be amusing, entertaining, and theatrically viable with *no* compromise in terms of the work's content, meaning, and effectiveness. She believes that performance artists in the future will develop the means to reach enormous audiences by utilizing electronic communication systems already in existence, like self-produced records, cable television, and alternative radio stations. At that point artists will be in the position to make direct contributions to a national dialogue rather than function as ineffective bystanders.

When Anderson first began doing public performances, she took an insular approach, which was pretty much expected of serious artists who worked within the exhibition context of museums and art galleries. She avoided recognition and acknowledgment of the audience, stared into space, and talked to herself rather than to the group watching her. She said at the time: "My idea of the perfect performance was analogous to a bad movie. At a bad movie you notice the popcorn under your feet, the height of the armrests, placement of the exit signs, etc. throughout the film. At a good movie you fall away . . . and at the end it's a mild surprise to find yourself sitting there." She now believes that these early pieces had a lot to do with the architectural ambiance of museums and galleries in the sixties: sparse, removed from the mainstream of life, and socially limiting.

Anderson also raises an interesting question about the cultural context and meaning of "boring" media art. Much of the performance and video art of the past decade and a half has been soundly criticized by the general public and popular press as boring and obtuse, while at the same time hundreds of articles in professional art journals have extolled the aesthetic virtues of this kind of radical art. It has been argued that this work represents fresh thinking and is in the process of developing a new visual language. To be sure, both sides — the viewing public and professional artists — have legitimate interests and concerns. The audience has turned to the work for unique creative and inspirational experiences; the concerns of the artists center on expressing their aesthetic perceptions about life. Entertainment has not been topmost in their minds; popular culture has always attended to that interest. Anderson began to conceive of a fusion of the legitimate desires, expectations, and needs of both parties. In this way performance artists *and* a potentially large and sophisticated audience could be satisfied and fulfilled.

As the diversity and interest of the artgoing public grew during the late sixties and seventies, the willingness to accept any feeble gesture as art diminished. Marcel Duchamps's belief that anything the artist does is art failed to impress many individuals who wanted artworks to be both meaningful and interesting. Many of the discoveries of the avant garde were incorporated into popular culture, raising the level of artistic consciousness in TV advertisements, popular music, and Hollywood films to new heights. Because of this assimilation the aesthetic distance between sophisticated commercial works and "high art" diminished significantly. If popular artforms could freely borrow from the avantgarde, why could the fine arts not learn from highly trained communication professionals in film, television, and recording? Postmodern performance artists like Laurie Anderson believed it was time to tap a source that greatly moved and affected millions: popular culture.

The oil embargo of the early seventies, and its resulting economic aftershocks, had a significant effect on the artistic climate in America. Obviously, social, political, and economic factors play an important role in the makeup of contemporary art. Like the fascinating sociological connection between women's hemlines and stockmarket trends—the shorter the skirt, the better the market—the art world appears to reflect the financial conditions of the times. Generally, artists are more cautious when confronted by the realities of rampant inflation, high unemployment, and a general feeling of uncertainty regarding the future. Audience acceptance appears to be a key issue in the eighties.

Nam June Paik has a theory about the nature of radical art in America's recent past. He feels that mass culture and popular art have been so refined and developed in this country that progressive artists—in the face of this formidable competition—have totally defined their work in opposition to popular culture. If films and television were directed towards light, fast-paced, easily understood entertainment, the artists would produce slow, difficult to understand, profoundly meaningful artworks; and if huge audiences of ordinary citizens were the target for the mass media, avant-garde artists would seek the approval of intellectuals and fellow artists. Paik feels that America's great wealth paved the way for an attitude of cultural arrogance that looked down upon sizable audiences and any art that had popular appeal. It was *aristocratic* to be concerned with the minutia of Formalist art; of course, this was an easy attitude to maintain when economic times were good and money was readily available. Highly specialized concerns also demonstrated artists' superiority over the masses whom they felt were only interested in network television and Disneyworld.

Over the last fifteen years, however, young people in America have taken advantage of educational opportunities in unprecedented numbers and today form a large, relatively sophisticated audience for many new forms of art. The old belief that only other artists were knowledgeable enough to appreciate serious work has faded. Avant-garde artists of the last decade discounted the premise of popular support for so long that they failed to see the emergence of a new audience—an audience willing to work with them. But this group has also demanded to be treated with some consideration and respect. The current climate of funding cutbacks and economic restrictions has changed the way many artists now view their audiences. Paik once explained that he did not write boring music or try to make boring video because he could not afford to: He came from a poor country and was himself poor.

Originally trained as an Egyptologist, Anderson grew up in Chicago, the daughter of a well-to-do paint manufacturer. Until the age of fifteen she was a serious student of the violin and achieved a great measure of technical virtuosity on the instrument; today she puts this ability to practical use in many of her performances. In her personal life as well as her art, Anderson is a far-ranging explorer and adventurer; one summer she tried to hitch-hike to the North Pole and got within several hundred miles by begging rides from bush pilots in the Alaskan Tundra. After college Anderson worked as a reviewer and critic for a journal well known for its theoretical and Formalist approach to artmaking. At this time in her life, she was obsessed with the immense beauty, popular appeal, and spiritual energy of Van Gogh's work and managed to introduce this concern into the context of almost every review. Her editor at the magazine called her in and complained about her excess of enthusiasm; he pointed out that it was not valid criticism to directly compare every contemporary artist to this acknowledged master. Anderson dutifully listened to the rebuff and began her next review with: "Unlike Van Gogh this artist . . ."

One of the role models for Anderson's work is the street musician. There was always something vital and engaging in street performances even if the musical standards were not terribly high. What made the difference was the context of the work—the street. Unlike the museum or art gallery's self-selective audience, here was the broadest possible range of individual backgrounds and responses.

Anderson's first performance event took place in 1972, when she wrote and orchestrated a concert piece for automobile horns, which was given at the Town Green in Rochester, Vermont. In 1974 she took to the streets—to put theory into action—and became an avant-garde street musician when she performed *Duets on Ice* (illustration) at public squares in Europe and America. Anderson froze the blades of a pair of skates into blocks of ice and appeared on several street corners in Genoa, Italy, under the sponsorship of the Saman Gallery, playing her violin until the ice blocks melted. In this piece Anderson accompanied a tape recording that came from a small playback unit hidden inside the instrument. Her musical program was intriguing, particularly for Italy; while the "ice skates" slowly melted, Anderson performed a continuous set of country and western songs for a slightly bewildered and vastly amused group of Genoans. *Duets on Ice* was also performed on street corners in each of the five boroughs of New York City, including a location near the entrance to the Bronx Zoo. The artist assumed, in a sense, the disguise of a street musician to enable her to reach members of an audience that performance artists rarely reached. Even the choice of thematic material—ice skates and cowboy music—was calculated to strike a note of popular awareness and recognition in the minds of the curious onlookers. No doubt seeing a young woman fiddling on ice blocks was an odd spectacle for Italians and Americans alike. It was not without its poignant side, as the ice slowly melted and drifted in time to the nostalgic, popular folk music played on the violin and recorder. Also elements of the situation were quite paradoxical; once the blocks melted, the skates were rendered even more useless without ice. At this point, the music stopped; and Anderson awkwardly walked away on the skates, leaving behind a puzzled knot of people, some of whom doubted what they saw.

In 1977 the artist developed her unique mix of popular and "serious" art further with an environmental-music piece at New York's Holly Solomon Gallery (illustration). Anderson hung illustrated song sheets on the wall and installed a commercial juke-

Laurie Anderson, *Duets on Ice* (photo courtesy of Holly Solomon Gallery, New York. Copyright by Paolo Rocchi, Genoa. Copy print by D. James Dee, New York).

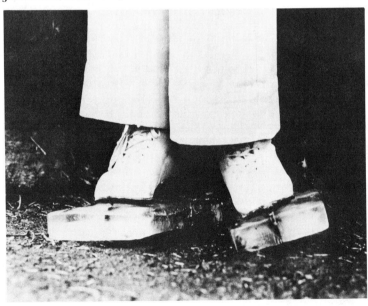

box that was loaded entirely with her own records. At twenty-five cents a shot, the listener heard a variety of catchy tunes with simple repetitive lyrics reminiscent of Mick Jagger. Statistics on frequency of play were automatically tabulated by the machine. Some of the most requested titles were: *If You Can't Talk About It, Point to It* (for screaming Jay Hawkins and Ludwig Wittgenstein); *Fast Food Blues* (advice to artists about marketing their work); *Unlike Van Gogh* (in response to her art critic days); *It's Not the Bullet That Kills You (For Chris Burden)*, a song in reference to Burden's late-sixties performance piece in which he had his arm grazed by a bullet; and *New York Social Life*, a satiric look at an artist's life in New York City.

In all of these songs the melodies and words were catchy and "top-ten" in tune: Some were done in standard three-chord rock format, others were influenced by the blues; most were humorous and decidedly clever. Anderson used simple language, pleasant tunes, and witty anecdotes to produce a body of engaging works that effortlessly mixed conceptual art with popular entertainment.

Although her music makes use of the hypnotic, repetitive, and simple phrasing of popular music, its thematic content is quite different. Anderson's strange songs inevitably offer subtle warnings about the dangers of contemporary life in corporate society. She uses the intrinsically attractive properties of popular commercial music as a knife — to cut through to the truth of the matter and expose the way in which we are all conditioned by ubiquitous forms of electronic communication. Anderson questions the identity of these anonymous "voices" (radio and television) and makes us think about what is really being said to us by them.

The artist feels that commercial songs, slogans, and advertisements form an information environment that totally surrounds and affects us. More than meets the eye is transmitted into us by these well-crafted communication missiles. We do not even have to consciously listen, they are always there entering and becoming part of us. *Handphone Table* (illustration), a sculptural sound installation at the Museum of Modern Art in 1978, metaphorically alludes to the phenomenon and enables us to perceive the way information enters our bodies and our minds in usually unrecognized ways.

For this piece Anderson constructed a simple pine table about five feet in length, and hid a tape recorder inside it. Two stools were placed at the far ends of the table; participants sat on the chairs, placed their elbows into rounded indentations, and covered their ears with their hands as if to keep sound out. But this procedure had the opposite effect; Anderson coupled two hidden sound-transmission nodules to the indentations from within the table. By a scientific phenomenon known as "bone conduction," music was relayed to the participants through body vibrations. The titles of the two pieces (a different one for each end) specifically referred to this effect: *And I Remember You in My Bones*, along with *Now You in Me Without a Body Move*.

But no decipherable language was heard in these "songs." One piece was assembled from repeated phrases of low-frequency organ music and the other featured plucked or slowly played passages on the violin. It was the nature of the audio experience, rather than the thematic content of the music, that was most remarkable; the sound seemed to originate from within the listener's body and mind. It was a sourceless and pervasive presence; the listeners *were* the music; it did not exist beyond them.

On a wall opposite the specially prepared table was a blurred photograph of a phantom listener with a legend beneath it that read: "the way you moved through me." Anderson's work enables us to experience first hand the unseen process through which information is unwittingly assimilated by our bodies and minds. This piece serves as proof that it is not necessary for us to "hear" things in an ordinary way to be affected by them; they are osmotically absorbed in a variety of ways.

Since 1979 Anderson has been developing an on-going epic performance series that operates under the umbrella title of

United States (Transportation, Politics, Money, and Love) (illustrations). Part I, *Transportation*, opened at the Kitchen — a New York performance space — where for the first time Anderson made use of a rock-concert format. *United States Part II* premiered October, 1980, at the Orpheum Theatre, an old vaudeville theatre in New York's Lower East Side, now the site of various "new-wave" events. Everything about this performance — even the "punk" musical hall — distinguished Anderson's work from the usually sparse and formalized work seen in traditional art-world contexts. This event also signaled Anderson's transformation from a midwestern college girl into an androgynous, new-wave punk. Her close-cropped, spiked hair and dark eyeshadow

Laurie Anderson, *Environmental Music Piece* (photo courtesy of Harry Shunk).

Laurie Anderson, *Handphone Table* (courtesy of Laurie Anderson).

made her appear powerful and frightening: a black angel of new music.

The stage that evening was literally jammed with lighting and complex sound equipment. Showing the sources of the sound is important to Anderson; rather than the faceless voices that emerge from our radios, Anderson exposes the technological origins of the sounds we are about to hear. A five-piece group accompanied her in *United States Part II* and provided the musical backup as well as visual effects. Projected above and to the left of Anderson during this performance was a gigantic schematic map of the United States showing clocks within the various time zones — all out of sync. The diagram bore a great resemblance to the kind of planning board that might be used by NORAD, the North American Air Defense Command, to monitor incoming missiles and bomber squadrons in time of war.

This fantasy was played out during the course of the program; a film of *Space Invaders*, a home video game in which the player tries to shoot down alien spacecraft, was projected onto the screen at one point.

Anderson attracts a diverse audience of

Laurie Anderson, *United States Part II* (photo © Chris Harris; courtesy of Holly Solomon Gallery).

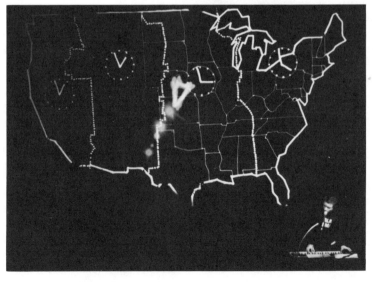

Laurie Anderson, *United States Part II* (photo by Paula Court, courtesy of Holly Solomon Gallery).

"new-wave" music fans, denizens of the contemporary art world, and hip culture buffs. During her 1981 tour she completely sold out the 1200-seat "Cinema" theatre in San Francisco. Anderson shared the bill with the writer William Burroughs, whose macabre, mad-cap novels have been rediscovered by a younger generation of "punks," and John Giorno, a New York sound poet. Although the reception for these writers was enthusiastic, it was clear that a majority of the audience had specifically come to see Anderson perform.

During the course of the evening Anderson used a variety of highly sophisticated audio devices, especially one called a harmonizer, which can totally transform the pitch, timbre, and range of the voice. This mechanism has been used for years by pop musicians to electronically modulate and thus substantially alter the original input signal in much the same way as the Paik–Abe synthesizer transforms video images. Through this instrument Anderson can effortlessly assume many different "voices," from a low-pitched, raspy man's voice to a strident "Alvin the Chipmunk" falsetto. Another device of which Anderson makes excellent use is a "black box" that multiplies her lone voice into the sound of a massed chorus.

Along with the fact that Anderson loves flashing diodes, switches, and calibrated dials for the sheer visual and conceptual wonder contained in them, she believes that working with this equipment is a means of taming and controlling machines that otherwise might intimidate and frighten us. In a recent interview she stated:

Another reason I use technology in performance is that it's a way of looking at information filters. The way that we receive information comes in so many varied ways. So for example, in this performance there's one sequence in which you hear a voice on tape and right after a live voice and then a written text. So you make this jump of the way you receive it hearing something pre-recorded, spoken live, and then receiving it through your eyes, so that there's some constant sort of jump.

One of the reasons I use filters to change my voice is I'm interested in a kind of corporate voice that might be compared to the writing in Newsweek *or* Time,

in which someone starts an article and then it's edited and re-edited and re-edited and the article finally comes out in Timese or Newsweekese and it's a corporate voice. It has someone's name signed on it, but it's in the style of that particular magazine. This corporate voice is a spooky voice, because it's a highly stylized voice. One of the things I'm trying to get at through these filters is to look at those kinds of—particularly American—voices that try to convince you there's a person behind it and there isn't. There's a corporation behind it. I like that technology can trick you and to call attention to the way it can trick you.[5]

Anderson's entrance into the massing of transistorized sound equipment at the Cinema was dramatic: She walked on stage simply dressed, carrying a wooden gavel. From the beginning there was the distinct feeling that this was to be a form of conceptual cabaret theatre rather than Formalist performance art.

Anderson casually approached the microphone and launched into what appeared to be an informal story about her recent tour. Her stage presence was smooth, confident, and well paced:

Lately I've been doing a lot of concerts in French. . . . I memorize it. . . . I mean my mouth's moving but I don't really understand what I'm saying . . . it's kind of like sitting at the breakfast table and it's early in the morning and you're sitting there not quite awake eating cereal and staring at the writing on the cereal box not really reading the words just sort of looking at them and suddenly for some reason you snap to attention and you realize that what you're eating is what you're reading but by then it's much too late. . . .[6]

The monologue is delivered with a self-assured, polished style that is designed to relax and prepare us for the event. Anderson is warming the audience up, people are obviously amused and prepared to have a good time. There is none of the overtly serious posturing and isolation from the audience that characterized performance art of the past decade. But beneath Anderson's well-modulated announcer's voice and humorous dialogue, a chilling undercurrent of subtle meanings and allusions emerges—a warning, in fact. Buried in the supposedly offhand story are coded messages that refer to the pervasive dangers inherent in our media-dominated society such as: the loss of personal identity we have experienced after having been conditioned by tens of thousands of hours of television and radio (". . . I mean my mouth is moving but I don't really understand what I'm saying . . ."); and the frightening realization that inevitably our total information intake determines who we are and what we become (". . . suddenly . . . you realize that what you're eating is what you're reading but by then it's much too late. . . .").

The mood starts to slowly shift; Anderson is toying with us; people are laughing but now it is an anxious laughter. She continues her story and tells us about French babies in strollers who are used as "traffic testers": "The mothers can't see what the traffic is like because of the parked cars so she just sort of edges the stroller out into the busy street. . . . the most striking thing is the expression on the babies' faces when they're stranded in the middle of traffic, banging those gavels that they all have . . . and they can't even speak English."

With this she abruptly ends the mono-

logue, goes over to an electronic board, raps the gavel several times while she adjusts some dials and succeeds in putting the sharp-pitched sound on a repeating tape-loop that immediately establishes the insistent rhythm of her first piece, *O Superman*. While this sound metronomically plays, Anderson goes over to a lectern with a small electronic organ mounted on top. Directly above and behind her is a projected circle of light originating from a projector in the orchestra pit (illustration).

Anderson's sense of staging and theatrical savvy is unerring. As *O Superman* musically unfolds, it succinctly establishes the whole mood and tone of the epic *United States* series. Although all of the sections in the work maintain a high-level of consistency, *O Superman* is clearly the most beautiful, evocative, and effective piece of the concert.

Just before she began the song the gavel sound slowly faded and was replaced by another loop of an electronic organ tone

Laurie Anderson, *O Superman* (courtesy of Laurie Anderson).

that maintained the previous rhythm. The first vocal sounds uttered by Anderson took us totally by surprise; instead of a solitary voice we heard a mass chorus of her "voices" singing a combination dirge and hymnlike refrain of, "ah, ah, ah, ah, . . . ah . . . ahhhh." By means of the complex electronic equipment and acoustic filters, she had transformed the sound of her voice into a group presence—a symbolic corporation.

After a few introductory bars Anderson named the characters in her musical drama by half-reciting, half-singing the invocation, "O Superman; The Judge; O Mom and Dad." Then, in a playful, chirpy, childlike, sing-song tone she sang as if a home telephone answering machine had been reached, "Hi, I'm not home right now but if you want to leave a message just start talking at the sound of the tone."

One of the most effective characteristics of Anderson's performance work is the way she mixes seemingly innocent humor with amorphous fear: "Hello, this is your mother," she sings in a Pollyanna tone, ". . . are you there? . . . are you coming home?" The passage appears to repeat itself. The next chorus starts out with the same "Hello" but there is a short pause, after which Anderson sings in a vaguely ominous tone: ". . . well, you don't know me but I know you and I've got a message to give you . . . here come the planes." With this last word the whole piece dramatically shifts in sound, meaning, and thematic scope. *O Superman* is clearly revealed here as a potent symbol for a certain concept of America: a country respected and feared throughout the world for its enormous military muscle, political leverage, and economic power. Anderson expresses this foreboding transformation by extending the word "planes" into the droning hum of a massed aircraft squadron. As this happens, her left hand, extended and flattened as if a child were playing "airplane" with it, lifts up and casts a weaving, plane-like shadow onto the lighted circle above her. The combined effect of the sound and the simple shadow image was sobering. No one was laughing now; an uneasy silence descended on the audience as she continued, "So you better get ready/Ready to go/Well you can come as you are/But pay as you go."

Anderson takes the saccharin catch phrases and advertising slogans that we inadvertently hear and read every day and makes us consider them in a new light. She warns that everyone is responsible for their actions and debts—even the wealthiest of countries—and suggests that a day of reckoning is fast approaching. The most frightening aspect of this work is its ambiguous, subtle meaning. Anderson recognizes that the most horrible and deep-seated fear is the one we cannot articulate—fear of the unknown.

A feisty and tough-sounding voice continues with, "OK, who is this really?" Another verbal persona—lower, "cooler"—follows, "And the voice said, 'This is the hand/The hand that takes.'" The shadow play—reminiscent of childhood dramas—not so innocently continues. Her hand is outstretched, palm up in the classic pose of a handout seeker. When the chorus is repeated a second time ("This is the hand/The hand that takes") Anderson instantaneously transforms the passive image of receiving into an aggressively violent image of taking—pointing her index finger to signify a gun. Because of the new visual imagery, the same words mean something entirely different now. Next, when she returns to the chant of "Here come the planes," the "gun" changes into a shadow plane and her voice once again hums with the sound of massed airplanes.

By now we are totally captivated and swept into the unfolding audio-visual saga of *O Superman*. Anderson's choice of ambiguous images extends the meaning of the piece into many realms. Through the simplest of means, she is able to conjure and fuse in our minds concepts of vague "authority," family, home, work, the military, and a broad and pervasive idea of the United States and its many ideologies.

Like the strong and violent contrasts that exist in America—hopeless poverty amidst untold wealth; cuts in social and educational programs alongside record defense spending—Anderson seesaws back and forth, exposing many of the inconsistencies inherent in the "messages" we receive through the media.

Now her voice assumes a false, sugary-sweet, public-relations tone as she wryly reassures us that: "They're American planes/Made in America/Smoking—or Nonsmoking?" The turnover rate between humor and trepidation is quicker now. Quite abruptly, she slips into a minor musical key and solemnly recites: "And the voice said/Neither snow nor rain/Nor gloom of night/Shall stay these couriers/From the swift completion of their appointed rounds." This legend is emblazoned over the pseudo-Greek architecture of the main post office in midtown Manhattan; no one living there can fail to notice this slogan because it runs the length of an entire city block. But it means something entirely different in the context of Anderson's song. Here it alludes to some kind of deeply ingrained but vaguely understood sense of mission on the part of the American people.

The song ends with a final verbal incantation and surrender to the immensity, power, and inevitable control contemporary technology extends over our lives: ". . . When love is gone there's always justice/And when justice is gone there's always force/And when force is gone there's always . . . Mom (Hi, Mom!)."

Finally, the organ plays a repetitive chord progression and Anderson plaintively sings, ". . . so hold me Mom in your long arms/So hold me Mom in your electronic arms/In your automatic arms. . . ."

In *O Superman* America is sublimated into an omnipotent, omnipresent, coyly benevolent "Mother" figure that is beamed at us through countless television shows, radios, magazines, newspapers, and popular songs. There is no escape, not even any choice; Anderson's voice speaks of a yearning to be held—in the long, electronic arms of America.

After this piece, the audience was totally in Anderson's power. She then went on to perform about a dozen other musical works, some keyed to prerecorded tapes and others involving various instruments: harmonizers, audiofilters, and electronic organs. Although the pieces ranged from stunning virtuoso performances on the violin to more episodic, narrative adventures, nothing else seemed to have quite the immense poetic power and magic of *O Superman*. This was Anderson's anthem and compelling leit motif.

Apparently, the enthusiasm of the audience is shared by many art critics, new music aficionados, and the general public; Anderson's popularity points towards the assimilation of Postmodern avant-garde works into the mainstream of popular culture. Surprisingly, an eight-minute, independent-label record of this song—only 5000 copies were originally pressed—found its way to England where John Peel, an adventurous BBC disc jockey, repeatedly played it on his popular radio show. Its enormous success surprised everyone. Even though it was unavailable in the United Kingdom at that time, other stations began to play it and soon the whole country was talking about *O Superman*. Naturally, curious record companies became interested in this piece, and within weeks Anderson signed a contract with Warner Brothers records and tapes. Over 100,000 records were quickly pressed by the company and shipped to England. Within a short time *O Superman* reached number two on the British pop chart.

In her most recent work Anderson continues to seek broader audiences and greater control over mass-media technology. In 1985 she released a feature film she made titled *Home of the Brave*, which played in movie theaters throughout the country. Through this film many people who had never seen her concerts could share the innovative Laurie Anderson experience. Much of the material in this film was adapted from her live concert performances, but many details were altered to relate to the special opportunities of the film medium. Exactly where Anderson will take this synthesis of popular culture and "serious" art is an interesting question. One thing is clear, however, there is a great deal of interest on the part of a younger generation with performance artists who leave behind the exclusivity of the art world and enter the marketplace with accessible and effective work. By gaining control over instruments of mass communication, the possibility now exists for both performance and video art to contribute to our culture in ways that were undreamed of in the past.

7 Spaces

Sculptural Events, Earthworks, Environments

A number of artists in the early seventies, primed by a bountiful economy yet deeply suspicious of prevailing artistic and cultural values, set out to rediscover through their work with environmental spaces some original characteristics of art buried by the "advances" of civilizations. Rejecting the coolness, overt rationality, and materialistic tendencies predominant in work of the sixties, they began to redefine the physical form and inner values of their art. All believed that the initial impetus of artmaking was not rooted in facile decoration, trivial amusement, or the acquisition of status and goods, but that it was founded on meaningful creative activity that put both the artist and audience in touch with the world in special ways. These individuals became increasingly disenchanted and detached from life; their inspiration came largely from early societies who cultivated close cultural relationships to the natural environment. Thus some of the newest forms of art were suggested by humanity's oldest concerns.

Although it is impossible to speak with certainty about the beginnings of art, anthropological evidence indicates that prehistoric societies believed that spiritually, through art, they could intervene in the processes of nature and favorably influence the course of events for the benefit of the community. In this way protoartists, or "shamans"—the first artistically active individuals known to us—contributed significantly to the physical–psychological well-being of the group. Early tribes of humans felt helpless in the face of natural phenom-

ena. The background of the world, with its "hidden things," was mysterious and inexplicable to them. Shamans were dedicated to developing their knowledge of the world to "explain" and subsequently harness the many interconnected environmental forces at work. Thus, in early cultures, artistic expression and creative acts played a primary role in the group's survival.

No better insight into the now-vanished beliefs and art activity of primitive hunter-gatherers exists than the account explorer Knude Rasmussen gives of his visit with the "Copper" Eskimos in 1926. Rasmussen described an amazing encounter with a present-day shaman who was called upon to tame a fierce storm that had raged for three days; food supplies were dangerously low, the tribe faced starvation. The whole community depended on the creative actions of the shaman to save them. Through ritual, or "individual acts of participation," the group hoped to discover the cause of the storm god's anger and halt the storm. A wide variety of artistic means was employed by the shaman during the course of a ceremony: pictorial representation and visual symbols, dramatized events, songs and group singing, ritualized movements, and hypnotic verbal incantations. This comprehensive approach combines many artforms that would, in our specialized society, be considered separate and unrelated disciplines. Above all, this approach demanded and received from the audience a level of involvement and participation unknown to our present culture.

The atmosphere in the ceremonial hut was serious but confident and hopeful; much was at stake: Tomorrow there would be no food left for the children. Late in the evening, after ritual sharing of food, songs, dances, and drama, the shaman seized a seemingly innocent man and fought him to submission, choking him until he lost consciousness and slumped lifelessly to the ground. In a dramatic and symbolic way the storm god was subdued and slain "in effigy." After this climax the shaman's work was concluded; the group, comforted and reassured, returned to their igloos through the howling storm.

"Sure enough," Rasmussen noted the following day we travelled on in dazzling sunshine over snow blown firm by the wind . . .[1]

Contemporary environmental artists believed that the broad-based creative activity of preindustrial societies was still a relevant artistic model and could effectively interact with many aspects of modern life, thus enlarging the role art can play in the world.

New kinds of "spaces" were necessary to contain and express this broadened view of art and life which would make it possible to experience in a fresh way the extraordinary organization and mystery of the world and the way in which we interact with it. Above all, this work signals a return to nature—not the idealized, rural "nature" of the last century—but an expanded view of the environment and an awareness of the role economic, psychological, cultural, and biological forces play in our lives. These artists function as modern alchemists transforming ordinary elements into magical events and structures that rejuvenate our psyches.

In these spaces old patterns of thought are replaced by new experiences that allow us the opportunity to perceive freshly the urban landscape, our relationships with plants and animals, the geologic history of the earth, the symbolic meaning of shelter, and changing socioeconomic relationships. Environmental artists have chosen to play direct roles *in* nature: to live, experience, and interact with it, not merely to represent it. Consequently, the real interest of these

individuals is in the entire "process" of artmaking, not only the finished product. Working with the natural environment, conceptual ideas, their bodies, and personal experiences, their goal is a broad visceral understanding of life and living. The traditional concern of art, visual representation, has been replaced by this expanded vision.

What these artists share in common is not their subject matter, materials, or methods of operation, but their interest in and commitment to an expanded view of artmaking and through it self-discovery. Out of the interaction of these elements personal, psychological, and physical spaces are created that are as unique as the artists who construct them. For instance, Christo is an artist who orchestrates vast sculptural events that explore, in experiential ways, interactions with political, social, and economic forces of our time. Robert Smithson's earthworks are meditations on geologic time and the history buried in the earth's strata. Smithson's work makes possible new perceptions about the contemporary landscape and how we might relate to it in a positive way. In the barren region north of Las Vegas, Michael Heizer lives and works on the largest grouping of environmental sculpture the world has ever seen. Eventually, this project, called *Complex I*, will cover three square miles of desert floor with eclectic structures that comment on the history of architecture and perhaps predict its future. A strong sense of social consciousness is evident in the work of Charles Simonds whose imaginary civilization of *Little People* exerts a very real and lasting effect on the lives of his neighbors in New York's Lower East Side. More than a decade ago sculptor Alice Aycock partially buried a small building on a Pennsylvania farm and uncovered in the process a flood of childhood memories recalling clubhouses, attics, and basements. Exploring this space conjures up deeply rooted feelings about shelter, personal security, and hidden fears. Perhaps the most formidable and challenging space remaining for us to explore in the world is the labyrinth of the mind. Dennis Oppenheim works in this uncharted area assuming the role of artist-shaman to reach and affect people with environmental works of compelling psychological power. The art of Jenny Holzer and Keith Haring is designed to work within heavily trafficked urban spaces; their work questions the social and political machinations of the world about us and makes us think about the cultural realities of modern life. Both employ forms of language to get their messages across: Holzer uses psychologically charged written statements; and Haring has developed a pictographic language of popular cultural symbols.

Based on a myriad of interests and concerns, this work challenges accepted notions of what art should concern itself with and asks an important question: What are the limits of artistic involvement? All of these spaces present us with experiences and feelings that expand our views of the world in special ways.

Many have chosen the role of generalist rather than specialist, exploring the unlimited world of the imagination and sharing their insights with us. As the environmental artist Hans Haacke explained;

The artist's business requires his involvement in practically everything. . . . It would be bypassing the issue to say that the artist's business is how to work with this and that material or manipulate the findings of perceptual psychology, and that the rest should be left to other professions. . . . The total scope of information he receives day after day is of concern. An artist is not an isolated system. In order to survive he has to continuously interact with the world around him. . . . Theoretically, there are no limits to his involvement. . . .[2]

Christo

In the fall of 1976, fifty miles north of San Francisco, Christo, a New York environmental artist, erected a sculptural project that vied with the great pyramids of Egypt in ambition, scale, and cost. *Running Fence* (colorplate), constructed of white nylon panels hung on supporting steel cables, interacted with the natural elements of landscape and sun to be transformed into an 18-foot-high curtain of light. It crossed 24 miles of northern California, from Cotati to the Pacific Ocean, skimmed the golden hills, zigzagged to avoid houses and barns, and accommodated itself to road crossings and groves of silvery green eucalyptus trees. The fence meandered in a tacking pattern, doubled back on itself through valleys, knolls, fields, and flocks of grazing sheep until it came to a high bluff overlooking the ocean, paused, and plunged straight down to disappear beneath the water.

The publicity surrounding this project was intense; press reports focused on the enormous cost, controversy, and legal battles, leaving us totally unprepared for the powerful visual presence of *Running Fence*. However, once the cloth panels were unfurled, the drama and mysterious wonder of this 24-mile curtain took hold and dwarfed the previous concerns. This was, in essence, one of the main concerns of the artist. Christo explained to the ranchers, whose property it crossed, that this undertaking was "Essentially for the enjoyment and beauty of it all": a disarmingly traditional statement from a highly unconventional artist. He reassured the residents that no mockery was intended: The project was meant to work in concert with and in celebration of the sensuous lush countryside it crossed.

Running Fence activated the landscape and was in turn amplified by its natural elements: the golden brown fields dotted with cattle, the sky, the wind, the small rural towns, the pickup trucks, fences, stores, white clapboard houses, barns, people, the twists and turns of the land. Christo challenged our notions of art and transformed the way we see. Natives of the region came away with a new appreciation and vision of the landscape they had been viewing all their lives. By juxtaposing this sculptural form with the environment, avenues of experience and perception were widened, making it possible for people to engage in fresh ways of thinking, feeling, and seeing.

By strict definition, Christo is not an "earthwork" artist like Smithson or Heizer, who carve artworks directly into the earth. His pieces function more as sculptural "events" involving large numbers of people working towards the completion of a project which is temporary, and once built, is dismantled, leaving behind only documentary photographs, working drawings, and memories of the experience. Christo insists that he makes *public* art by involving a great number of people in the most diverse way possible. These projects, which harness the collective efforts of many people, focus on issues that concern us all: the individual's role in a complex and socially interdependent world, the limits and controls of technology, and the effects of law and government on our lives.

Christo, born Christo Javacheff in Bulgaria on June 13, 1935, grew up in a family that was active in scientific as well as artistic circles. His father was the owner of a chemical-manufacturing company which introduced him to the worlds of science, business, and finance, areas of specialization Christo was able to link together in his mature work. A socialist education in Bulgaria prepared him to address, through his art, a variety of social issues that confront our world.

While a student at the Fine Arts Academy in Sofia, studying painting, sculpture, and set design, Christo was disappointed in the school's emphasis on Social Realist art and rigid political ideology. However, he did enjoy the trips he made with fellow students into the countryside to engage in government sponsored "agit-prop," a form of politicized environmental art. Propaganda slogans were painted on rock outcroppings; and along the route of the Orient Express, the group would cover unsightly mounds of hay and old farm machinery with tarpaulins to create a better impression of Bulgaria for foreign travelers. The collective work process and mysterious presence of the shrouded hay bales made a lasting impression on Christo and was to influence profoundly the future direction of his work. *Running Fence*, in fact, was inspired by the artist's memory of the Orient Express slowly making its way through the Bulgarian countryside.

The study of set design in Prague under Emile Burian (a disciple of Brecht) first exposed Christo to Modern art. Paintings by Kandinsky, Matisse, and Miro were stored in the basement of the city museum, hidden from the general view of the public. Christo had the opportunity to view these works and was greatly impressed with their bold, imaginative approach. Realizing that his destiny lay in the artistic freedom of the West, Christo managed to leave Eastern Europe by covering himself with medical supplies in a train bound for Vienna. Thus, Christo's amazing career based on concealment began with a covert escape. After a disappointing semester of study at the conservative Vienna Fine Arts Academy, he moved to Geneva where he supported himself painting fashionable society portraits. With money saved from this commercial venture, he was ready for Paris, center of the European art world.

It was here, in the late fifties, that Christo met Pierre Restany, an influential art critic, and Daniel Spoerri, a leading artist of the emerging "Nouveau Réalisme" movement. These "New Realists" were concerned with the presentation of actual objects rather than representations of them in paint or bronze. They believed the era of illusion was past and that the artist's new role would be that of a fabricator and presenter of real objects and experiences. Christo's associations with these people and their theories formed some of the basis for his future work.

Recalling his student days covering bales of hay and old farm machinery, Christo quickly established his reputation as a packager of a wide range of objects. Bottles, cans, furniture, motorcycles (illustration), oil drums, trees, human bodies, even a container of air became transformed by his twine and cloth wrappings.

In some respects his early works are similar to those of other New Realist artists such as Arman and Spoerri. But Christo soon distinguished himself with large-scale works such as the Cologne waterfront piece of 1961, in which he enlisted the help of dockworkers to cover existing stacks of oil drums and rolls of paper with tarpaulins as part of his one-man show at the Haro Lauhus Gallery. The workers were surprisingly cooperative and helpful; they asked Christo questions about his work, told him about their own jobs, and expressed a gen-

Christo, *Wrapped Motorcycle*, 1962 (photo by Shunk-Kender).

uine interest in his ideas. Engaging these people in a dialogue about art was the most exciting aspect of leaving the confines of the gallery and doing on-site work. In this important but short-lived piece, Christo established the social-environmental-temporal framework that was to form the basis for his mature work.

In 1964, he moved to New York where he became fascinated with the architecture of Lower Manhattan, particularly the commercial storefronts that were filled with product samples and goods. Christo fabricated a series of full-sized sculptural replicas of these storefronts (illustration) that were made of the same materials as their real-life counterparts. His windows, however, were empty or covered with cloth or paper that prevented viewers from seeing what was inside. Thus Christo effectively subverted their usual function; we are left with the impression of an architectural wrapping without contents. The paper or cloth in these windows covers a great deal of the glass but not all of it, allowing us a limited view of the interior space. Lights are installed inside these empty windows calling further attention to the denied function of display. Unlike his wrapped

packages whose contents are discernible despite the covering, these pieces focus on emptiness, creating a subtle mood of desolation and loneliness. No goods can be seen, no activity is taking place. In a materialistic, product-consuming, possession-conscious society, these windows antithetically present us with feelings and space rather than "things," thus forcing us to consider our cultural roles as consumers. Frustrated by our attempts to see goods inside, we experience the space between ourselves and the window in a new way. Christo has, through these works, replaced illusion with contemplation, banal goods with provocative thought. By placing a full-scale building fragment inside another building (the gallery) he had disrupted our normal timespace perception, and has created a Surrealistic and dreamlike effect. This work wraps us in a unique architectural space and in the process transforms *us* into the missing contents of the storefront windows.

Christo's projects involving architectural concealment became increasingly larger until, in 1969, they reached their logical conclusion: the packaging of an entire building. With 10,000 square feet of brown

Christo, *Four Store Fronts*, 1964–65 (photo by Ferdinand Boesch).

Christo, *Museum of Contemporary Art Chicago, Wrapped*, 1969 (photo by Shunk-Kender).

C U R R E N T S

tarpaulin and 4000 feet of manila rope, professional building contractors from the Art Institute completely wrapped the Chicago Museum of Contemporary Art (illustration). The transformed building drew much attention from a public barely aware of the museum's existence. Some people wanted to know if remodeling was going on. Was it done to conserve heat? People laughed at it and some defended it, but no one ignored this event. The resulting publicity served, in fact, to support Christo's belief that contemporary art could be an effective instrument of social dialogue and change. This project came about at a time when art's traditional role was being questioned and when museums were unsure of their responsibility to the public. Jan van der Marck, Director of the Museum, was criticized by the public and some colleagues, but he maintained his beliefs and staunchly defended the project. In the preface to the exhibit catalogue he states:

With the whole idea of a modern museum and its usefulness somewhat up for grabs, Christo's packaged monument succeeds in parodying all the associations a museum evokes: a mausoleum, a repository for precious contents, an intent to "wrap up" all of art history.[3]

Despite mixed reviews from local newspaper critics, Christo was acknowledged to have generated more thought about contemporary art than any other exhibitor in the area.

As part of the Museum's "package" deal Christo agreed to van der Marck's invitation to wrap the exterior if he could also construct a floor piece inside. *Wrapped Floor* (illustration) made use of 2800 square feet of rented drop cloths covering the entire first floor of the museum, creating a sea of swirls and folds suggestive of water. Far from being randomly placed, the subtly stained sheets were carefully arranged by Christo into a baroque composition of great

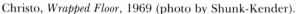

Christo, *Wrapped Floor*, 1969 (photo by Shunk-Kender).

complexity before being tied down with string to the supporting columns. All the furniture removed from the floor was piled and covered with the same material, creating a lumpy package at the far end of the gallery. Finally, as a finishing touch, Christo wrapped the exit sign and pay telephone with translucent plastic sheets. Upon entering, some unsuspecting visitors, anticipating a show of painting or sculpture, would see the entire space covered with painters' drop cloths and announce to their friends, "The exhibit hasn't opened yet." Students found the exhibit meditative and visually soothing, and returned often to sit on the floor and contemplate this unusual space. In *Wrapped Floor* Christo interfaced the real with the unreal and through "concealment" offered a unique experience to thousands of Chicago residents.

As early as 1966, Christo had contemplated working directly in nature on a scale that would exceed his urban architectural projects. *Wrapped Trees* (illustration) was a proposal to wrap with plastic the tops of live trees along an avenue in Paris, France. Securing permission from the city of Paris proved impossible so the project was dropped. The idea of wrapping large areas of ground can be traced to a series of proposals Christo submitted to New York's Museum of Modern Art in 1968. One of his projects was to wrap the entire Abby Aldrich Rockefeller sculpture garden in a plastic fabric covering: pools, trees, statuary, steps, and the garden wall that parallels West 54th Street. Blanketing large areas of ground presented a distinct challenge to Christo. He was interested in extending his control not only over urban landscapes but over natural ones as well. In 1969, his dream of creating enormous outdoor landworks was realized in a project that succeeded in wrapping a mile of the Australian coast.

Christo, *Wrapped Trees,* project for the Avenue des Champs Élysées Paris. Collage 1969 (photo by Bernard Rouget).

Wrapped Coast (illustration), installed at a site called Little Bay near Sidney, utilized an impressive 1 million square feet of open-weave synthetic fabric and 35 miles of rope to become Christo's first large-scale environment. The origins of this piece are interesting. In 1968 Christo met John Kaldor, an art-collecting textile sales representative from Australia who was in New York City on business. Kaldor purchased a small wrapped piece and after returning home wrote the artist a letter suggesting that he visit Australia to give a few lectures and mount an exhibit. Christo quickly replied he would like to come and execute the most ambitious project so far—wrapping a section of the Australian coast. Kaldor, explaining that such a project would help the cultural climate in the country, persuaded numerous civil servants and elected officials to support this project and obtain the necessary permits. The execution involved use of an open-weave cloth that would not harm indigenous insect life, and workers logged 17,000 hours constructing this enormous project that remained on view only four weeks. At a preinflation cost of only $110,000 (some of the materials were donated), Christo achieved worldwide attention planting his artistic concepts firmly in the minds of millions of people.

Following this triumph, in 1972 Christo hung an orange-colored veil across a 1200-foot-wide canyon in Rifle Pass, Colorado, enclosing the natural space rather than wrapping it. *Valley Curtain* (illustration) involved more people and money than any previous endeavor of Christo's and set the stage for even greater undertakings; it was,

Christo, *Wrapped Coast*, Little Bay, Australia, 1968–69 (photo By Shunk-Kender).

however, up only twenty-eight hours before it was ripped apart by high winds.

The $850,000 cost of this project made it necessary to employ new methods of fund raising. Christo and his French wife, Jeanne-Claude (the financial and organizational genius behind Christo's schemes), formed "The Valley Curtain Corporation." Christo donated drawings, scale models, montages, and older works to the corporation, which in turn sold them to museums and dealers. The money raised by these artworks went towards the financing of the *Valley Curtain* project; when the work was finished, the corporation would balance its books, show no profits, and therefore pay no federal tax.

With the difficulty and challenge of *Valley Curtain* behind him, Christo took some time off, regrouped his forces, and began work on *Running Fence* (illustrations), a project he had been thinking about for years.

After traveling more than 6000 miles up and down the West Coast of the United States looking for a site, Christo chose the rolling countryside 50 miles north of San Francisco as an ideal area for his 24-mile curtain. Golden grasslands, victorian houses, weathered barns, and broad valleys offered a rural tableau of picturesque serenity. The climate was perfect, little rain was forecast for September (the target date of the project), and daytime temperatures averaged in the 70s. In this area a gap in the Coastal mountain range allows the fog that usually hangs over the beach to penetrate inland for miles, creating dramatic visual effects that would enhance the fence. The location, with its proximity to millions of people in the San Francisco Bay area, would also provide Christo with the audience necessary for the proper celebration of this "sculptural event."

From the beginning, *Running Fence* had

Christo, *Valley Curtain*, Rifle, Colorado, 1970–72 (photo by Shunk-Kender).

Christo, *Running Fence*, Sonoma and Marin counties, California, 1972–76 (photos by Wolfgang Volz).

been embroiled in what Christo optimistically calls "process." Local artists and citizens, suspicious of this foreigner with a seemingly unlimited supply of money and influential friends, formed a "Committee to Stop Running Fence." The local newspapers could not resist attacking Christo, implying that somehow they were all being "conned" by this fast-talking artist, even though no public money or grants were used to finance him. But Christo, rather than viewing these obstructions as negative and undesirable, believes this controversy creates an interpenetration of "real life" and art so that our experience of both is heightened. The publicity that surrounds his work is used as a catalyst to set into motion fresh thinking about our lives and the motivating forces that surround us. Christo is uninterested in abstract, Formalist art that functions detached from the reality of contemporary life. Effective art, according to him, must seriously engage in a dialogue with the significant forces of our time. In an interview with art historian Jonathan Fineberg, Christo stated:

We live in an essentially economic, social, and political world. Our society is directed to social concerns of our fellow beings . . . that of course is the issue of our time, and this is why I think any art that is less political, less economical, less social today, is simply less contemporary.[4]

Christo's belief that obstacles are an important part of his artistic process was soundly put to the test by the inordinate opposition he encountered in the *Running Fence* project. The California Coastal Zone Conservation Commission, a state agency designed to monitor and protect the valuable coastline, requires a special permit for any permanent construction done in its jurisdiction. The permit is granted only if the applicant proves that no permanent damage will be done to the environment. Initially the Commission gave its permission, with the understanding that *Running Fence* was a temporary structure, but protest groups appealed the decision and after hearing testimony from a retired biologist, Dr. Joel Hedgpath, that harm would be done to the marine ecology by spectators,

the agency voted nine to three to deny the construction permit. Christo proposed an alternative viewing area at Dillon Beach, one mile south of the fence site, but permission was never granted. This forced him to defy the Coastal Commission and erect the ocean section without official approval. San Francisco newspapers accused him of contradicting his avowed belief of working within the "system" and deplored this defiance of the law. The day after the article appeared, Christo defended his action and explained to Calvin Tomkins:

Illegality is essential to the American system, don't you see? I completely work within the American system by being illegal, like everyone else, if there is no illegal part the project is less reflective of the system. It's the subversive character of the system that makes it so exciting to live here — one reason anyway. And no make believe, remember. We challenge, and we pay the consequences.[5]

The Coastal Commission was adamant about those consequences and went to the Marin County Superior Court requesting the maximum penalty: $500 for each day construction takes place without a permit and a $10,000 fine for disobeying the law. But considering *Running Fence's* $3 million budget, the threat of these fines amounted to no more than a hand slap.

Christo's problems were not limited to the Coastal Commission alone; other groups filed complaints and lawsuits that led to seventeen public hearings and required a small army of lawyers to defend the project in three sessions of the California Supreme Court. "The Committee to Stop Running Fence" was outraged when the counties approved Christo's proposal without requiring an Environmental Impact Report (EIR) because it was seen as a temporary project and thus exempt from normal requirements. Filing suit in the Sonoma County Superior Court, this group finally received a ruling that an EIR was in fact necessary. This action postponed the project for a year but Christo was consoled by the knowledge that his was the first work of art to require an Environmental Impact Report. It took eight months and $39,000 to produce this 265-page report that concluded in language that seemed to echo Christo's thoughts perfectly: "The only large-scale irreversible change may very well be in the ideas and attitudes of people. . . ."

In September of 1976, after workers had embedded the 21-foot-long poles 3 feet into the ground, and cable was threaded, top and bottom, from one support to the next, 165,000 yards of nylon were fastened to the cable by steel hooks. Skilled construction workers had labored for months planting the poles and stringing cable, before 350 local college students moved in and hung the cloth sections in three days.

In a curious way the fence restated many of Christo's projects of the past: Where it enters the Pacific it is reminiscent of his Australian coast project; at certain stretches through small valleys it evokes memories of the Rifle Pass curtain; the panel sections were delivered to the site in bundles that recalled his early wrapped packages, and when they unfurled and were hung they would sometimes briefly wrap the bodies of the students who positioned them. Once the fence was up, weekends became Happenings for the thousands of people moving in cars across the northern California countryside. Christo's team of student workers, dressed in yellow T-shirts with the *Running Fence* logo silk screened on them, kept traffic flowing and prevented people from trespassing on private property. Highway Patrol cars were among the participants of this sculptural event and repeatedly urged those who had stopped to move along. No accidents occurred, and it never proved necessary to close down the curtain, according to prior arrangement, because of traffic congestion or fire hazard.

After a two-week run, the panels were taken down, posts removed, anchors driven beneath the earth, holes filled, and bare spots reseeded. No trace of the fence remained. Ironically, one of humanity's largest works of art had a short lifespan and left without a trace. Many ranchers, by now staunch supporters of Christo, lamented the removal of the fence and wished it could remain in place. Others, whose perception was altered by the project, said it was just as well, as they had gained a new appreciation of the region.

Ultimately, it was faith in people that moved this fence 24 miles and overcame hostile, reactionary elements in the community that would have opposed any enterprise that dared to venture outside strict bounds of conventionality. This intense, gaunt, determined artist had, along with his wife, successfully overcome every obstacle to see their grandiose dream completed. They had self-financed a multimillion dollar operation by the use of an ingenious corporate system and the sale of original art. Through sincere and determined efforts to explain their ideas, they had won the support and cooperation of many landowners. Three years of legal battles, countless meetings, public hearings, and an obstructive maze of petty bureaucratic rules did not discourage them from realizing their goal. Christo offers a world sadly in need of meaningful social activity the opportunity for people to join in the collective enterprise of constructing "useless" projects, to play with concepts foreign to our minds, and to discover ways of seeing and feeling that even the artist could not entirely anticipate. As Christo puts it:

You cannot say that a work of art is representing only what you think it is representing. The "Running Fence" and "Valley Curtain" and previous projects have this very broad relation. The ranchers in California or the cowboys in Colorado understood that the work of art was not only the fabric and steel and cable, but there was the hills, the wind, the fear; they really cannot separate all these emotions. The work of art was that life experience of a few months or years. It is rewarding to see that the ranchers' appreciation of art was in a complex way. That is, not only a formal way, and not only a human way, but all these pieces put together; that they can find that a part of "Running Fence" is their cows, and the sky, and the hills and the barns and the people.[6]

Following Christo's successful collaboration with the rolling hills and valleys of Northern California, he then turned to the lush green islands and inland waterways of Florida's Biscayne Bay. *Surrounded Islands* (illustration), completed in May of 1983, was another in Christo's series of large-scale environmental artworks. This piece took its inspiration from the natural beauty of Miami's Biscayne Bay, a large body of water dotted with a string of small green islands. After two years of planning and study, eleven islands in the bay were chosen to be surrounded with bright pink "skirts" of polypropylene fabric. Seen from the air these islands looked like gigantic water lilies floating in a blue-green pond. Christo noted that this color was chosen because "it is in harmony with the tropical vegetation of the island, the light of the Miami sky and the colors of the shallow waters of Biscayne Bay." Although Christo's work may seem like a radical departure from art of the last hundred years, this piece was reminiscent of Monet's famous water-lilly paintings which blended reflections of water with pastel coloration.

Once again the concept and technical considerations behind this project stood out in sharp contrast to the formal aesthetic beauty of the completed work. Visually it

Christo, *Surrounded Islands* (aerial view) (copyright: 1980–83 Christo/C.V.J. Corporation/Photo © 1983 Wolfgang Volz).

was spectacular and as attractive as anything Christo has done. Although Christo supplies the vision for this environmental artwork, like his other large-scale projects it was a truly collaborative endeavor: The *Surrounded Islands* project employed four consulting engineers, a building contractor, 430 workers to install the fabric, two lawyers, a marine engineer, marine biologist, two ornithologists, and a mammal expert to make sure no manatees (unusual aquatic animals that make their home in the bay) were ensnarled in the underwater cables or cloth.

Using aerial surveys of the islands, Christo and his team of experts plotted and cut a total of 79 individual cloth patterns out of which the 11 skirts would be constructed. To prevent the fabric from sinking in the water, 12-inch-diameter booms were assembled in sections and attached to the outer perimeter of the pink skirts. Holding the booms and fabric in place against wind and tide were 610 anchors which were deeply embedded in the limestone seabed of the shallow bay.

The impact of *Surrounded Islands* on the public was considerable. Typically not everyone appreciated it and some questioned whether it was even a work of art. But, it did spark lively debate and often provided an excuse to have a party along the waterfront to celebrate Christo's temporary addition to the natural beauty of Miami's environment. Thousands of people viewed it from boats, planes, and from the shore. Like *Running Fence, Surrounded Islands* inevitably affected people's appreciation of the local environment. By the end of its two-week run many individuals, skeptical at first, became Christo converts.

Christo's next major project took place in the heart of a densely populated urban environment—Paris, France. In 1985 Christo successfully completed another project he had planned and worked toward for years: wrapping the Pont-Neuf (illustration). Few places in Paris (or all of Europe for that matter) could have provided a more spectacular urban setting than Pont-Neuf—a bridge that joins the Left and Right Banks and the Île de la Cité. Certainly there

ABOVE: Christo, *The Pont-Neuf Wrapped* (copyright: C.V.J. Corp./Christo 1985. Photo by Wolfgang Volz).

RIGHT: Christo, detail of *The Pont-Neuf Wrapped* (copyright: C.V.J. Corp./Christo 1985. Photo by Jeanne-Claude).

S P A C E S

are few places that could have matched the historic overtones of this site since bridges of one sort or another have graced this river-crossing area for over two thousand years.

Although the logistical figures for this project do not quite match physically larger works such as *Surrounded Islands*, they are impressive enough: 440,000 square feet of woven polyamide fabric was used to cover the bridge; the fabric was held down by 42,900 feet of rope (illustration) and secured by 12.1 tons of steel chains encircling the base of each tower, 3 feet under the water. In all, a crew of 600 professional workers were employed by this temporary work of art. While it was up, crews of 40 monitors worked in shifts round the clock to maintain the project and provide information for the hoards of people who thronged to see it.

Because of the prime urban location more people had a chance to physically interact with this wrapping than any other Christo project of the past. A total of 3 million people walked over Pont-Neuf during its 14-day life — 220,000 in the first 10 hours alone.

Not only was this an artwork one could walk on but, from a distance, it was also a visually stunning addition to the Paris skyline. After much consideration Christo chose a light-yellow fabric that was complimentary to the surrounding stone buildings and responded to the changing light. Seen at dawn the bridge became pale and ghostlike; but as the sun rose and made its way across the sky the color of the wrapping changed accordingly. In the late afternoon, oblique rays of sunlight filtering through the atmosphere transformed the fabric into a glowing orange-yellow.

One might suspect that Christo's temporary transformation of Pont-Neuf represented a deviation from its presumably uneventful existence. Such is not the case. This particular Parisian landmark has undergone many profound structural changes in its long life. Begun under Henri III, the bridge was finally completed during the reign of Henri IV in 1606. Christo's brief embellishment was but a small footnote to its extraordinary personal history — change it seems is very much a part of its tradition.

By making use of such a famous and historic structure, Christo was able to provide a working example of how contemporary artists could use elements of the past to express new sensibilities of the present. In this way Christo's Pont-Neuf project was similar to the work of many contemporary artists today as they seek to combine new visual forms and languages with traditional values of the past.

Because of the long lead time and the social ramifications of Christo's environmental art, typically he always has several projects underway simultaneously. One of these proposals may well prove to be the most difficult undertaking of his career: wrapping the Reichstag in Berlin with 60,000 square yards of white nylon cloth (illustration). One might assume that the construction problems would be relatively simple for a team that has wrapped a mile of the Australian coast and erected a 24-mile fence; however, the powerful political implications of such a work might prove to be too much for this intrepid couple. Their attempt to wrap the Reichstag deliberately provokes controversy and arouses strong feelings on both sides of the Iron Curtain. The reasons are clear. Built in 1871, the Reichstag functions — in a divided Germany — as the symbol of former unity.

Christo believes that wrapping the Reichstag would propel this building (and Berlin) into an aesthetic spotlight, calling attention to the international tensions, political beliefs, and economic forces that swarm around it. This project, with its direct political overtones, reinforces Christo's idea that significant contemporary art should be directly involved with the social concerns of our fellow human beings. Christian Democrat Karl Carstens, president of the West German parliament, recently vetoed the plan; Christo, however, refuses to admit defeat and continues to rally support for the project. Prospects for completing this project seem dim despite help from many important political and cultural figures, including Willy Brandt, the city's former mayor. And yet what is most compelling here is the fact that Christo's interest in social dialogue and process, even at this incomplete stage, lends a power to this piece that few finished artworks can claim.

There is no doubt that Christo's projects are very concerned with the creation of visually beautiful experiences. But more important, his works go far beyond the realm of traditional aesthetics to disturb people with their compelling mystery, to force people to question the social responsibilities of the artist, and to offer fresh insights into both everyday life and artistic experience.

Wrapping the Reichstag or constructing a 24-mile nonutilitarian fence seems to represent a complete and irrevocable break with art of the past. Even the challenging work of Malevich and Duchamp, seemingly incomprehensible during its time, has now assumed a respectable place in museums and art-history textbooks. Regardless of how hermetic and difficult these paintings seemed, they were above all art *objects*, fashioned out of traditional art materials such as canvas and paint. In works of art in which

essentially no image or historical forms can be perceived, such as sculptural events or earthworks, the confusion seems total. Some of the reasons for the development of these hybrid forms can be attributed to the rapid transformation our society has undergone over the last century. A founding belief of Modernism was that as our world changed through technological and social reform, the form and role of art would appropriately respond to these innovations. Because of the invention and widespread use of photography, television, and film, a flood of mechanically reproducible images inundates our society. Illusion and representation, once the exclusive domain of the artist, have been taken over by commercial interests to sell and promote everything from politicians to automobiles. Art, once bound to the image, has been released by these developments to pursue other objectives and goals; artists could now use their work to express philosophical ideas and thus engage in a kind of visual research. The theoretical scientists, with their concern for

Christo, *Wrapped Reichstag*, project for Berlin. Collage in two parts 1980 (photo by Eeva-Inkeri).

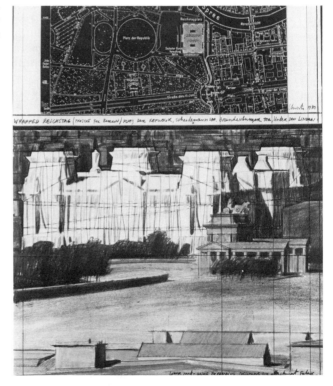

"pure" knowledge, became role models for artists who were dissatisfied with existing modes of expression.

Robert Smithson

Robert Smithson, an artist engaged in this speculative form of thinking, viewed his art as a form of specialized knowledge of the world. Ecosystems, archaeology, geology, evolution, and organic processes were studied and incorporated into the work of this highly reflective artist and writer. His earthworks were meditations on the transforming role art could play in the understanding of the environment and ourselves. He questioned the idea of evolutionary "progress" and made us realize there was always a price to pay: Something was gained and something was lost. Smithson was an innately optimistic individual who saw ugliness and decay as positive elements awaiting the artist's transformative vision. His concept of art included a new role for the viewer: directly experiencing the earth and contemplating the contemporary landscape. Smithson's dissatisfaction with the gallery system and the type of art shown in it was one of the motivating factors which led to his involvement with land art. He believed that most contemporary work was overly specialized and myopically self-centered, and lacked the intellectual scope and resources to engage people in a fresh dialogue with the world. A new, more public art was necessary according to Smithson — not more large-scale traditional sculpture, but work that forced issues and provided new realizations about ugliness and beauty, progress and destruction. Robert Smithson's art was public in the sense that it was accessible to anyone who thought about and confronted the ideas and experiences it contained. Consequently his art, like knowledge, could never be owned, only understood. His earthworks functioned as services, bringing us new perceptions and feelings about the earth, industrial development, and our relationship to society.

As a boy growing up in Passaic, New Jersey, Robert Smithson built a small museum in his basement to house a collection of minerals and shells. Rocks held a particular fascination for him and he remembered always being on the lookout for fossils. The earth and its buried contents has occupied Smithson's thinking ever since. By the time he entered high school, art had gradually replaced his interest in science. New York City was only fifteen miles away and during his senior year he attended Saturday classes there at the Art Students League; this brief schooling was all the formal art education Smithson was to receive. After six months in the army and a hitch-hiking trip around the country he moved to Manhattan where he befriended a number of well-established artists and started working on his own. Smithson's art gradually evolved into a Minimalist format and by 1964 he was exhibiting with the Dwan Gallery and was represented in the important *Primary Structures* show at the Jewish Museum. Smithson was not altogether happy with this success: Like some of his friends, he was discouraged by the limitations of Minimalist art and its Reductivist tendencies. His participation in the New York art scene dwindled and he spent a lot of time reading and thinking about crystalline structures and organic processes. Abstraction — the kind he had been practicing in his early work — then seemed meaningless and far removed from reality. Nature, with its complexity of processes and meanings, offered more possibilities. Recalling his boyhood interest in geology Smithson wrote:

The strata of the earth is a jumbled museum. Embedded in the sediment is a text that contains limits and boundaries which evade the rational order, and social structures which confine art. In order to read the rocks we must become conscious of geologic time, and of the layers of prehistoric material that is entombed in the earth's crust.[7]

Smithson's sources expanded to include not only books on crystalline structures but texts on geology, mythology, nature, and chemistry. What emerged from this program of reading and reflection was the realization that what really interested him was the ongoing process of a work of art, not merely the finished product.

Frederic Law Olmsted, a turn-of-the-century landscape architect and the designer of Central Park, became one of Smithson's heros. In a provocative article, entitled "Frederic Law Olmsted and the Dialectical Landscape" written for *Artforum* magazine, Smithson introduced Olmsted as America's first earthwork artist. Olmsted's view of nature was highly original for its time: Whereas most of his contemporaries saw nature as a distant, nostalgic, idealized, static entity, he conceived of it as a vast, dynamic, sometimes frightening process, in which humans played an important part. The urban park was assigned an important role by Olmsted in the expanding aesthetic, social, and political process of America. Central Park was designed to be a democratic mediator between the nostalgic landscape of the past and the harsh industrialized life of the present. Open and available to everyone for a multitude of uses, the park was not a "thing to look at" but a process to enter into and experience.

Today Central Park looks like a natural section of land that was fenced off 100 years ago and preserved while New York's concrete high rises rose around it. Nothing could be further from the truth. In 1885 it was the site of a human-caused disaster area — the result of indiscriminate tree cutting by early settlers. Before Olmsted's transformative vision, it was a muddy, rubbish-strewn area overrun with goats and ramshackle squatters' huts. Early photographs reveal an urban wasteland reminiscent of strip mining regions in West Virginia. The "Greensward" plan, as the park was first called, involved moving 10 million horse carts of earth, constructing an extensive underground water system for the creation of a lake, putting in miles of paths, and planting acres of grassy meadows. Its slight elevation above street level and surrounding groves of trees separate it from the city but leave an openness that invites entry and allow for the widest possible social interactions. The Metropolitan Museum of Art is housed in Central Park but few people either recognize the relationship or realize that the park itself is a form of environmental art, expanding our awareness, heightening our sense of "place," and allowing rediscovery of the self through contemplation and play. Olmsted's view of the park's role in a changing social and physical environment helped form the theoretical basis for Smithson's development of earth art.

The first pieces Smithson did after his self-imposed exile from the art world were in direct response to his readings about natural process and his interest in working directly with nature. He referred to these pieces as "non-sites" and each one usually consisted of a topographical map mounted on the wall with a rock-filled metal bin directly below it. Earth and minerals were gathered from the area shown on the map and transported to the gallery; a connection was set up between the map or "concept" and the nonabstract reality of the natural objects.

Geologic time and entropy (simply, the measure of the disorder, or randomness, of a system) were concepts that formed the basis for much of Smithson's later work. Order and disorder, disintegration and reconstruction were understood as basic to the workings of the universe and life itself. He saw in this fundamental law of thermodynamics a direct relationship to our perception of time, the consciousness of our lives, and the inevitability of our death. In 1967 Smithson wrote in *Artforum*:

I should now like to prove the irreversibility of eternity by using a jejeune experiment for proving entropy. Picture in your mind's eye the sand box divided in half with black sand on one side and white sand on the other. We take a child and have him run hundreds of times clockwise in the box until the sand gets mixed and begins to turn grey; after that we have him run anticlockwise, but the result will not be the restoration of the original division but a greater degree of greyness and an increase of entropy.[8]

Spiral Jetty (illustrations) combines Smithson's fascination with geological sites and his broad interpretation of entropy to produce a work of complex meaning and compelling beauty. The spiral shape becomes a potent symbol for life itself, continuously expanding outward while simultaneously shrinking inward. His choice of Utah's Great Salt Lake as a site developed out of an earlier piece executed in California's Mono Lake. This unique body of water in the high Sierra supports no fish or reptile life yet at certain times of the year its surface swarms with millions of insect larvae that develop into flies. The contradiction of lifeless water and teeming surface life stirred Smithson's imagination and led to further research on unusual and specialized organisms. In a book called *Vanishing Trails of Atacama* Smithson read about Bolivian salars, or salt lakes, that are filled with microorganisms that can tolerate the high mineral content and turn the water tomato-soup red. Bolivia seemed too far away to be practical, and Mono Lake lacked the ele-

Robert Smithson, *Spiral Jetty* (photos by Gianfranco Gorgoni/Contact).

ment of color; a friend suggested that Smithson investigate Utah's Great Salt Lake. Smithson placed a call to Ted Tuttle of the Utah Parks Department for more information about this special phenomenon. Tuttle recalled in this interview that the northern part of the lake, beyond Lucien Cutoff, turned a vivid red at certain times of the year and gave Smithson the names of several people in the area who knew the lake.

One of these individuals was a well driller named Charles Stoddard who had attempted to homestead the lake on Carrington Island in 1932 but was forced to abandon the site when he was unable to find fresh water. Stoddard took Smithson to Little Valley on the east side of the Lucien Cutoff and it was here, in the abandoned artificially constructed harbors, that Smithson saw his first red water. Access proved impossible at this location, however: "Keep Out" signs were posted everywhere and hostile ranchers warned Smithson to get off their property. Stoddard suggested Rozel Point as a possible alternative. The dirt road leading there slowly made its way through an immense landscape unlike any Smithson had seen before. As the road neared the water, salt flats became visible, ringing the lake and trapping bits of wreckage from abandoned oil rigs nearby. According to Smithson, the whole area had a look of "Modern prehistory." Black tar pools dotted the landscape just south of the point, luring people unsuccessfully for the past forty years to attempt to tap this site for oil. Smithson was intrigued with the decaying machinery and the abandoned hopes it signaled. An area one mile from the oil seep, where the water comes up to the shoreline, was finally selected as the site. Standing on the shore looking out over the quiet, mirrorlike lake, Smithson perceived the land, sky, and water as a rotary that continually folded back on itself in the immense Utah wilderness. The spiral configuration would encompass all of these elements in a sweeping way, both visual and symbolic.

A twenty-year lease on the property was secured and in April of 1970, a front-loader,

two dump trucks, and a tractor were sent to the site to begin work. Smithson, realizing that few people would ever visit the site, wanted to make a film that would document the process and communicate his ideas to a greater audience. The Ace Gallery in Los Angeles agreed to help fund this project and sent a movie-camera operator to record the beginnings of *Spiral Jetty*. Early footage shows trucks loaded with earth and basalt carefully backing to the edge of the water and dumping their loads at the prepositioned marking stakes. Occasionally a truck would get stuck in the soft mud and the rest of the day would be spent filling in this soft spot. There was fear that these sections of the jetty would sink but the entire structure held.

Smithson recalled his feelings and perceptions of this place in his article "Spiral Jetty," published in *Arts of the Environment*:

No ideas, no concepts, no systems, no structures, no abstractions could hold themselves together in the actuality of that evidence. . . . On the slopes of Rozel Point I closed my eyes, and the sun burned crimson through the lids. I opened them and the Great Salt Lake was bleeding scarlet streaks. My sight was saturated by the color of red algae circulating in the heart of the lake, pumping into ruby currents, no they were veins and arteries sucking up the obscure sediments. My eyes became combustion chambers churning orbs of blood blazing by the light of the sun. All was enveloped in a flaming chromosphere. I thought of Jackson Pollock's Eyes in the Heat (1964; Peggy Guggenheim Collection). Swirling within the incandescence of solar energy were sprays of blood. My movie would end in sunstroke. Perception was heaving, the stomach turning, I was on a geologic fault that groaned within me. Between heat lightning and heat exhaustion the spiral curled into vaporization. I had the red heaves, while the sun vomited its corpuscular radiations. Rays of glare hit my eyes with the frequency of a Geiger counter. Surely, the storm clouds massing would turn into a rain of blood.[9]

In 1971, following the successful completion of *Spiral Jetty*, Smithson was invited to do a piece in an international art exposition at Sonsbeek, The Netherlands. The park site, because of its formal European design, proved unsuitable to his ideas; consequently museum officials searched Holland for a more appropriate location.

Emmen, a post-World War II planned community, was finally chosen because of its geologically unique quarry at the edge of town. Glacial movements had created strange multicolored layerings of soils that set the area apart from the otherwise tame landscape of Holland. The wild appearance of the exposed strata provided Smithson with the proper setting for his ideas. Placed against this visually rich landscape, *Spiral Hill* (illustration) was created by bulldozing the black earth into a large cone-shaped mound set on the edge of the lake. A white sand path slowly circled the cone, ascending to the top. Still green water reflected the orange oxide cliff across the pond and introduced a mirroring effect. The artist's enthusiasm for the site was contagious and soon Dutch workers and townspeople were actively involved in the process.

The second piece at the quarry, *Broken Circle*, was built by dredging sand from the lake bottom, the way dikes are built, rather than bringing in landfill, the way *Spiral Jetty* was made. Smithson was extremely interested in these Dutch "earthworks" that were responsible for reclaiming so much land from the sea and felt a strong connection with his own ideas. A large glacial rock, already in place on this lake's edge, became the central locus of this piece. Smithson dug a semicircular canal inland, mirrored by the lake and set off against the multicolored hill in the distance. The citizens of Emmen were so impressed with his project they voted to allocate public funds to preserve and maintain it, affirming the democratic goals of his art.

Recalling Olmsted's "reclamation" of the Central Park site, Smithson was convinced that his art could function in a similar way with areas devastated by industrial processes. He wrote to several strip mining companies submitting proposals for projects and reminding them of their responsibility to the land. Finally, with the help of friends in New York, he succeeded in getting the Minerals Engineering Company of Denver interested in his ideas. At their Creede, Colorado, facility great quantities of rock are crushed and chemically processed to extract the ore; the residue is then flushed out with tons of water. This waste eventually flows into large bodies of water called "tailing ponds." The company needed a new pond because the old one was filling up;

Robert Smithson, *Spiral Hill* (courtesy of the Estate of Robert Smithson).

therefore the proposal to create an earthwork out of a tailing pond seemed reasonable and cost effective. Smithson envisioned this work's continuing for twenty-five years, incorporating 9 million tons of tailing, and creating a pond 2000 feet in diameter. After years of negotiations, planning, and fund raising, just when it seemed possible to effect industrial processes through art, the project was abruptly canceled for financial reasons. Frustrated, but still firmly believing in the idea that art could play an important role in our society, Smithson took a short vacation in New Mexico to relax. Through a chance meeting with a friend there, he learned of some shallow desert lakes in the Texas Panhandle that might relate to his abortive *Tailing Pond* project. Excited by the possibilities of these bodies of water, Smithson and his wife, Nancy Holt, arrived in Amarillo, a prosperous agricultural and cattle center in northwest Texas, and discovered Tecovas Lake. This 8-foot-deep body of water is part of a unique irrigation system called the Keyline that was

constructed by an innovative Australian engineer, P. A. Yeomans. It ingeniously avoids expensive pumping stations by utilizing the natural downhill slope of the land to distribute water for irrigation. Stanley Marsh, a landowner on Tecovas Lake, was contacted by Smithson and was intrigued by the idea of an earthwork on his property. Aerial surveys of the lake were undertaken to chart its shape and size, and working drawings were made with the aid of these photographs. Smithson and Holt waded the shallow lake and staked out a preliminary shape for the piece, which was tentatively called *Amarillo Ramp* (illustration). Unsatisfied with the scale of the first attempt, changes were made and an aircraft hired to view the modifications. On July 20, 1973, while its passengers were observing the artwork from the air, the single-engine plane stalled, went into a dive, and crashed, killing Smithson and everyone on board. Nancy Holt felt the piece should be completed as a memorial to her husband. Soon after the funeral she returned to

Robert Smithson, *Amarillo Ramp* (photo by Gianfranco Gorgoni/Contact).

Amarillo with two of Smithson's friends and resumed the work. Although there is some controversy as to whether it really is a valid Smithson piece, Holt claims she was informed of the initial planning and that it remains true to Smithson's ideas and disposition.

Years after his death, Smithson's ideas continue to exert a profound influence on our perception of art and nature. Artists are still discovering and responding to this rich body of work. Smithson's legacy leaves us with an enlarged sense of the world, one that is complex, dynamic, and mutable. His meditations on art and the landscape called for a remembrance of origins. He wrote:

We have a memory of things forgotten in the distant past, prehistoric time; in other words there are things that we can't remember but there are traces left behind.[10]

Smithson's work makes us remember and takes us to those places where remote futures meet distant pasts.

Michael Heizer

Religious figures and spiritual savants have always been attracted to the desert. In the early seventies, perhaps in reaction to the confusion and noise of contemporary life, artists rediscovered in these spaces a quiet and austere environment conducive to self-discovery, meditation, and work. Nevada meets all of the geographical requirements of the historical deserts but coexists alongside Las Vegas, a technological dream (or nightmare, depending on your point of view) of immense proportions. Perhaps this paradox is what made the region attractive to contemporary artists. Certainly the range of experiences in this region is unique.

Approaching Las Vegas by car at night may be as close to a religious experience as we can find today. The endless miles of wasteland and hypnotic drone of the car prepare us to receive this modern miracle of neon and plastic. Its halo of light can be seen in the distance reaching up for miles in the clear, by now cooling, desert air. Millions of people make pilgrimages here every year: Tourists with beatific looks faithfully feed the slot machines, high rollers from Los Angeles place thousand-dollar bets at the luxurious gambling tables. The temperature never changes, an even 68 degrees; daylight is banned and with it a sense of the changing rhythms of the day. Time stops, the moment is frozen, and a serene feeling pervades the air. This is a special place, one that greatly affects perceptions and attitudes.

Michael Heizer, an earthwork artist, works in the dry desert valleys and eroded hills outside of Las Vegas. He likes the feeling of space and scale in this vast wasteland, the sense of time experienced walking around on it, and the challenge of pitting oneself against a hostile and unyielding environment. The contrast between Las Vegas and the desert is enormous; instant divorce rubs shoulders with a frightening timelessness. The Las Vegas desert is a long way from New York's art world and its highly commercial overtones. This distance, severity, and distinct sense of place is very attractive to artists who feel the limits and constraints of the gallery world.

It was obvious to artists working in these regions that the desert offered more than a different set of spatial conditions; it also seemed to make possible a new perception of spirituality that could be effectively incorporated in their work. Heizer once told Calvin Tomkins in an interview, "One of the implications of earth art might be to remove completely the commodity status of a work of art and allow a return to the idea of art as . . . more of a religion."[11] By the late sixties many artists were questioning not only the validity of the gallery-museum system but the integrity and meaning of the forms themselves. Overly refined and highly finished machinelike forms shifting from one white room to another seemed an unrewarding activity to artists like Michael Heizer. The artist recalled: "After I'd been in New York for a couple of years I saw art here as really dead. A lot of things had been killed off at the same time, because the forms themselves were coming into question. I was lucky because I wasn't really involved with all that; I was prepared to deal with the new situation."

In reaction to the materialistic concerns and spiritual emptiness of this period, Heizer turned to the space of the desert as a way out of this dead end. Influenced by the American Indian societies he learned about from his father (an archaeologist who specialized in Pre-Colombian cultures) and aware of the harmonious relationship that existed between their art and the land, Heizer found a new direction to follow. In the winter of 1968, he traveled to Nevada to begin work on a piece called *NESW* (the name represented a connection between the Northeast of New York and the Southwest site of his first earthwork). No photographs are available, and the artist is reluctant to talk about this early, perhaps premature, earthwork. Heizer returned to his studio excited by these new experiences and continued to paint; about this time he met the well-known art collector Robert Scull, who visited his loft to view his paintings but soon became more interested in the desert work.

Scull, a man who liked to be in the collecting vanguard, felt these pieces were important and financed Heizer's return to the Nevada desert that summer.

A dry lake bed north of Las Vegas became a new arena for the extension of Heizer's painterly interest in simple geometric forms. Heizer created *Circular Surface Planar Displacement* (illustration) by using his motorcycle like a gigantic mechanized pen to inscribe a series of linear marks in the soft desert floor. Seen from an aerial perspective the shallow tire cuts formed a series of small circles (200 feet in diameter) tangent and adjacent to a large circle (400 feet in diameter). Walking alongside the tracks on a dry lake bottom, however, does not allow us this perspective. What we perceive at ground level is a series of mysterious, overlapping motorcycle tracks forming gently curving arcs. In this respect *Circular Surface Planar Displacement* is reminiscent of ancient South American Indian

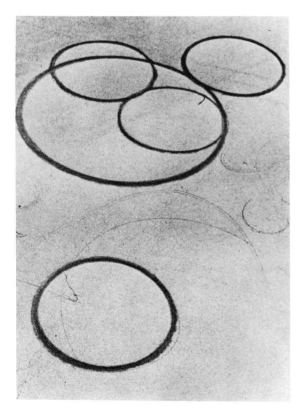

Michael Heizer, *Circular Surface Planar Displacement* (courtesy of Michael Heizer).

landworks that were formed by scraping the dry desert floor and are visible only from an airplane. Under the patronage of Scull, Heizer produced a series of pieces that involved displacing or moving large amounts of rock and earth great distances. The geologic history of this desert region provided the inspiration and impetus for *Displaced, Replaced Mass* (illustration). Millions of years ago, before the Sierra mountain range was thrust 9000 feet upward, all of the land in that desert region existed at the same elevation. Heizer dynamited several 68-ton granite boulders out of the mountains, transported them 60 miles by truck, and placed them in concrete-lined holes he constructed on the desert floor. These rocks, displaced by millions of years of slow geologic upheaval, were thus returned to their original elevation.

After the intense activity in the summer of 1968, which saw the completion of nine "depressions" in the Nevada desert, Heizer was offered a show at the Dwan gallery during the fall of 1969. As part of the agreement Heizer was to execute a major earthwork in conjunction with the indoor exhibit. That fall the artist left for the de-

sert and returned two months later with photographs of an immense piece called *Double Negative* (illustrations). The photographs revealed two enormous cuts in the desert floor that displaced a total of 249,000 tons of earth. Mormon Mesa, located 80 miles from Las Vegas, was the site for this project, which has become Heizer's most famous earthwork. The mesa is difficult to find even with the best directions; it can be reached only by jeep or four-wheel drive vehicle and usually entails traveling for hours through intense heat. The arduous approach is matched only by the overwhelming size and effect of the work itself.

No sign or marker announces *Double Negative's* presence, just the immensity of two large, 30-foot-wide and 50-foot-deep trenches dug from opposite sides of a canyon. One of these notches bores into the cliffside for 1200 feet; its counterpart, directly across the divide, is 400-feet long. Heizer pitched the ramplike entrance at a 45-degree angle to make it too steep for cars but passable on foot; it is in walking down this trench that the piece envelops us and reveals its size. Bulldozer marks, graffiti, and beer cans are apparent as we walk

Michael Heizer, *Displaced, Replaced Mass* (courtesy of Michael Heizer).

CURRENTS

its full length; at the edge of the mesa, the Virgin River is visible flowing undisturbed far below. The 240,000 tons of displaced earth and rock pushed into the canyon are barely visible, swallowed by the vast desert landscape. It is this contradiction of scale, large by human standards but dwarfed by the endless wasteland, that permeates *Double Negative*. In certain respects Heizer feels it is one of the smallest pieces he's done, considering its insignificant size in relation to the scale of the mesa.

There is a painterly aspect to this earthwork. The deep cut into the desert floor reveals a rich palate of earth colors: warm ochers, browns, rust reds, and streaks of green where rain has leeched out the minerals. Heizer believes one needs to spend twenty-four hours with the piece in order to fully experience the subtle colors and constantly changing conditions of light. But there is a dangerous and foreboding quality to this region that is reflected in the scale and presence of *Double Negative*, and most

Michael Heizer, *Double Negative* (courtesy of Michael Heizer).

visitors from the city are content to spend a few hours walking around this robust work and then to retire to a safe, air-conditioned motel for the night.

Many such contemporary earthworks are related to the earliest earthworks, prehistoric cave art. Both are inaccessible, remote, make use of indigenous materials, and respond to natural features of the site. Both have a mysterious, mute presence that goes far beyond purely visual effects. Like Smithson, Heizer went forward by returning to the beginnings of art and rediscovering old feelings buried in the earth.

Unlike the cave pieces, however, there is little in the contemporary works of the "handcrafted" look we normally associate with archaic art. In the construction of earthworks, hands now play a new role and manipulate the levers of gigantic bulldozers to control hundreds of horsepower. These machines are extensions of the human body, performing similar functions but accomplishing them at great savings of time and money. Heizer likes the feeling of power he gets moving and altering the earth, relating it to oil and mining magnates and the wealth they derive from such activities. He is not particularly interested in financial power, though, and in the past has taken a dim view of art as a medium of economic exchange: "I want to get rid of the 'parasites' in art. With this thing (earthworks), you can't trade it or speculate with it the way they do with traditional pieces." Recently, after several successful painting shows in New York galleries, Heizer may have revised his view of "parasites" to include some beneficial varieties. Today he claims he never intended to destroy the gallery system and the art object, but was simply exploring a new area and process. Despite his distaste for monetary speculation, someone had to be found to provide the financial assistance to make his desert projects possible. Heizer's primary patron was gallery owner Virginia Dwan, who raised the $27,000 to buy the square mile of land surrounding the site of *Double Negative*. Attempts to sell the piece to some European dealers in 1971 proved futile when Heizer,

disgusted by the financial dickering, canceled the deal.

Because most earth art takes place in deserts and other regions far removed from population centers, few people have actually seen an earthwork. Although Heizer disclaims the effectiveness of photographs to convey any important aspect of his work, documentation plays an important role in presenting these works to a greater audience. The artist believes that a photograph cannot possibly convey the experience of standing 50 feet below the desert floor enveloped in its geologic past or the stillness and sense of time one feels in that special place. Nevertheless, documentary photographs are important adjuncts to the earthworks' becoming contemporary icons by verifying their existence and narrating the Herculean processes that formed them.

Heizer has singlehandedly moved thousands of tons of earth and rock in *Double Negative* to carve what would seem to be an immutable and lasting monument to his vision. But, acutely aware that no artwork lasts forever, he is quick to acknowledge the slow, unrelentless changes nature imposes on his work. Heizer accepts these changes and considers them important factors in the life of his art: the fluctuating light of the day and season, the eroding action of wind and rain, even visitors who carve graffiti on the earthen walls. This particular piece is expected to remain visible for 100 years or so, but eventually the land will be restored to its natural state. After the artist imposes his vision on the landscape, a greater force takes over and has the final say in its destiny; it is in the integration of these forces—art and nature—that *Double Negative* excels.

To ensure that future works will not be destroyed by "progress" Heizer had raised the money to purchase land in a desert valley north of Las Vegas on which he hopes to build the most ambitious permanent sculptural project of our time, *Complex of the City* (illustration). This is no ordinary complex of streets and dwellings, but a metaphorical city of earth and concrete forms resembling a modern Stonehenge.

Michael Heizer, *Complex of the City* (courtesy of Michael Heizer).

After years of planning, digging, pouring concrete, and living in the desert alongside his construction, Heizer completed the first section in 1976. *Complex I*—a huge, elongated mound of earth with concrete sides framed by steel and concrete columns—is symbolic architecture with no utilitarian function. From certain viewpoints the columns line up to resemble Constructivist sculpture of the 1920s. Incorporating 250 tons of earth, 210 tons of concrete, and 25 tons of steel, *Complex I* is only a small section of an epic "city" that will eventually cover 3 square miles of desert land.

To make this project possible, Heizer has set up a corporation and has borrowed money to pay for earth-moving equipment and materials. As the sections are completed, he intends to offer them for sale to collectors or museums. The buyer gets the work itself and the land on which it is built, but agrees in the contract not to move the piece. In this roadless, dust-choked Nevada Valley where he lives with a small crew of workers and few distractions, Heizer has plenty of time to speculate on the future

of this project, and he has wryly observed that it may not be for our time, but for the future.

Charles Simonds

The political unrest caused by racial disturbances and the Vietnam War movement forced many artists in the 1960s to reconsider the role art could play in social intervention and reform. A major question was what form would this art take? Direct, heavy-handed approaches usually produced ineffective "propaganda" that was politically ineffective and artistically dull. Charles Simonds, however, is one artist who successfully combines social action with highly personal imagery.

During the seventies Simonds wandered through the streets of New York's Lower East Side, building and abandoning miniaturized clay dwellings of the *Little People*, an imaginary civilization whose social customs, religious beliefs, and building techinques he has thoroughly worked out in his mind. At first, Simonds was taken for a harmless "crazy" but through the years his ideas and persistent work have affected the consciousness of the neighborhood, and today he is an active leader of community groups such as the "Lower East Side Coalition for Human Housing" and the "Association of Community Service Centers." Watching him build these diminutive structures, talking about the Little People's lives, and seeing the destruction of the dwellings when he departs have prompted many local people to become more active in the planning and preservation of their own neighborhoods.

The connection between people and their environment, and Simonds' own personal artistic activity, is very important to the artist. Simonds feels the art world exists within a limited and narrow framework compared to the broad and highly energized world of New York City's streets. He believes that the art world is only a fragment of the real world and that art should participate in important social concerns that are usually beyond the scope of traditional art activity. People who have never seen an artist at work can stop, watch him, talk with him about his ideas, and sometimes help with the construction. Simonds feels that by leaving the traditional art world he has gone from a drab and lonely "prison" into a rich and exciting world of enlarged possibilities and broad social concerns.

Since 1971 Simonds has constructed over 250 tiny clay dwellings and has written extensively about the Little People and their mythological civilization. The buildings represent a form of metaphorical architecture that explores the psychological dimensions of a society's beliefs and customs. Miniaturized dwellings of an imaginary civilization provide the perspective and framework for studying the social structures of our own society and for examining the meaning of our personal lives.

Simonds' early work was obsessed with a return to origins by the use of his body in primitive, ritualistic ways. *Birth*, a 16-mm film made in 1970, symbolically heralded the emergence of the Little People and established a personal vocabulary of myths relating to Simonds' body and the earth. This film shows Simonds slowly emerging from a pool of muddy clay, half-man, half-earth, in a primordial "birth" sequence. In another film, *Landscape/Body/Dwelling*, Simonds buries himself in the earth and constructs a group of fantasy buildings on his body. Tiny places of worship, ritual dwellings, and gardens blend into his form, which merges into the soft earth. It is impossible to distinguish where one material begins and another ends. In an interview with Lucy Lippard, Simonds explains his feelings about these films.

"Landscape/Body/Dwelling," is a process of transformation of land into the body, body into land. I can feel myself located between the earth beneath me (which bears the imprint of my body contour) and the clay landscape on top of me (the underside of which bears the other contour of my body). Both "Birth" and "Landscape/Body/Dwelling" are rituals the Little People would engage in. Their dwellings in the streets are part of that sequence. It's the origin myth — the origin of the world of man and of the people.[12]

The ending of *Landscape/Body/Dwelling* is startling. Simonds, who had been previously lying down, slowly rises, creating a cataclysmic geological event of microscopic proportions in which buildings, gardens, and places of worship disintegrate into a formless mass of clay as a human being emerges from the earth. The exquisitely fragile, temporal, and mythic aspects of Simonds' work are expressed quite vividly in this particular scene.

In the late sixties, after graduating from the University of California at Berkeley and moving to Manhattan, Simonds renounced his student work in welded sculpture to concentrate on a series of highly personal fetishistic works. Fashioned out of blood, urine, and saliva placed in transparent connecting tubes, these works formed a room-sized environment. Relationships were set up between biological growth, personal ritual, and an interest in "dwellings." These constructions, established at an early stage, are the major themes of Simonds' mature work. In an interview with Daniel Abodie, Simonds talks about his early work and its sources:

Basically it was . . . an elaborate project that transformed my loft into "stations" using clay, my body (hair and fluids), fantasy images, and art historical and architectural image quotations as a means of creating a fantasy history of a thought. The ingredients were in all different scales; there were fragments of "the colossal dream," small biological specimens, childlike paintings, shadows made with hair, broken and bandaged timbers, tadpoles in various stages of growth, sculptured reliefs of a voyage to Cythera, figures with birdheads, sacrifical columns, fragments of a large stairway covered with plasticene. . . .[13]

The Little People and their fictional civilization grew out of personal examination of the self, fantasy, literature, and ritual. Simonds scrutinized our lives and material objects, returning them to us in the form of tiny sculptural cities that become condensed, metaphoric symbols for many civilizations and many ways of life. The act of miniaturizations is an economical device that enables us to see an overall view of a particular time and place. The whole work is visible at once, making it possible to contrast it in scale and content to the real bricks and buildings of New York City.

These tiny dwellings become gateways between the oppressive reality of a poor neighborhood and a mythic past with its half-forgotten dreams and magical beliefs. They allow us to enter another world, to take part in ceremonies, customs, and beliefs that are foreign yet familiar to us. We are given a new perspective on the city of the present by being given an opportunity to look at a symbolic city of the past. Our own time scale is greatly enlarged by Simonds' work. Seeing these crumbling remnants of the past lends hope to the possibility that significant change could occur in the present. Depressing neighborhoods are no longer seen as immutable and unchangeable. Simonds' imaginary view of the past leads to a hopeful vision of the future. What is built into these personal fantasy worlds that Simonds constructs is the realization that through persistence, faith, and coordinated effort, positive change is possible in our lives.

Over the years three distinct groups of Little People have evolved to populate Simonds' fantasy world: the circular people, the spiral people, and the linear people—each group having its own rituals, beliefs, dwellings, and visual characteristics. Simonds develops literary and philosophical programs that fit the visual characteristics of these people. The endless repetition of human activity is expressed by the circle people; building upward on the self-consumptive decay of the city are the spiral people; and the road, traveling, searching, and migrating relate to the linear people. Simonds has appropriated visual aspects of these shapes (circle, spiral, and line) to express philosophical attitudes relating to habitation and the way we live. Each distinct group builds with its own architectural style, using clay, sand, rock, and small twigs. Simonds acknowledges that the origin of these structures goes back to Pueblo Indian buildings he saw on childhood visits to the Southwest with his parents, who were Viennese-trained psychoanalysts. Simonds' older brother worked on realistic clay sculpture at home, and for as long as Simonds can remember he was always involved with clay, a material the artist views as one of the primal materials of art.

Simonds obtains the red clay he uses in his construction from a clay pit in Sayreville, New Jersey, where he has befriended one of the workers. A scoop from the "front loader" supplies all the free material he needs for quite a long time. The Sayreville facility has been in operation for over 100 years and was the main provider of raw materials for the manufacture of bricks used in many of New York's oldest buildings. Simonds finds this historic connection appropriate to the nature of his work, which seeks to combine an awareness of the past with contemporary concerns. Finding and gathering materials is not so much a chore as an adventure; meeting people and sharing his interests with them is enjoyable and provides a connection to the world usually missing in studio art. Besides the pleasure of the hunt, using easily obtainable materials frees him from economic pressures and is consistent with his process of building and freely abandoning these tiny structures to the whims of the city.

In the early 1970s, after filling his loft

with dwelling places of the Little People, Simonds recalled the happy day when he was inspired by the thought of building these structures outdoors. He talked a friend into letting him use the second-story window ledge of his building, and the first group of Little People set up housekeeping in the canyons of New York. This early group, called *Cliff Dwellers* (illustration) occupied walls and window ledges of the SoHo district in lower Manhattan.

Soon, dwellings were constructed on street level among gutters and against the bases of buildings. The people who occupied these sites were referred to as the *Herdsmen*. Simonds invented a complex fantasy life for these two societies; *Cliff Dwellers* were fierce warriors and hunters who periodically swept down and raided the peaceful *Herdsmen*, an agriculturally based tribe living on the "plains" of the street.

SoHo is a neighborhood of New York that basically hosts two types of people: workers in the light manufacturing factories and artists who have converted industrial spaces into living and working areas.

Simonds found that the local artists responded to his dwellings in a predictable and generally uninteresting way. They were usually interested in placing him and his work into categories that easily fit their frames of reference. The response of the workers, however, was more open, enthusiastic, and quite unpredictable. Because of this, Simonds knew he needed to find a more coherent neighborhood where the Little People could move about and interact freely with a wide variety of people. The Lower East Side, a cohesive community with a lively street life, seemed ideal; soon the Little People had migrated there and set up habitations among the primarily black and Puerto Rican inhabitants.

The first dwelling in this new area was constructed on Avenue C, one of the busiest streets on the Lower East Side. Simonds recalls that it was a joyous, exciting day; small children climbed over him, people stopped to talk, junkies eyed him suspiciously, and a local store owner bought him coffee—for New York City a miracle in itself.

Over the years Simonds has built these

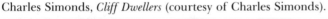

Charles Simonds, *Cliff Dwellers* (courtesy of Charles Simonds).

tiny dwellings all over the neighborhood: in vacant lots, cracks in the wall, and gutters. His method of operation is consistent; he rides through the neighborhood on an old delivery bicycle (so beatup he claims he never has to worry about it being stolen) until he finds a suitable site. He then unpacks his tools and building materials and begins work. Mornings are ideal: The mood of the street is peaceful and enough time is available to slowly and methodically develop the complex structures. Simonds never works more than a day on a single site, preferring to realize his vision within the span of an eight- to ten-hour workday.

Simonds prepares the brick or concrete surface by smearing it with wet red clay, and then places his tiny bricks on this foundation. He mass produces these building blocks by rolling out a slab of clay about one-sixteenth of an inch thick and carefully slicing it with a sharp knife into the prefabricated rectangles. Using long-handled industrial tweezers to handle the bricks, he dips them into a diluted solution of Elmers glue and presses them into place. The final design of the structure is always determined by irregularities of the site, but most of the buildings are generally rectangular and reminiscent of Southwestern Pueblo architecture (illustration).

Simonds gets a great deal of pleasure out of sharing his ideas and experiences with people. Children form the bulk of his audience, staring quietly at first, but later asking many questions about the Little People and their ways. Usually the children end up helping Simonds, fetching materials, handing him tools, and sometimes making their own additions to his structures. When the artist abandons the finished dwelling at the end of a workday, the adults and children who saw him devote so many hours to its construction are bewildered. They find it hard to believe that someone could put so much work into a project, without pay, and leave the construction to such an uncertain fate. The older children, aware of the beautiful fragility of Simonds' work, zealously guard it from the younger ones interested in playing war or blowing it up with firecrackers. But no matter how many

Charles Simonds, *Dwelling, Houston Street, New York City* (courtesy of Charles Simonds).

people provide protection, eventually the dwellings are left unguarded and get destroyed.

These constructions, built for the enjoyment of everyone in the neighborhood, resist being moved. Because of the materials and methods used, any attempt at personal ownership results in their certain destruction. Only a few buildings, done in inaccessible places, survived for any length of time; the record holder is a dwelling that lasted for five years. This fragility makes us think about the physical permanence of New York City itself and the culture it houses; the destruction of Simonds' works is a reminder that violence is a part of everyday life for the inhabitants of the Lower East Side. Simonds accepts this and views the dwellings as part of a continuous, on-going narrative. It is the story of a people moving through the city determined to survive despite the constant presence of danger. When local residents see the artist's determination to tell the Little People's story over and over again, their own will to survive and improve the condition of the ghetto environment in which they live is strengthened. Although disturbed at first by the quick destruction of the street dwellings, Simonds now feels he has built up a sizable population in the minds of people who watch him work and are moved by the story of the Little People. In the long run, this imaginary existence might be more substantial than a continued and on-going physical life.

To view these dwellings on a formal aesthetic basis is a mistake, according to Simonds. He feels that these structures are expressions of a people's beliefs, attitudes about nature, and their relationship to the land. They question the economic and social aspects of territory and private ownership in an area where virtually no one owns the property on which he or she lives. "The earth the Little People live on is free," observed Simonds, who through his work calls attention to the potential of public land and neighborhood projects that could reactivate this space for the benefit of all. The notion of "reclaiming" vacant city lots appeals to Simonds as much as it did to Robert Smithson and his mentor Frederick Law Olmsted. By the use of political and artistic means, the real world is acted upon and new visual, economic relationships are set up between individuals and the land. Simonds envisions building environments recalling ancient forms to which people can relate in complex psychological ways, unlike most contemporary parks, which are based on nineteenth-century forms and attitudes. Trash-filled lots on the Lower East Side represent wasted public spaces that need to be activated and transformed through political, economic, and aesthetic means. Simonds believes that his art is not an end in itself but a way to effect his own consciousness and that of the people with whom he works.

Because of the Little People, their philosophical relationship to the land, and his work with neighborhood community-action groups, Simonds has developed a keen interest in the economic and political life of the Lower East Side. Working in the street and talking to the residents about their life in the area has made him aware of their situation and what could be done to improve it. In an interview with Lucy Lippard, Simonds observed:

The earth that the Little People live on is very free. It just appears under them, and they can nestle up to it or make their bricks from it or bury things in it; they can wander about and be wherever they want to be on the earth. The park reflects the same thoughts in real space, the difference being that in the city, land is worth money, and political power is needed to free it. I'm interested in finding out how a capitalist society, the city bureaucracy, and the communities have articulated a piece of land; what are the wrinkles in that system which allow for that land to have another life, a different function, a different way for people to relate to it, and a different form.[14]

La Placita (illustration), a park project Simonds worked on in 1975, developed out of his expressed concern for people and the land on which they live. An abandoned city lot was transformed into a community playground by constructing large, gently curving mounds of earth that created a sensuous, almost primitive environment that could be enjoyed by adults as well as children. In an all-too-pervasive world of concrete and asphalt, Simonds evoked the forgotten presence of the earth lying dormant beneath the city. *La Placita* expresses the same childlike wonder and archaic feeling for the past that his miniature dwellings do, but allows people to engage in a full-scale participatory experience. A similar park was constructed in Cleveland by contacting the residents of an apartment building and soliciting their ideas about what could be done with a nearby vacant lot. Viewing his role in the project as a facilitator for other people's ideas, Simonds was excited by their enthusiastic participation which eventually led to the building of a playground, barbecue pit, and garden. His art functioned as a catalyst, transforming their imagination and practical needs into a useful reality.

Growth House (illustration), constructed in 1975 at Artpark, an outdoor facility for environmental artists, is another full-scale project that expresses the artist's ideas about decay, renewal, and the symbolic–social meaning of shelter. Simonds built this structure out of what he called "growth bricks," burlap bags filled with earth and seeds. Its circular form reflected the seasonal cycles of nature that are analogous to our own cycles of birth, maturation, and death. *Growth House* started out in the winter as a barren, sand-bagged structure and gradually transformed itself into a leafy green dwelling that blossomed and produced a wide variety of vegetables by autumn. Alan Saret, an environmental artist living at Artpark, harvested Simonds' artwork and lived on its produce for months. At the center of the structure is an unchanging "navel"-like dome of earth surrounded by the radiating growth bricks reflecting their seasonal cycle. *Growth House* was designed to function as a continuing organic process, renewing itself with fresh vegetation each spring, but because of time limitations placed on use of the building site, it was only up for five months.

An earlier project at Artpark, called *Niagara Gorge*, was created in 1974 by enlisting the help of local residents to excavate the remains of a nineteenth-

Charles Simonds, *La Placita* (courtesy of Charles Simonds).

century railroad tunnel and construct a full-sized, three-story dwelling and ritual area. Simonds related this project to local Iroquois legends that believed that a race of little people carved the cliffs in this region of New York State. On the river bank below the tunnel, fifteen cairns (conical heaps of stone usually raised as memorials) were constructed, ranging in size from a few feet to 8 feet high. This was the "ritual area" where Simonds reenacted a performance of *Landscape/Body/Dwelling* as a culmination to the project. Both the miniature and large-scale projects share a concern for the philosophical ideas of habitation, how people live, and how their beliefs and values affect the structures they build. Local residents who helped excavate the abandoned tunnel experienced through this project an awakening of interest in the history of their region and a new appreciation of the land. For Simonds this social awareness is part of the deeper content of his work, which expands to encompass people and their lives in a variety of ways.

Cities in America are rife with contemporary ruins, boarded-up buildings awaiting "renewal," vacant storefronts in economically depressed areas, and freeway spans that were stopped in midair when protest groups convinced city rulers that their harm far outweighed their value. For example, the Stanley Tankel housing project, at Breezy Point, New York, was an

Charles Simonds, *Growth House* (courtesy of Charles Simonds).

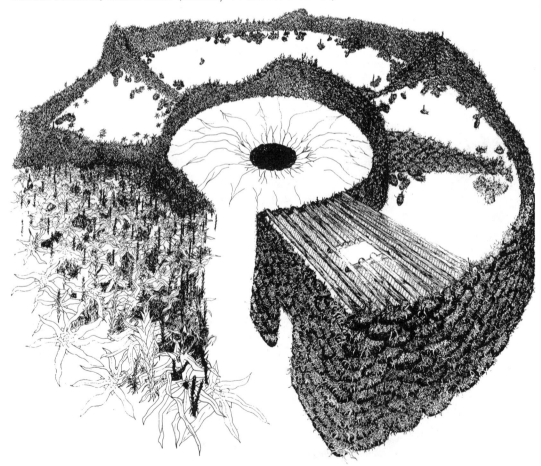

unfinished, abandoned, fifteen-story monument to the folly of bureaucrats and planners. Simonds proposed that the dying complex be transformed into a combination hanging garden and youth hostel by planting wisteria throughout the skeletal structure. Simonds uses gardens and vegetation, with their implied optimism, in his large-scale projects as metaphors for the rebirth that is possible through "reclamation." *Hanging Gardens* (illustration) would transform an embarrassing eyesore into a living structure, thus becoming a catalyst for real urban renewal and positive social change.

In Simonds' recent proposal entitled *Floating Cities* (illustration), he foresees the need for rearrangeable housing and services that would respond to the realities of an economically changing world. Work spaces, living quarters, and stores would float on water and be easily shifted and repositioned according to the needs of a utopian society in the future. Potential for redirection and change would be built into the structure of this project, to avoid the

stagnation evident in the inflexible and slowly changing cities of today.

Writing is important to Simonds. It serves as a definer of his ideas and a catalyst for his actions. In 1975, Simonds' book, *Three Peoples*, was published, describing in great detail the Little People's social customs, work habits, and religious beliefs. These writings form the theoretical basis for the clay "dwellings" and the mythical people who inhabit them (illustration). In this book Simonds establishes the philosophical basis for their existence and describes their journeys in lyric and symbolic terms:

Their dwellings make a pattern on the earth as of a great tree laid flat, branching and forking according to their loves and hates, forming an ancestral record of life lived as an odyssey, its roots in a dark and distant past. . . .

. . . an old woman came back from her journey into the past with a vision: what everyone had believed to be a life following an endless line was really part of a great unperceived arc that would eventually meet itself. At that point everyone would join their ancestors in a great joyous dance. . . .[15]

Charles Simonds, *Hanging Gardens* (courtesy of Charles Simonds).

Charles Simonds, *Floating Cities* (courtesy of Charles Simonds).

Charles Simonds, *Dwellings* (photo by Rudolph Burckhardt; courtesy of Charles Simonds).

Simonds uses the circle, spiral, and line to express three distinct philosophies; every dwelling site becomes a specific chapter in the life of these nomads and undergoes its own cycle of birth, life, and death. Looking into these recently abandoned dwellings places us somewhere between the distant past and the immediate present. When we come upon them suddenly on the street they make us pause; we are lost for a time in pleasurable contemplation about their lives and travels. Peering inside the tiny rooms one sees remnants of extinguished fires, storehouses of food, clay water jars, and what seem to be ritual artifacts. It looks as if they have recently abandoned the structure but their presence remains, buried in the details of the construction. Simonds' work is a constant reminder of the eroding effects of time and the relentless cycles of life and death that envelop all of us.

Charles Simonds often says he is the only Little Person but through his persistent efforts a fantasy civilization coexists alongside the grim reality of life in Lower Manhattan. Entrusted to city dwellers, Simonds' constructions are almost always destroyed by them, but their story is kept alive through oral traditions and the haunting images they conjure in our minds. Working on the street, talking to friends and local residents, Simonds thinks a lot about the way we live and how our beliefs shape the world. Through his dwellings and stories of the Little People, he stirs our imaginations and affects our thoughts about ourselves and the world. Simonds' constructions take their meaning and direction from people and their responses to his fragile, poetic world (illustration). Even the wanton destruction of these dwellings serves as a stimulant to make people conscious about their environment and what could be done to im-

Charles Simonds, *Dwelling, Houston Street, New York City*; courtesy of Charles Simonds).

prove it. In this way Simonds' work is political, serving to activate and channel people's energy through community organizations towards the realization of projects like *Placita Park*. "The reality of the park," Simonds observed, "is the result of two fantasies — mine and theirs, which met through the Little People."

What Simonds achieves through his work is the fusion of personal expression with social conscience. Most artists, working in the isolation of their studios, have difficulty relating their ideas and concerns to people outside of their profession. Simonds' art, with its rich blend of individual fantasy and social implications, avoids the alienation many artists experience with their work. Working on the street, talking to people who stop, and watching their reactions to the Little People has provided Simonds with a responsive audience and support structure few artists enjoy. It is this human factor that energizes his work and firmly plants it within the mainstream of life.

Perhaps the single most rewarding experience to come about through Simonds' work on the streets was this story a man once told him: The man was on his way to kill someone with a knife, saw Simonds on the street working on one of his "dwellings," became fascinated with the story of the Little People, watched him work for quite a while, and consequently never completed his mission. That story means a lot to Simonds.

Alice Aycock

In the early seventies, while Smithson, Heizer, and De Maria constructed enormous earthworks and presented a new concept of sculpture as "place," artists Alice Aycock and Dennis Oppenheim utilized certain aspects of architecture for new directions and possibilities. They reacted against the simple, primary-form, nonliterate sculpture of the late sixties that functioned in an "abstract" and isolated way from its environment. Inspired by site-specific work, Aycock and Oppenheim constructed quasiarchitectural works (both

indoor and outdoor) that explored the idea of shelter from several points of view: psychological, literary, and experiential. These rooms and architectural spaces could be walked into and experienced in many ways; they extended the boundaries of art far beyond the perceptual. When we entered those rooms we left the outside world and, in a sense, entered the artist's mind. *Spiral Jetty* claimed as much physical space as possible, even reaching out, as Smithson believed, into the universe; Aycock and Oppenheim, however, are interested in exploring the more interior and intimate psychological space of the mind. Bruce Nauman, an environmental artist from California, constructed an artwork in 1970 that illustrates this interest in an inner "mind-space" quite clearly. For an exhibit at the Los Angeles County Museum Nauman constructed a full-sized room, visually empty but housing many hidden speakers evenly distributed throughout the space. A voice on audio tape played continuously through the speakers, commanding us repeatedly, with constant variation of intonation and pitch, to "Get out of this room, get out of my mind." This insistent voice was omnidirectional and unchanging in volume no matter where one stood in the room; it came from nowhere and everywhere, its target and source was our minds. After a few minutes in this disconcerting room one began to feel that by entering this space one had, in fact, entered the artist's mind.

Aycock and Oppenheim make use of such architectural spaces as containers for personal, psychological, and literary ideas. There is an element of theatre connected to these roomworks, which function as empty stage sets awaiting our presence to activate and bring them to life. They exert an effect on us also. Through them we enter the artist's mind and our imagination is stirred; we are literally and figuratively taken to places we have never been before, and experiences unfold that anchor us in a unique time and space. Through this process we discover new things about ourselves and leave full of new ideas and perceptions.

Alice Aycock has been producing large outdoor and indoor constructions since she built her first full-sized structure in 1972. Her work exists as meta-architecture embodying her view of our time as one of imbalance, anxiety, and fear. Personal history and childhood experiences form much of the basis for her work. Aycock remembers when she was a child growing up in Pennsylvania:

Those times when there is no clear point of view, those moments when the world is out of joint, topsy-turvy, upside down. I remember an experience I had as a child in an amusement park. There was a long wooden barrel about 15 feet in diameter and 50 feet long which revolved. People were supposed to walk through the center of the barrel while the barrel turned around them. Wall was continuously becoming floor, floor becoming wall and wall becoming ceiling. I went inside and I couldn't walk through it. I began to scream.[16]

Aycock incorporates the disjunction, uncertainty, and fear that she experienced at that amusement park into her environmental works. Doors that open to solid walls, ladders that abruptly end in space, shafts that drop away menacingly, low roofs, narrow passageways, chimneys, all create powerful emotional effects. She is interested in creating spatial experiences that relate to the idea of architectural "shelter" and develop out of the complex matrix of human behavior. Aycock has done extensive research on the history of cities and dwellings, studying the way these structures responded to emotional and utilitarian needs. For instance, she is interested in the way medieval walled cities developed out of a need for personal protection and access to a central marketplace. Wall construction was expensive, and resulted in towns with narrow streets and small interior living spaces. Emotional needs and economic forces combined to produce these particular spaces, which in turn affect our behavior. Aycock is fascinated by architectural "behaviorism" and the way dwellings reflect and express our beliefs, fears, and customs. She cites an interesting example of this human element in the case of the forest-dwelling Mbuti pygmies. Their camps have no preplanned shape and are constantly changing as these nomadic people move in and out. The Mbuti do not build according to the orientation of the compass or sun; they are, however, very concerned with the positioning of their doors. If a neighbor is in high regard, the door faces in the direction of the neighbor's hut. If the neighbor falls out of favor, the door is blocked off and another is opened in the opposite direction. Although most people see architecture as primarily fulfilling a need for shelter, Aycock is interested in the social symbolism and human drama inherent in its forms.

While looking through the *World Book Encyclopedia* for the definition of "magnetic north," Aycock accidentally came across a circular plan for an Egyptian labyrinth originally designed as a prison. This illustration became the genesis for her first full-sized outdoor structure, *Maze* (illustration), constructed in July, 1972, on a farm near New Kingston, Pennsylvania. After discovering the Egyptian plan in the encyclopedia, Aycock continued her research into this form and eventually incorporated aspects of fourteenth-century English mazes, American Indian stockades, and Zulu "Kraals" into this piece. *Maze* was, essentially, a large twelve-sided wooden structure, roughly 32 feet in diameter and 6-feet high, that became somewhat of an attraction for the local teenagers who would drive their cars through the field, enter the maze (they tore down some barriers to make access easier), and perform their own adolescent rituals of building fires, drinking, and smoking marijuana. Aycock believes that *Maze* symbolically and spatially represents "the path" of life itself, with its complexities, false starts, and inherent terrors. Notes from her journal describing a trip to the Southwest provide further clues to the work's background and meaning:

Originally, I had hoped to create a moment of absolute panic—when the only thing that mattered was to get out. Externalize the terror I had felt the time we got lost on a jeep trail in the desert in Utah with a '66 Oldsmobile. I egged Mark on because of the landscape, a pink and gray crusty soil streaked with mineral washouts and worn by erosion. And we expected to eventually join up with the main road. The trail wound up and around the hills, switch back fashion, periodically branching off in separate directions. Finally, the road ended at a dry riverbed. We could see no sign of people for miles. On the way back, I accused Mark of intentionally trying to kill me.[17]

For Aycock, the terror and fright of being lost in the maze is balanced by the rediscovery of direction made possible through the experience.

Participatory experience plays a large role in Aycock's early work. She eclectically incorporates spatial aspects of classical Greek architecture, caves, temples, tombs, and bunkers into artworks that provide rich, exploratory situations for the participant. These spaces are only known by moving one's body through them; they are environmental sculptures that must be literally entered and experienced directly, not merely looked at.

In 1973, a year after *Maze* was completed, Aycock returned to the Gibney farm in Pennsylvania and constructed *Low Building with Dirt Roof (for Mary)* (illustrations) on a small hill in the middle of a large field. As we approach the entrance, the earthern roof seems to form an artificial horizon. Aycock was interested in setting up associations between the curve of the mound and the curve of the earth. There is a lonely, isolated look to this piece, reminiscent of a settler's house on the prairie, abandoned and almost completely reclaimed by the land. The normal relationship of house to ground is questioned by this work. Did it sink into the soft accepting field? Was the earth engaged in a process of swallowing it whole? Or over the years did the field slowly grow up to meet the house, taking it over bit by bit? Above all, there is an organic unity to this piece that suggests a close symbolic relationship between nature and art. The

Alice Aycock, *Maze* (courtesy of Alice Aycock).

usually clear separation between building and land has vanished, they have become one. *Low Building with Dirt Roof (for Mary)* is also about transition: land reclaiming itself, rotting wood turning into soil, and a house transforming into the earth. Poetic and nostalgic references are made to abandoned structures found throughout America: New England's rickety barns, Midwestern sod shelters of land-rush days, and the free-standing chimneys of Appalachia.

In a sense, all that remains of this house is the attic. Attics and cellars are important to Aycock because they are less functional spaces of a house and are usually storage areas for memorabilia. Freed from the utilitarian demands made on other areas of the house, they are the most contemplative and meditative spaces. Aycock recalls memories of the attic in her family's home:

About fifteen years ago I visited the house in which my great-great-grandparents Benjamin and Serena lived and where my great-grandfather Francis was born. It was a small wood frame house. I climbed up alone into the attic where they slept and stood under

Alice Aycock, *Low Building with Dirt Roof (for Mary)* (courtesy of Alice Aycock).

Alice Aycock, detail of *Low Building with Dirt Roof (for Mary)* (courtesy of Alice Aycock).

the rafters. In the yard was the family cemetery. I remember the tombstone of Catherine, who died at age three. Years later, I dreamt that my brother Billy came for me and took me to that same wooden house set into the hills of Greece like a tholos tomb. I climbed the stairs again and behind a screen, a young girl, whose face I could not see, lay dead.[18]

Alice and her mother built *Low Building with Dirt Roof (for Mary)* with stones gathered from the field and wood from a partly collapsed building on the property. The interior dimensions of the building range from a maximum 30 inches at the entrance to a low of 12 inches where the pitched roof begins at the far ends (illustration). You must assume a hands-and-knees posture as you enter and be prepared to crawl on your stomach to explore it all. The juxtaposition of large, open field and cramped dark, interior space is enormous. Entering this small, damp room is like entering the earth itself. When we emerge into the bright sunlight after crawling through this atticlike space we experience an equally exhilarating effect. Aycock has confronted spatial effects similar to this at historic ruins and sites throughout the world. These often-deserted, always mysterious places function as catalysts for her imagination. She has written:

Because the archeological sites I have visited are like empty theatres for past events, I try to fabricate dramas for my buildings. To fill them with events that never happened, to allude to function, a function they never had.[19]

Aycock's next major construction, *Walled Trench/Earth Platform/Center Pit* (illustration) done in 1974, was inspired by drawings she saw of Rhodesian circle-pit dwellings that offered Aboriginal tribes protection from animals and enemy raiding parties. This piece calls to mind shaft graves at Mycenae and the golden treasure contained in them. A central shaft in *Walled Trench/Earth Platform/Center Pit* draws us into the earth, promising if not treasure, at least discovery and adventure. This experience makes us remember our fascination as children with primitive structures such as forts, caves, and pits. Some of Aycock's shelter, stripped of an ordinary architectural utility, become psychologically unique and reassuring places, conjuring up memories and feelings not felt since childhood.

Walled Trench/Earth Platform/Center Pit is

Alice Aycock, *Walled Trench/Earth Platform/Center Pit* (courtesy of Alice Aycock).

a shaft within a pit, formed by three concentric concrete-block walls, and reaching a depth of five feet. It is also situated in a low-lying area on the Gibney farm, between two small hills. *Low Building with Dirt Roof* is visible on a hilltop beyond *Walled Trench/Earth Platform/Center Pit*. The raised inner platform is filled with earth excavated from the surrounding trench, and is solid except for the tunnel that bores through it and connects the outer trench to the central shaft. There is no obvious point of entry to this piece. The open central pit is inviting but the surrounding trench presents a physical barrier. Aycock teases and invites us to leap across to gain entry through the central shaft. Although the distance across the trench is not great, only 52 inches, it is enough to present some danger or risk. There is always the possibility of not leaping far enough and hitting your head on the concrete blocks or over-reaching and falling into the center pit. Aycock, who describes herself as a person who takes few risks, has jumped only once, preferring the less risky method of lowering herself into the outer trench and scaling the inner wall to reach the earth platform and central shaft. Once on the central earth platform, the well bottom is visible 5 feet below; it is punctured by a tunnel opening that invites us to lower ourselves and explore the inner chamber. The artist recalls childhood stories of a well, of a child falling into a well, and writes that her response to this piece is "half terror/half love." Aycock skillfully synthesizes myth, history, and personal experience to create a structure that objectifies a vast range of feelings and memories buried within us.

Artists interested in site-specific environmental works have always had to deal with the problem of accessibility and use. Securing permission from property owners to make use of their land for environmental artworks has not been an easy task. This is particularly true around high-population centers, where a potentially large audience resides. Artpark, previously mentioned in regard to Simonds' works, is a project of the New York State Parks Department, and was set up to respond to the special needs

of these artists. In 1974, a section of land along the Niagara Gorge, in Lewiston, New York, was set aside as a laboratory workshop for people engaged in large outdoor projects that explore ideas about art and the environment. Every summer artists have been invited to make use of the natural features of the location to realize projects that for lack of space or opportunity could not be accomplished otherwise. The grounds are used on a rotating basis; at the end of the season the work must be removed or destroyed to make room for a new group of works. It was at Artpark, in 1977, that Alice Aycock built one of her largest outdoor pieces to date: *The Beginnings of a Complex . . . Excerpt Shaft No. 4 Five Walls* (illustration). Towering over the Upstate New York landscape were five parallel wooden walls, 28 feet high. Many familiar aspects of Aycock's work were contained in this piece: ladders, tunnels, shafts, doorways, and windows leading to more door-

Alice Aycock, *The Beginnings of a Complex . . . Excerpt Shaft No. 4 Five Walls* (courtesy of Alice Aycock).

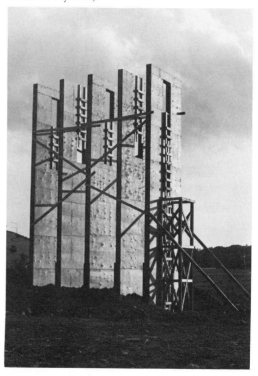

ways and windows. The usual function of these familiar architectural forms is subverted in Aycock's work; they bother, confuse, and delight, making us wonder and think about their original use. Aycock was interested in imparting a medieval look to these towers, one that was suggested by the strangely deserted site. It reminded her of landscapes seen in paintings and illuminated manuscripts of the Middle Ages. She wrote regarding the piece

I wanted to give the feeling of a Bosch landscape like in Temptation of St. Anthony — *pervert architectural conventions. Start with a landscape and make it strange.*[20]

Aycock intended to construct artificial hills in the park to help illustrate her medieval vision, but lack of time prevented her from doing so. *The Beginnings of a Complex . . . Excerpt Shaft No. 4 Five Walls* also has the look of an incomplete movie awaiting the finishing touches before the actors and bit players scale the walls, assume their positions, and wait for the cameras to roll.

After the construction of these large outdoor pieces, Aycock moved away from participatory experience in her work and assigned the spectator a new role. In *Studies*

for a Town (illustration) at the New York Museum of Modern Art, museum guards prevented us from physically exploring the elliptical plywood construction on display. Denied actual access and physical experience, we are left with a heightened sense of expectation that engages our imaginations and makes us explore it with our minds. Although it remains visually open and inviting, we cannot enter the wedgelike opening or walk the precarious stairways. This piece simultaneously invites our entry and denies it, recalling the contradictory themes of fear-security, attraction-repulsion, and open-closed spaces found in Aycock's outdoor works.

This architectural sculpture almost fills the small museum gallery. A passage around it seems open, but attempts to circumnavigate it prove futile when it comes to a dead end three-quarters of the way around. The recurrent themes of frustration and contradiction reinforce the nonfunctional and absurdist elements of Aycock's work. Her aesthetic power is rooted in the poetic subversion of an errant functionalism that has pervaded architecture for the last sixty years. Freed from the need to shelter and house, her constructions, with their rich mixture

Alice Aycock, *Studies for a Town* (courtesy of Alice Aycock).

of personal experience and historic reference, returns to us an experience of "shelter" that is more concerned with emotional overtones than with utilitarian needs.

During a visit to California in 1976, Aycock visited a Hollywood soundstage where the set for *Hello Dolly* was housed. She was intrigued by the sight of New York City at the turn of the century juxtaposed against modern-day Los Angeles. Seeing this exterior theatrical set functioning indoors reinforced Aycock's current interest in building large-scale theatrical environments inside museum and gallery space.

How to Catch and Manufacture Ghosts (illustration), installed at the John Weber Gallery in 1979, is not only a theatrical set with props but also a stage for performers who activate it at a certain time every day. Based on a Tantric diagram of the location of stars, the piece consists of a circular wooden platform flanked by two semicircular wings. Mysterious and magical objects are placed provocatively throughout the set. A glass jug bobs in a tray of water. Wheels and cranks are hung from the ceiling and a soap-bubble blower sits next to a large white bird in a cage. The lighting is dramatic and evocative; an air of expec-

tancy hangs over the construction and there is a feeling that something is about to happen or has just happened. Once a day, gallery attendants enter and bring this tableau to life. One person sits on the step slowly blowing bubbles and another assistant transfers the bird from the cage to the bottle. The words of a mysterious "N.N." are written on the gallery wall: "That is how I go home sometimes. Somebody at home — my mother — dreams of me and then I am at home with a broom in my hand helping her. . . . Once I was putting on my shoes at home. My feet and shoes began to lift themselves up in the air and I thought I would be lifted off the ground." As we read this script, bubbles, like the "ghosts" to whom they allude, drift silently through the gallery, reinforcing the fantasylike atmosphere of this concrete dream.

All of Aycock's constructions provide the opportunity for exploration and imaginary habitation. Spaces are created that allow free-flowing associations, experiences, and feelings to blend together in a felicitous way. Her work successfully combines architectural references, personal memories, and skillful execution to create metaphorical structures for the mind, body, and spirit.

Alice Aycock, *How to Catch and Manufacture Ghosts* (courtesy of Alice Aycock).

Dennis Oppenheim

One of the themes that consistently runs through the work of process and environmental artists is the desire to impart meaning to their lives through their art activities. Isolating, meaningless activities that do not serve the real needs of the artists and their audiences are rejected. One of the most important needs today is a psychological sense of well-being to counter the fragmenting effects of contemporary life. Dennis Oppenheim is an artist whose art activity functions as a therapeutic tool, putting him in touch with difficult feelings and helping him to overcome personal problems. He is also able to share these insights with a greater audience when his work is shown publicly. Although classical psychological literature has tended to view art as a sublimation of neurosis, art therapy in recent years has proved to be a valuable tool in separating neuroses from art and to see this activity as beneficial and healing. Although Oppenheim does not practice "art therapy" in any direct way, there are interesting analogies between the nature and content of his mature work and its "therapeutic" value that are worth noting. Jack Burnham, a highly respected scholar and art critic, delves into this question of art activity and emotional well-being in an article titled "The Artist as Shaman." Burnham views the recent work of some artists as essentially shamanistic in nature. In "primitive" societies, shamans were highly perceptive individuals who functioned as spiritual leaders and concerned themselves with the psychological problems of the community. Burnham suggests that Dennis Oppenheim is a kind of contemporary shaman, an individual whose life has come to a crisis and through "artistic" activity gains control over powers that have been dominating him. Through a form of psychodrama, the traditional shaman undergoes ritualized trials that force him to confront himself and his past, perhaps exploring early stages of infantile development and family relationships. Having survived these experiences, the previously troubled individual is transformed, strengthened, and emerges as a changed person with remarkable powers. In a certain sense what is experienced is a mystical reenactment of a psychic "death" and a consequent spiritual "rebirth." After this initiation, the shaman–artist possesses a new-found ability to reach others; using song, dance, theatre, and poetry, he bridges the gap between worldly and spiritual problems to provide a valuable service to the community. Although it would be misleading to imply that Oppenheim has consciously directed his work towards shamanistic ends, the similarities are worth noting.

Oppenheim's early work was involved with data systems and forms of process art done directly in the landscape. In the mid-sixties he approached a Pennsylvania farmer and asked if he could direct the course of his harvesting machinery through a 200-foot by 500-foot field of wheat. Permission was granted and the artist's instruction was to simply make a cut directly down the center of the field, creating a landscapular art-work that transcended the spatial and thematic limitations of a gallery. Oppenheim, like many of his fellow artists working with the environment and nature, envisioned art interacting in a greater way with the world. Audiences would also be enlarged, they reasoned, when art that was previously confined to gallery and museum spaces confronted people in accessible, outdoor settings.

Oppenheim's interest in information systems and what he called "data transfers" led to a conceptual "snow-work" he constructed on the U.S.–Canadian border in 1967 (illustration). The annual ring growth pattern of a tree cross section was enlarged to 150 feet in diameter and "transferred" to that northern border site by shoveling away the snow on both sides of a river. In these early works Oppenheim was fascinated with unusual juxtapositions of concepts, scale, and geographical locations. Another "data-transferral" project that never materialized consisted of a scheme whereby an old World War I trench plan would be excavated and recreated in a section of northern New Jersey.

While Oppenheim was lecturing at the

University of Wisconsin art department in 1970, he proposed an environmental project called *Maze* to be constructed on a nearby farm. The inspiration and source for this piece was a classic psychology learning experiment that timed laboratory mice as they made their way through a complex maze. Oppenheim's scheme envisioned an outlandish enlargement of the original plan, substituting cattle for mice and constructing the maze out of bales of hay. Measuring 500 feet by 1000 feet, this piece parodied the miniscule laboratory setup on a grand scale. Corn was placed in the maze as the reward and cattle stampeded through the enclosure eating and presumably learning as they progressed. Oppenheim was interested in the systems operant in this piece and drew a direct analogy between "digesting" the system by learning from it and literally digesting the alfalfa of the enclosure. A few hundred feet from *Maze* was a white, 100-foot-square shape called *Decoy*. Oppenheim suggested to viewers, totally confusing them, that *Decoy* was the real piece and that *Maze* was only an arbitrary off-shoot. This interest in setting up seemingly logical systems and then disrupting them is a quality that figures greatly in Oppenheim's work. Many of his pieces function in a curious state between logic and illogic, forcing us to constantly reconsider what we are seeing. Admittedly, there is a disturbing quality to this interruption; Oppenheim fuses the surface order of the world with an undercurrent of mysterious, unexplainable phenomena. His work has the ability to make us question what we know and what we think we are seeing. Oppenheim refutes the old saying "seeing is believing" and insists in his pieces that believing is really seeing. There is a firm refusal on his part to be categorized, pinned down, or easily explained. His diverse output surges with enormous energy, sometimes obsessively repeating itself, often breaking away on a new tangent. It is impossible to categorize his work. To do so would be to deny the essential qualities of his thinking. Oppenheim consciously employs diversity as a means of continually breathing new life into his actions and art. In a recently pub-

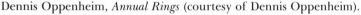

Dennis Oppenheim, *Annual Rings* (courtesy of Dennis Oppenheim).

lished article entitled "Catalyst, 1967–1974," he transcribed the ideas and activities directly from his notebook. The results are a dizzying compendium of language, ideas, materials, and processes that continue breathlessly with no punctuation, paragraphs, or clear indication of where one proposal ends and the other begins. Oppenheim has always kept notebooks. The earliest were filled with traditional sketches and plans; later, when concepts began to play a prominent role in his work, language was used as a way of giving form to ideas. Cryptic, disjointed, and confusing, the notes nevertheless reveal the complex thinking that flows beneath the surface of his work. They are notes to himself, incomplete recipes, a stream-of-consciousness collage of raw ideas:

void — buy a product from a consumer
shelf — photograph space — ask merchant to
leave space for specified time — EN
FORCE VOID — exhibit product
sterilized surface — glass
list of antidotes
plug in
gallery engages in alternative business
during show
6 reservation: airlines, bus, train, restaurant, sports
*   event, pre-need burial*
plot-tape-recorded telephone calls
infected zone-Milan-kill traps/rat
poison/burned grass-in progress
fish channels
branding irons-brand floors/trees/grass
brand a mountain
proper channels
processing list/new materials/end product
change/one month's variation/bend
pass over/bow/installation pit
catalyst vs. pigment[21]

Language is used in Oppenheim's work as a container and generator of ideas; projects flow out of the experience of language itself and the notion of taking language literally. Certain idiomatic expressions are taken literally by Oppenheim, and lead to finished pieces, such as: "take my word for it," "the wrong frame of mind," and "spinning a yarn." Quite often a written statement will accompany a visual piece, not as an explanatory footnote but rather as a linguistic element effecting its meaning and final appearance. Oppenheim's broad appropriation of language is a declaration that nothing is beyond his manipulation and incorporation. The similarities between his use of language and the "verbal magic" practiced by shamans in primitive cultures is interesting. Both seek power and control by "naming"; to know and call out the name of a person or object is believed to exert control over these things.

Photography plays an important role in Oppenheim's work, both as a recorder of visual ideas and as a means of expressing time and duration through sequence. In *Reading Position for Second Degree Burn* (illustration), the artist placed an open book over his chest, had a "before" photograph taken, then exposed himself for hours to the sun. The "after" photograph, with the book removed, documents the painful results of sunburn. The artist's body is used as a photosensitive "canvas" to record the effects of natural phenomena like the sun. Oppenheim makes a direct connection between the light-sensitive qualities of both skin and film. An important concern is the time factor involved in this process. The "before" and "after" photographs establish a time scale and communicate it through color change, contrasting the white, unexposed skin with red, burned area.

Polarities (illustration), completed in 1972, was a turning point in Oppenheim's career, and directly relates to Burnham's idea about shamanism in contemporary art. When Oppenheim's father died in November of 1971, Oppenheim returned to the family home and went through his father's personal papers and sketches looking for the last drawing his father had done before his death. Oppenheim saw significance in this last sketch and identified the area that "was the last movement of his hand across the page." He juxtaposed this diamond-like shape with the first drawing done by his daughter Chandra, and plotted these two shapes — enormously enlarged — with flares on the ground in Bridgehampton, New York. After sunset, the flares were ignited and Oppenheim flew overhead in a small plane (flying was believed to be a common shamanistic feat) photographing the flare-lit configuration. Oppenheim combined the last mark his father made before his death with the first scrawls of his young daughter in a work of art that comments on the genetic continuity of life. This artistic event can be read in shamanistic terms, as an action that provides meaning to otherwise meaningless, incomprehensible, and tragic elements of life.

Most of Oppenheim's recent environmental works operate outside of logical, rational, and explicable systems. They become reflections and meditations on the quiet horror of contemporary existence. No answers are given, no way out shown, there is no easy way to deal with these strangely juxtaposed elements; they force us to make new connections, discover new feelings, and confront aspects of life usually hidden.

Adrenochrome is a powerful piece documenting Oppenheim's "recovery" from an earlier mental collapse. Installed at the Sonnabend Gallery in 1973, it consists of three elements: a large pyramidal base with one gram of oxidized adrenochrome (a chemical compound connected with some forms of mental illness) placed on the top, a projected slide of the crystal enlarged 500 times, and a continuous tape recording of a letter from Oppenheim's doctor outlining his mental condition and medical needs. By objectifying his past medical history in a public way, Oppenheim allows himself to reach a new awareness about a difficult period of his life and to share these perceptions with an audience. The formal organization of this piece is in direct opposition to the psychological state it depicts. Although it refers to a confused and disoriented state of mind, it presents this story in an extremely ordered and controlled way. The fearful beauty and geometric order of

Dennis Oppenheim, *Reading Position for Second Degree Burn* (courtesy of Dennis Oppenheim).

Dennis Oppenheim, *Polarities* (courtesy of Dennis Oppenheim).

the enlarged crystal is juxtaposed against the large pyramid holding at its apex a precious ounce of this difficult-to-isolate psychoreactive chemical. The sound of the letter being read drones on, reminding us of the serious medical implications behind this mysteriously beautiful display.

In *Wishing Well* (illustration), constructed in San Francisco in 1973, Oppenheim installed a conveyor belt that emerged through a wall and extended to a container of water. An endless line of pennies were deposited by the moving belt into a sunken well. As the coins regularly fell into the bucket, the artist, on a tape-recorded sound track, expressed his wish to pass through solid matter:

I want to be able to sink downward, I wish I could do this, I want this well to act in favor of this, I wish these pennies could act as sparks, to charge the eventuality of this. I know where I want to stand. I know

the location to be in this city, near by. I'll stand there as often as I can. It's not far away. My shoulder blades will be pressed against a building, a thick wall. I'll be looking out toward the street. The heels of my shoes will be against the concrete and planted firmly on the sidewalk. It's not as if this material begins to soften or relax, but the first sensation is similar to that. I can feel the pavement begin to surround the soles of my shoes. At the same time, my shoulder blades have gone at least 1 inch back, as my feet are further engulfed by the concrete. I can feel my shoulders dragging downward, as they sink deeper into the wall. My feet are now in the pavement, not quite up to the beginning of my ankles. . . . My thighs are well into the ground, my neck is practically engulfed now, only my chin remains. My thighs are well into the ground, my feet passing through difficult forms of rock and gravel, dampness. My chest and stomach become the only visible portions, as the sidewalk surface is now level with my waist.[22]

Oppenheim contrasts his verbal fantasy of standing somewhere in the city and slowly

Dennis Oppenheim, *Wishing Well* (courtesy of Dennis Oppenheim).

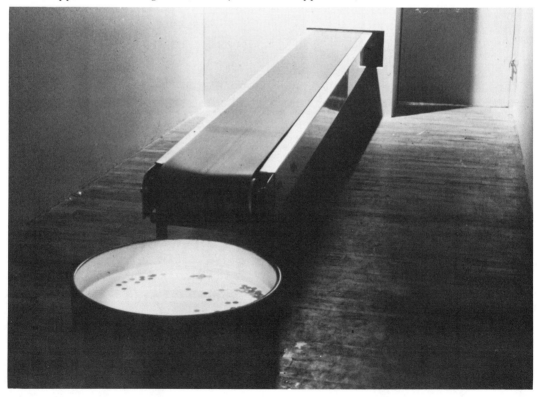

being engulfed by it with the reality of penny after penny dropping into a bucket. He transforms our childhood memories of making a wish while throwing a penny in a well into a magical act. The steady motion of the slowly moving belt merges with the voice to create an event that occurs in one's mind. Sound track and moving belt, working in concert, create a powerful mental image of a man slowly being swallowed into the pavement.

In many of Oppenheim's recent environmental pieces there is an underlying atmosphere of doom. *Echo* (illustration), installed at the John Gibson Gallery in January of 1976, makes use of four synchronized film projectors, one focused on each wall, that fill the room at intervals with the image of a gigantic hand slapping the wall. A room-reverberating sound accompanies the visual action; what becomes anxiety provoking is the unpredictability of just where and when the hand will strike. Oppenheim confronts us with an imaginary but powerful fear. No place is safe. All four walls of the room are projection surfaces, so no matter where you stand at some time the hand will fall on you. Another Oppen-

heim installation at John Gibson's that conveys an ominous sense of dread is *Attempt to Raise Hell* (illustration). A small steel-headed puppet in a green suit sits quietly in close proximity to a suspended bell. At 100-second intervals, an extremely powerful electromagnet is switched on briefly and forces the head to lunge forward, striking the edge of the bell with great force. The sound resonates for over a minute, stopping just before the next cycle. It is a frightening, demented piece that defies rational explanation. Oppenheim obviously wants us to experience this work in a psychological as well as a physical way. In the exhibit notes he quizzically wrote, "The sound fills the room as well as the mind." *Attempt to Raise Hell* invites the uninitiated to participate in a strange contemporary ritual that alternately amuses and horrifies.

In other environmental works Oppenheim creates situations that describe the workings of redundant psycholgocial traps. By the presentation of repetitiously obsessive acts he describes and comments upon the alienating aspect of our individuality in contemporary society. His use of puppets symbolically represents isolated individuals

Dennis Oppenheim, *Echo* (courtesy of Dennis Oppenheim).

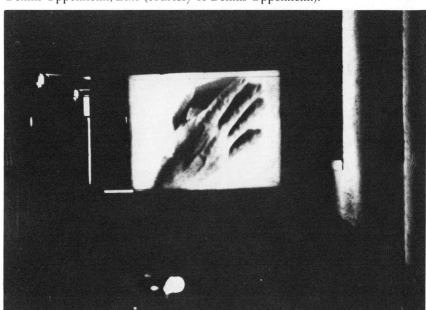

and at the same time the artist himself. These mechanically activated humanistic forms are able to perform feats that would be impossible for Oppenheim to do, such as slamming his head against a bell and dancing tirelessly for days. Oppenheim also avoids the anxiety and stage fright involved in doing a live performance piece but retains much of the effect.

Oppenheim's *Theme for a Major Hit,* incorporating a motorized marionette, was shown in New York for the first time in the spring of 1974. On a small, round stage, a 30-inch-high puppet wearing a grey suit with a white shirt and tie springs to life with the playing of a two-hour recorded song Oppenheim wrote: "It Ain't What You Make, It's What Makes You Do It." Under a theatrical spotlight the jerky movement of the puppet and droning sound of the tape combine to create a disconcerting mood that is difficult to forget. The movements are lifelike enough to convey the feeling that perhaps there is a form of android life buried in this small figure. The face, which resembles Oppenheim's, is frozen in a grim, expressionless posture, condemned to participate in this senseless ritual forever. Oppenheim's use of puppets calls to mind their employment in early theater and primitive magic. The classical Greek stage exploited the visual effects of large masks to underscore the meaning of the spoken word; primitive tribes still make use of lifeless effigies to exert power over the living in voodoo ceremonies. Although Oppenheim conceives of and plans these pieces with great care, they are not essentially concerned with conceptual or formal problems. His goal is the awareness of deep and complex feelings that have been buried for thousands of years.

Anyone who has attended lectures or given them (this includes almost everybody in our society) can immediately relate to the familiar scene presented to us in Oppenheim's *Lecture Piece No. 1* (illustration). There is, however, a disturbing aspect to this otherwise ordinary tableau. Everything is greatly reduced in size. Forty 12-inch-high chairs and two tiny motorized marionettes, one lecturing in front of a podium and one solitary "student" in the last row, greet the visitor. A sparse institutional look pervades, the details are nondescriptive, a rug runs down the center to the speaker's stand, which is flanked by two rows of simple, ordinary chairs. The lips of the 18-inch puppet move in synchronization to a 20-minute stereo sound track that continues in an un-

varied tone. By drastically changing the scale of this unremarkable place Oppenheim creates an otherworldly atmosphere that mixes, in a curious way, banal features with unsettling effects. The "lecturer's" voice is clear and understandable, but this event, although physically close, seems to be taking place at a great distance and in another dimension. Lips moving mechanically, the vacant seats (a teacher's nightmare), a solitary listener who cannot hear all contribute to an overwhelming feeling of emptiness that permeates the room.

In these thought-provoking installations Oppenheim transforms familiar aspects of life into extraordinary dramas. We witness events that challenge and transform our perceptions, and force us to enter new regions of thought and experience feelings that are usually beyond us. Behind the manipulation of objects and complex ideas lies the central issue in Oppenheim's work: an introspective search for meaning in a sometimes difficult and confusing world.

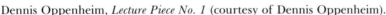

Dennis Oppenheim, *Lecture Piece No. 1* (courtesy of Dennis Oppenheim).

Jenny Holzer

In the late seventies a group of artists emerged whose exhibition space was more likely to be the grimy walls of subway stations and city buildings than the pristine white interiors of uptown gallerys. Although they were not interested in the remote rural sites that appealed to their predecessors, the earthwork artists, they did share some of their sensibilities: Like them, these new environmental artists were involved in reaching a different public with work that did not fit easily into gallery formats. The city was their proving ground, and their art for the most part was overtly political and socially conscious.

Jenny Holzer first became known when people noticed the buildings of lower Manhattan plastered with posters containing strange printed messages. Called "Truisms,"

Jenny Holzer, *Truisms*, 1978 (courtesy of the Barbara Gladstone Gallery and Jenny Holzer).

ages other than simple, black italic type set against white paper (illustration).

The verbal content of these "Truisms" is both curious and disturbing. It consists of a collection of anarchistic and mundane sayings that make satiric use of the kind of anonymous babble one hears being used in casual social encounters. These Truisms total several hundred in number and are systematically presented in alphabetic order. The first Truism reads: "A LITTLE KNOWLEDGE GOES A LONG WAY." Innocent enough as a starter, but as we read on — presumably standing on a sidewalk in lower Manhattan — we experience diabolical shifts in meaning and effect. Verbal information is presented to us as a series of contradictory half-truths and half-lies. Clearly separating the "truths" from the "lies" becomes an impossible task. For example:

Morals are for little people
Murder has its sexual side
Myths make reality more intelligible
Noise can be hostile
Nothing upsets the balance of good and evil
Occasionally principles are more valuable then people
Often you should act like you are sexless

Holzer has purposefully arranged these quips in a nondevelopmental way — they have no beginning and consequently, no end. They are viewed by Holzer as discrete entities, whose linkages are formed by our own perceptions and experiences rather than by the author's design. In this sense they mimic and mock the bits and pieces of verbal information that we glean everyday in our visual environment from advertisements and commercial signs.

Unlike the vague, anonymous jingles of most advertising — "be a part of it," "escape," "for treasured moments," "because your job is making decisions," etc. — Holzer's have bite. Many of them comment on subjects which are taboo for advertising because they are unpleasant realities: insanity, sexual disillusionment, and economic inequity. Advertising presents us with a dream world where, if one could believe the beautiful images and reassuring slogans, everyone is entitled to an attractive sexual partner, a new car, and a salary in six figures. Holzer's Truisms collide with these hopeful but for the most part fictitious dreams by announcing: "PEOPLE WHO GO CRAZY ARE TOO SENSITIVE"; "ROMANTIC LOVE WAS INVENTED TO MANIPULATE WOMAN"; and "STARVATION IS NATURE'S WAY."

Contradiction is one of Holzer's most powerful literary devices. She blatantly and purposefully uses half-truths in order to point out the complexities of meaning in the modern world and how language is manipulated by powerful economic interests. For instance:

CHILDREN ARE THE CRUELIST OF ALL
CHILDREN ARE THE HOPE OF THE
    FUTURE

EVERYONE'S WORK IS EQUALLY
    IMPORTANT
EXCEPTIONAL PEOPLE DESERVE SPECIAL
    CONCESSIONS

The tensions created by these paradoxical verbal mediations force us to confront our own conflicting desires, beliefs, and perceptions of the world around us. Holzer's statements often blend elements of truth and falsehood, sometimes within the same statement. In many ways her art is a form of visual poetics and often uses linguistic ploys of literature. Similar tensions and questions about the meaning of "truth" are incorporated into Mark Strand's poem *Elegy for My Father*, which takes the form of a dialogue between a son and his dead father:

Why did you travel?
BECAUSE THE HOUSE WAS COLD.
Why did you travel?
BECAUSE IT IS WHAT I HAVE ALWAYS
    DONE BETWEEN SUNSET AND SUNRISE.
What did you wear?
I WORE A BLUE SUIT, A WHITE SHIRT,
    YELLOW TIE, AND YELLOW SOCKS.
What did you wear?
I WORE NOTHING. A SCARF OF PAIN
    KEPT ME WARM.

Who did you sleep with?
I SLEPT WITH A DIFFERENT WOMAN
    EACH NIGHT.
Who did you sleep with?
I SLEPT ALONE. I HAVE ALWAYS SLEPT
    ALONE.
Why did you lie to me?
I ALWAYS THOUGHT I TOLD THE TRUTH.
Why did you lie to me?
BECAUSE THE TRUTH LIES LIKE
    NOTHING ELSE AND I LOVE THE
    TRUTH.
Why are you going?
BECAUSE NOTHING MEANS MUCH TO ME
    ANYMORE.
Why are you going?
I DON'T KNOW. I HAVE NEVER KNOWN.
How long shall I wait for you?
DO NOT WAIT FOR ME. I AM TIRED AND I
    WANT TO LIE DOWN.
Are you tired and do you want to lie down?
YES, I AM TIRED AND I WANT TO LIE
    DOWN.[23]

Holzer's art is ultimately political in the way it questions and comments upon the ideological beliefs of our time. Rather than focusing on individual issues she presents us with an introspective social discourse that sidesteps the idea of answers and focuses on the power of language to assume authority. Perhaps the most significant issue of her Truisms is not which statement is true or false but who is making the statement? Is it the same smooth, accentless, authoritarian voice that speaks to us from the inside of televisions and radios?

In 1980 and 1981, Holzer and artist Peter Nadin collaborated on a series of works called "Living." Instead of printed posters, Holzer and Nadin fabricated bronze plaques with three-dimensionally embossed statements on the surface. They were similar to the kind of metal markers historic societies are fond of placing in front of a house that an important political figure once lived in. Holzer's commentary was much more provocative, however, as she outlined the potential of downward economic mobility that was looming during the recessionary years of 1981–1982 (illustration).

Recently Holzer has become interested in the prospect of reaching even more people than before. Using the animated spec-

AFFLUENT COLLEGE-BOUND
STUDENTS FACE THE REAL
PROSPECT OF DOWNWARD MOBILITY.
FEELINGS OF ENTITLEMENT
CLASH WITH THE AWARENESS OF
IMMINENT SCARCITY. THERE IS
RESENTMENT AT GROWING UP AT
THE END OF AN ERA OF PLENTY
COUPLED WITH REASSESSMENT
OF CONVENTIONAL MEASURES OF
SUCCESS.

Jenny Holzer and Peter Nadin, "Living" series, 1981–1982 (courtesy of the Barbara Gladstone Gallery and Jenny Holzer).

tacolor lightboard high above Times Square, she has presented her messages simultaneously to thousands of passers-by in a program presented by the Public Art Fund, a nonprofit organization (illustration). This means of presentation seems a perfect match for Holzer's concept of the public street as an exhibition arena for her art. But she is also aware of the strict limitations this animated sign imposes:

I realize that people's attention span, especially if they are on their way to lunch, might be 2.3 seconds and so I try to make each statement have a lot of impact and stand on its own. That still doesn't mean that I don't try to make a whole program that works. I want the statements to reinforce each other so the whole is more than its parts. I just try to make the parts okay,

too. I also want to supply people with a cheap electronic retrospective of my work.[24]

The context in which her work appears enhances the effectiveness of Holzer's art. Seen on the street instead of in a book or gallery it stands a better chance of catching us unaware and affecting us with its exposé of the fraudulent potential of language.

Another artistic venue that interests Holzer is broadcast television. She reasoned, Why not use the means of the medium to comment on itself? Following a recent internship with a station in Hartford, Connecticut, Holzer became excited about the prospect of developing 30- and 60-second spots that could be aired on broadcast television. One of her discoveries was the surprising affordability of air-time at that local Hartford Station: "You can buy 30 seconds in the middle of Laverne and Shirley for about 75 dollars, or you can enhance the CBS Morning News for a few hundred dollars. The audience is all of Connecticut and a little of New York and Massachusetts."

The prospect of Holzer reaching so many unsuspecting "art patrons" as they sit bleary-eyed with a morning cup of coffee watching the news is humorous; certainly the prospect of using broadcast television has far-reaching implications for future media artists. By harnessing these new avenues of visual communication, she has claimed a new territory and redefined opportunities

Jenny Holzer, *Truism: Your Oldest Fears Are the Worst Ones*, 1982 (courtesy of the Barbara Gladstone Gallery and Jenny Holzer/ photo by Lisa Kahane).

CURRENTS

for artists interested in socially conscious forms of art.

Keith Haring

The presence of graffiti in urban restrooms, subways, and alleys in large cities such as New York is such a common sight that, for natives at least, it evokes no particular response — it remains an annoying nuisance like uncollected garbage or dirty streets. For the most part, graffiti appears to be a primitive, semiliterate system of undecipherable scrawls and scribbles done by teenage vandals acting out territorial-marking urges. Some critics, however, believe that graffiti represents the aesthetic expression of a disenfranchised underclass who are openly hostile to a system that has locked them out. As such they tend to view graffiti as the work of contemporary urban folk artists — most of whom come from ethnic backgrounds and have no formal art training. Keith Haring is one prominent graffiti artist who does not fit this profile. He is white, middle-class, has attended art school, and now shows and sells his work in fashionable New York galleries.

Like Holzer, Haring is interested in exploring the audience found in well-trafficked urban spaces. But Haring's choice of venue is the vast number of underground tunnels in the New York Subway system. Few regular travelers on the trains are unfamiliar with his distinctive white-chalk images drawn on the black paper which covers unsold advertising space (illustration). These simple, linear drawings are done rapidly in a style that is both economical and visually rich.

Over the years Haring has developed a large yet stylistically coherent repertoire of popular images that include: crawling babies, flying saucers, dancing men, ringing phones, and barking dogs. Haring strives for concise, emblematic images that are immediately recognizable — they take on the appearance of corporate trademarks for New Wave sensibilities. These diagrammatic images represent Haring's awareness of how much contemporary life is affected by the realities of technological progress.

Haring explains the importance certain communication devices hold for him: "The TV and telephone have significance in my drawings just as everyday things that have so much power over our lives. People who grew up in the modern world tend to take these things for granted, without stopping to think that they are new and that the world existed without them for so long."

Haring was born in Kutztown, Pennsylvania — an area known for its folk-art hex signs on barns — and came to New York in 1978 to enroll in the School of Visual Arts. After working on a series of language-based performances and videos, he became frustrated with the limits of conventional word usage and began to develop a "pictographic" language based on the repeated use of signs and symbols. Haring recalls:

The earliest subway drawings were done with black Pilot markers over Johnny Walker Red ads depicting a snowy landscape with train tracks. I would usually

Keith Haring, drawing in subway (courtesy of Tony Shafrazi and Keith Haring/photo by Tseng Kwong Chi).

draw a crawling baby or a dog, or combinations of the two, getting zapped by a flying saucer. They were done on the same scale as the landscape. Whenever I would see one of those ads, I'd do that drawing over it. Then, around January, when the black paper (used to cover unrenewed ads) came along, I started doing chalk drawings on that paper. Chalk is a great medium—clean, economical, fast; just stick it in your pocket. I lived near Times Square and worked downtown, so I was on the subway at least twice a day. By moving from the East Village to Times Square, I increased my audience by tens of thousands of people.[25]

Many of Haring's subway drawings appear to have themes of power and force. Lines radiate from a variety of images which can be read as representing emanations of sound, mental telepathy, illumination, or a vague sense of power mysteriously inherent in the object itself. *Untitled*, 1981 (illustration), is a drawing that illustrates this feature of force radiation. Two half-bird, half-human creatures are shown firmly holding what appear, for lack of a better term, to be "power rods." Between them a smaller representation of a human figure is shown with arms outstretched as if worshiping or reaching for these charged artifacts. All of the human figures are featureless and undifferentiated in terms of sex and age. There is a humorous, whimsical side to these simple drawings yet they clearly become more than superficially

amusing cartoons. Themes of subjugation, fear, and capitulation to an unknown authority of great power appear over and over again. Some subway storyboards feature human figures falling down from the sky while a barking dog, with its head upraised in a howling position, stands on the ground wildly gyrating. Another one reveals a human figure ceremoniously holding aloft a large, triangular-patterned snake whose body passes through a circular opening in the person's abdomen. Mysterious lines of "force" emanate from the figure's hands and the opening (illustration). Many of Haring's drawings are like this one, combining disarmingly amusing images within dark scenarios.

Recently Haring has emerged from the underground to exhibit in a variety of museum and gallery settings. To enter these new show spaces, he has had to adapt his tactics to produce more permanent, portable forms of art. Working on colored plexiglass, paper, and sheets of vinyl plastic (illustration), Haring continues to refine his visual images on portable (thus saleable) surfaces. One of the difficulties he faces, however, is maintaining the visual and psychological tension of his early subway drawings. Since drawing on a subway sign is a misdemeanor in New York, Haring always worked with one eye on the lookout

Keith Haring, ink drawing, untitled (courtesy of Tony Shafrazi and Keith Haring).

Keith Haring, drawing in subway (courtesy of Tony Shafrazi and Keith Haring/photo by Tseng Kwong Chi).

for Transit Police. But the urgency of this situation seemed to sharpen his control and this factor no doubt helped shape these mysterious, terse drawings. Also missing in the gallery work is the sense of surprise and discovery one feels upon finding one of these illegal artworks in the underground. To some extent Haring misses this risky and adventurous activity (in one year the New York Transit Police arrested nearly 800 "graffiti artists"); but commercial success also has it rewards, and like rock music this genre seems to attract mainly young practitioners.

In the long run Haring's work must be viewed as a recent extension of Pop-art sensibilities. Having been born in the late fifties the artist sees himself as "one of the first babies of the space age. I grew up on T.V. I feel more that I'm a product of Pop, rather than a person who is calling attention to it." In line with this view, Haring has given Vivienne Westwood, an avant-garde fashion designer, permission to reproduce his black line drawings for a line of clothing she is producing. Soon transplanted Haring drawings may be seen on people's backs as they blithely walk past one of his white-chalk drawings on blacked-out advertising spaces. Somehow this curious overlapping of art and life, culture and commerce, seems quite appropriate to Haring's aesthetic.

All of the artists profiled in this chapter have effectively challenged accepted notions of visual format, thematic content, and even the physical materiality of the artwork

Keith Haring, untitled, 1982 (courtesy of Tony Shafrazi and Keith Haring/photo by Ivan Dalla-Tana).

itself to extend the bounds of thought and, as Hans Haacke explained, to continuously interact with the world around them. Through its multidisciplinary nature, this work expresses the realities of our increasingly complex and interdependent society; contemporary images, ancient myths, and half-forgotten feelings are brought forward and presented to us by these unique spaces and events. Much of the work examines our beliefs and actions, how we live and what our values are—experiencing these environmental artworks makes us think about where our culture has come from and where it is going.

Notes

Chapter 1

1. Sam Hunter, *Art and Business: The Philip Morris Story* (New York: Harry N. Abrams, Inc., 1979), p. 36.

2. Hunter, *Art and Business*, p. 45.

3. Beverly Pepper, "Space, Time and Nature in Monumental Sculpture," *Art Journal*, Spring 1978, p. 251.

4. Pepper, "Space, Time and Nature in Monumental Sculpture," p. 251.

5. Clara Weyergraf, *Richard Serra: Interviews Etc. 1970–1980* (Yonkers, NY.: Hudson River Museum, 1980), p. 168.

Chapter 2

1. Calvin Tomkins, "Raggedy Andy," in *Andy Warhol* (Boston: New York Graphic Society, n.d.), p. 12.

2. Jonas Mekas, "Notes after Reseeing the Movies of Andy Warhol," *Film Culture*, No. 33, 1964.

3. Andy Warhol, *The Philosophy of Andy Warhol (From A to B & Back again)* (New York: Harcourt, Brace, Jovanovich, 1975), p. 22.

4. John Arthur, *Richard Estes: The Urban Landscape* (Boston: Museum of Fine Arts, and New York Graphic Society, 1978), p. 26.

5. Arthur, *Richard Estes*, p. 42.

6. Arthur, *Richard Estes*, p. 27.

7. Arakawa and Madeline H. Gins, *The Mechanism of Meaning* (New York: Harry N. Abrams, Inc., 1979), pp. 4–5.

8. Arakawa and Gins, *The Mechanism of Meaning*, p. 77.

9. Calvin Tomkins, "Romare Bearden," *The New Yorker*, November 28, 1977, p. 53.

10. *Art In America*, December 1982, p. 65.

11. Kristin Olive, *Arts Magazine*, November 1985, p. 84.

Chapter 3

1. Print Council of America, 527 Madison Avenue, New York City, 10022.

2. Constance Glenn, *Jim Dine, Figure Drawing, 1975–1979* (New York: Harper and Row Publishers, Inc., 1980).

3. Richard Field, *Jasper Johns: Prints, 1970–1977* (Middletown, CT: Wesleyan University Press, 1978), p. 15.

4. Carter Ratcliff, *Pat Steir Paintings* (New York: Abrams, 1986) p. 89.

Chapter 4

1. Robert Morris, *Robert Morris: Mirror Works 1961–78* (New York: Leo Castelli, Inc., 1979), pp. 4–7

2. John Gruen, "Jackie Winsor: Eloquence of a Yankee Pioneer," *Art News*, March 1979, p. 59.

3. Gruen, "Jackie Winsor," p. 61.

4. Lucy Lippard, "Jackie Winsor," *Artforum*, February 1974, p. 58.

5. Gruen, "Jackie Winsor," p. 60.

6. From the statement by Jackie Winsor that appears in *Jackie Winsor*, edited by Kynaston McShine. Copyright © 1979 The Museum of Modern Art, New York. All rights reserved. Reprinted by permission.

7. Gruen, "Jackie Winsor," p. 57.

8. Lucas Samaras, "A Reconstituted Diary: Greece, 1967," *Artforum*, October 1968, p. 27.

9. Kim Levin, *Lucas Samaras* (New York: Harry N. Abrams, Inc., 1975) pp. 45–46.

Chapter 5

1. John Szarkowski, *Mirrors and Windows: American Photography Since 1960* (New York: Museum of Modern Art, 1978), p. 18.

2. Joel Meyerowitz, *Cape Light: Color Photography by Joel Meyerowitz* (Boston: New York Graphic Society, 1979).

3. *Diane Arbus* (Millerton, NY: Aperture, Inc., 1972) p. 3. © 1972 The Estate of Diane Arbus.

4. Allan Porter, "Imaginary Interview with Ralph Gibson," *Camera*, April 1979, pp. 20–27.

Chapter 6

1. Caroline Tisdall, *Joseph Beuys* (London: Thames and Hudson, 1979), p. 10.

2. Tisdall, *Joseph Beuys*, p. 34.

3. Tisdall, *Joseph Beuys*, p. 16.

4. Robert Keil, "Nam June Paik — Video Philosopher," *Artweek*, July 5, 1980, p. 4.

5. "Interview with Laurie Anderson," *Impressions*, Spring 1981, pp. 14–15.

6. Transcribed by the author from the Laurie Anderson Concert, Cinema Theatre, San Francisco, Fall 1981.

Chapter 7

1. Knud Rasmussen, *Across Arctic America, Narrative of the Fifth Thule Expedition* (New York: G. P. Putnam, 1927), p. 440.

2. Jack Burnham, *Great Western Salt Works: Essays on the Meaning of Post-Formalist Art* (New York: George Braziller, 1968), p. 30.

3. Jan Van Der Marck, *Wrapped Museum* (Chicago: Museum of Contemporary Art, 1969).

4. Jonathan Fineberg, "Theater of the Real: Thoughts on Christo," *Art in America*, December 1979, p. 98.

5. Calvin Tomkins, "Christo," *The New Yorker*, March 28, 1979, p. 43.

6. Fineberg, "Theater of the Real," p. 96.

7. Nancy Holt, ed., *The Writings of Robert Smithson: Essays with Illustrations* (New York: New York University Press, 1979), p. 89.

8. Holt, ed., *The Writings of Robert Smithson*, pp. 56–57.

9. Holt, ed., *The Writings of Robert Smithson*, p. 113.

10. Philip Leider, "For Robert Smithson," *Art in America*, November 1973, p. 82.

11. Calvin Tomkins, *The Scene: Reports on Post-Modern Art* (New York: Viking Press, 1976), p. 139.

12. Lucy Lippard, "Charles Simonds," *Artforum*, February 1974, p. 36.

13. Daniel Abadie, *Temenos* (Buffalo, NY: Albright-Knox Art Gallery catalogue, n.d.), p. 7.

14. Lippard, "Charles Simonds," p. 38.

15. Allan Sondheim, ed., *Individuals: Post-Movement Art* (New York: E. P. Dutton, 1977), p. 300.

16. Alice Aycock, *Work 1972–1974*.

17. Sondheim, *Individuals*, p. 106.

18. Aycock, *Work 1972–1974*.

19. Margaret Sheffield, "Alice Aycock: Mystery Under Construction," *Artforum*, September 1977, p. 63.

20. Ellen Thalenberg, "Site Work, Some Sculpture at Artpark 1977," *Arts Canada*, October–November 1977, p. 20.

21. Sondheim, *Individuals*, p. 251.

22. Sondheim, *Individuals*, p. 258.

23. Mark Strand, *The Story of Our Lives* (New York: Atheneum, 1973).

24. Jeanne Siegel, "Jenny Holzer's Language Games," *Arts Magazine*, December 1985, p. 68.

25. Barry Blinderman, "Keith Haring's Subterranean Signatures," *Arts Magazine*, September 1981, p. 164.

Timeline

Developments in Art from 1940 to the Present

1940

Established American painters	Stuart Davis, Charles Demuth, Arthur Dove, Marsden Hartley, Edward Hopper, Georgia O'Keeffe
Beginnings of Abstract Expressionism and the New York School of painting	William Baziotes, Arshile Gorky, Adolph Gottlieb, Hans Hofmann, Robert Motherwell, Jackson Pollock, Mark Rothko, Clifford Still
Sculptors working in America	Alexander Calder, Marcel Duchamp, Elie Nadelman, Isamu Noguchi, Jose de Rivera, David Smith
European artists who moved to New York in the early 1940s	Andre Breton, Max Ernst, Fernand Leger, Matta, Piet Mondrian, Yves Tanguy
Artists working in Europe	Balthus, Georges Braque, Alberto Giacometti, Henri Matisse, Joan Miro, Pablo Picasso
Photographic artists	Ansel Adams, Bill Brandt, Henri Cartier-Bresson, Walker Evans, Dorothea Lange, Man Ray, Edward Weston

EVENTS

1940 Battle of Britain

1941 Germany invades the USSR

1943 Jackson Pollock's first one-person exhibition

1944 Rome and Paris liberated

1945 Franklin D. Roosevelt dies, Truman becomes president, Germany surrenders

1949 People's Republic of China founded

BIBLIOGRAPHY

Alloway, L. *Topics in American Art Since 1945* (New York: Norton, 1975).

Anderson, W. *American Sculpture in Process: 1930/1970* (Boston: New York Graphic Society, 1975).

Read, H. *The Philosophy of Modern Art* (London: Faber and Faber, 1964).

Rosenberg, H. *The Tradition of the New* (Chicago: University of Chicago Press, 1983).

1950

Artists working in America	Joseph Cornell, Willem DeKooning, Philip Guston, Grace Hartigan, Jasper Johns, Edward Kienholtz, Alfred Leslie, Richard Linder, Joan Mitchell, Louise Nevelson, Robert Rauschenberg, Ad Reinhard, Larry Rivers, Jack Tworkov
English "Pop Artists"	Richard Hamilton, Eduardo Paolozzi
England	Francis Bacon, Henry Moore, Ben Nicholson
"CoBrA" group (Copenhagen, Brussels, Amsterdam)	Pierre Alechinsky (Belgian), Karel Appel (Dutch), Asger Jorn (Danish)
France	Jean Fautrier, Hans Hartung, Pierre Soulages, Wols
Spain	Antonio Tapies
Italy	Alberto Burri

EVENTS

1950   Korean War begins

1953   Dwight D. Eisenhower takes office as president, Korean War ends

1954   Supreme court rules segregation unconstitutional

1956   Hungarian uprising

1959   Fidel Castro comes to power in Cuba

BIBLIOGRAPHY

Ashton, D. *The New York School: A Cultural Reckoning* (New York: Penguin, 1979).

Ritchie, A. C. *The New Decade: Twenty-two European Painters and Sculptors* (New York: MoMA, 1955).

Sandler, I. *The New York School: The Painters and Sculptors of the Fifties* (New York: Harper & Row, 1978).

1960

American Pop Art	Robert Indiana, Roy Lichtenstein, Claes Oldenburg, James Rosenquist, Andy Warhol, Tom Wesselman
Minimal art (painting)	Alexander Lieberman, Brice Marden, Agnes Martin, Dorothea Rockburne, Robert Ryman
Primary Structures (sculpture)	Carl Andre, Larry Bell, Ronald Bladen, Eva Hess, Donald Judd, Robert Morris, Tony Smith
"Happenings"	Jim Dine, Red Grooms, Allan Kaprow, Claes Oldenburg, Robert Whitman
Post-Painterly Abstraction	Richard Diebenkorn, Helen Frankenthaler, Al Held, Morris Louis, Kenneth Noland, Jules Olitski, Frank Stella
Optical art	Joseph Albers, Richard Anuszkiewicz, Larry Poons, Bridget Riley, Victor Vasarely
Earth works and environmental art (late sixties)	Christo, Walter De Maria, Michael Heizer, Richard Long (English), Dennis Oppenheim, Robert Smithson
Photographic artists	Richard Avedon, Harry Callahan, Bruce Davidson, Robert Frank

EUROPEAN ARTISTS

Sculptors	Anthony Caro (England), Edwardo Chillida (Spain), Marino Marini (Italy), Arnaldo Pomodoro (Italy)
"New Realism" (France)	Arman, Yves Klein, Jean Tinguely
Germany	Joseph Beuys

EVENTS

1961	John F. Kennedy takes office as president, Russians orbit first man in space
1962	Cuban missile crisis
1963	Assassination of President Kennedy
1967	Six-Day War between Israel and Arab countries
1968	Student unrest in Europe and U.S.
1969	Richard M. Nixon takes office as president

BIBLIOGRAPHY

Battcock, G., ed., *The New Art* (New York: Dutton, 1966).

Burnham, J. *Beyond Modern Sculpture: The Effects of Science and Technology on the Sculpture of this Century* (New York: Braziller, 1968).

Coplans, J. *Serial Imagery* (Pasadena, CA: Pasadena Museum of Art, 1968).

Russell, J., and **Gablik, S.** *Pop Art Redefined* (New York: Praeger, 1969).

1970

Figurative and "Super Realism"	Robert Bechtel, Chuck Close, Richard Estes, Janet Fish, Audry Flack, Duane Hanson (sculpture), David Hockney, Alice Neel, Philip Pearlstein
Conceptual Art	John Baldessari, Jan Dibbets, Hans Haacke, Douglas Huebler, Joseph Kossuth, Sol LeWitt
Sculptors	Magdalena Abakanowicz, Louise Bourgeois, Barry Flanagan (England), Nancy Graves, Red Grooms, Robert Irwin, Richard Serra, Joel Shapiro
"New Image" artists	Nicholas Africano, Jennifer Bartlett, Robert Moskowitz, Donald Sultan, Joe Zucker
"Pattern and Decoration" artists	Robert Kushner, Kim MacConnel, Rodney Ripps, Robert Zakanitch
Personal Directions	Mel Bochner, Nancy Holt, Bruce Nauman, Edward Ruscha
Video-Performance	Vito Acconci, Chris Burden, Peter Campus, Juan Downey, Gilbert and George (England)
Photographic artists	Robert Cumming, Bill Owens, Stephen Shore, Eve Sonneman
England	Peter Blake, Gilbert and George, Allan Jones, R. B. Kitaj, Bruce McLean

EVENTS

1970 U.S. invades Cambodia, four students killed at Kent State by state militia

1972 Watergate burglary in Washington, D.C.

1973 War in Vietnam ends

1974 Richard M. Nixon resigns presidency over Watergate scandal

1976   U.S. celebrates its bicentennial

1978   Civil war in Nicaragua

1979   Soviet invasion of Afghanistan

BIBLIOGRAPHY

Marshal, R. *New Image Painting* (New York: Whitney Museum of American Art, 1978 [exhibition catalogue]).

Meisel, L. K. *Photo-Realism* (New York: Abrams, 1980).

Meyer, U. *Conceptual Art* (New York: Dutton, 1972).

Robins, C. *The Pluralist Era: American Art 1968–1981* (New York: Harper & Row, 1984).

Sondheim, A., ed., *Individuals: Post-Movement Art in America* (New York: Dutton, 1976).

1980

| American artists | John Ahearn (sculpture), Siah Armajani (sculpture), Ashley Bickerton, Ross Bleckner, Richard Bosman, Eric Fischl, Leon Golub, Jeff Koons, Barbara Kruger, Sherry Levine, Robert Longo, Allan McCollum, Elizabeth Murry, Tom Otterness (sculpture), Judy Pfaff, Martin Puryear (sculpture), Kenny Scharf, Julian Schnable, Gary Stephen, Philip Taaffe, Mark Tansey, Robert Therrien |

EUROPEAN ARTISTS

Germany	George Baselitz, Rainer Fetting, Jorg Immendorff, Markus Lupertz, A. R. Penck, Sigmar Polke
England	Lucian Freud, Antony Gormley, Howard Hodgkin, Bill Woodrow
Italy	Sandro Chia, Francesco Clemente, Enzo Cucchi

EVENTS

1980   Ronald Reagan defeats Jimmy Carter for the Presidency

1982   Falkland war (Britain and Argentina), economic recession hits the U.S.

1984   Reagan reelected by a landslide

1986   President Ferdinand Marcos of the Philippines swept from office

1987   "Black Monday"—stock market drops 508 points on October 19

BIBLIOGRAPHY

Art of Our Time: The Saatchi Collection (London and New York: Lund Humphries, 1984). Essays by M. Auping, P. Carlson, L. Cooke, H. Kramer, K. Levine, M. Rosenthal, and P. Tuchman.

Foster, H., ed., *The Anti-Aesthetic: Essays on Postmodern Culture* (Port Tounsend, WA: Bay Press, 1983).

Hertz, R. *Theories of Contemporary Art* (Englewood Cliffs, NJ: Prentice Hall, 1985).

Lucie-Smith, E. *American Art Now* (New York: Morrow, 1985).

Present

Index

Abstract Expressionism, 36–37
Anderson, Laurie, 241–52
    artworks:
        Duets on Ice, 244
        Environmental Music Piece, 246
        Handphone Table, 245
        Home of the Brave (film), 251
        O Superman, 249–51
        United States (performance series), 246
    William Burroughs and, 248
    education, 243
    John Giorno and, 248
    Holly Solomon Gallery and, 244
    Museum of Modern Art and, 245
    Van Gogh and, 243
Arakawa, Shusaku, 60–68
    artworks:
        Ambiguous Zones of a Lemon, 64–65
        The Mechanism of Meaning, 63, 64, 65
        Mistake, 61
        Panel No. 3, 65–66
        Untitled 1969, 60, 61, 62
    Madeline Gins and, 63, 64
Arbus, Diane, 186–92
    artworks:
        *Child with a Toy Hand Grenade in Central Park,
            New York City 1962*, 192
        Identical Twins, Roselle, New Jersey, 190
        *A Jewish Giant at Home with His Parents in the
            Bronx, New York 1970*, 187
        *Teenage Couple on Hudson Street, New York City
            1963*, 190
        *Woman on a Park Bench on a Sunny Day, New
            York City*, 187
    Museum of Modern art and, 191
    Venice Biennale and, 191
Art Students League, 36
Aycock, Alice, 292–99
    archeological sites and, 296
    architecture and, 293
    artworks:
        *The Beginnings of a Complex . . . Excerpt Shaft
            No. 4 Five Walls*, 297
        How to Catch and Manufacture Ghosts, 299
        Low Building with Dirt Roof (for Mary), 295

        Maze, 293–94
        Studies for a Town, 298
        Walled Trench/Earth Platform/Center Pit, 296
    Hollywood and, 299
    John Weber Gallery and, 299

Bearden, Romare, 69–77
    artworks:
        Baptism, 74
        Jazz Savoy, 71
        Memories, 75
        Of the Blues, 69
        The Return of Odysseus, 75
        Watching the Good Trains Go By, 70
    Cordier and Ekstrom Gallery and, 69
    education, 71
    George Grosz and, 71
    Nanette Rohan and, 73
    "The Spiral Group" and, 73
Beuys, Joseph, 216–31
    artworks:
        Celtic + -, 227
        Chief, 222
        Crossing the Rhine, 230
        Eurasianstaff (film), 228
        How to Explain Pictures to a Dead Hare, 223, 224
        I Like America and America Likes Me, 228, 229
        Iphigenie/Titus Andronicus, 225, 226
        Vacuum ↔ Mass (film), 228
    coyote, 229–230
    Düsseldorf Academy of Art and, 221
    "Fluxus" and, 222
    Free International University and, 230
    Peter Handeke and, 225–26
    Ewald Mataré and, 221
    Nam June Paik and, 221
    Performance piece, Middelburg, Holland, 221
    politics, 223
    René Block Gallery and, 222, 228
    Schmela Gallery and, 223
    shamanism and, 221
    study of zoology, 220
    university studies, 220
    visit to America, 228

Bogosian, Eric, 215
Borofsky, Jonathan, 168–73
  Art Students League and, 72
  artworks:
    Counting, 169
    Five Hammering Men, 172
    I Dreamed I Was Taller than Picasso, 171
    Installation Piece, 1980, 170
    *2,845,317 Green Tilted El Salvador Painting with
      Chattering Man*, 173
  concept art and, 169
  Documenta exhibition and, 171
  Venice Biennale and, 171
Burton, Scott, 35
  artworks:
    Viewpoint, 34

Cabaret Voltaire, 215
Calder, Alexander, 5
  artworks:
    La Grande Vitesse, 5
    Rooftop Painting, 7
Christo, 225–70
  artworks:
    Four Store Fronts, 258
    Museum of Contemporary Art Chicago, Wrapped,
      258
    The Pont-Neuf Wrapped, 267
    Running Fence, 255, 263
    Surrounded Islands, 265–66
    Valley Curtain, 262
    Wrapped Coast, 261
    Wrapped Floor, 259
    Wrapped Motorcycle, 257
    Wrapped Reichstag, 269
    Wrapped Trees, 260
  Emile Burian and, 256
  Coastal Commission and, 264
  education, 256
  Environmental Impact Report and, 264
  Jonathan Fineberg and, 264
  John Kaldor and, 261
  New York City and, 257
  Nouveau Réalism and, 256
  Pierre Restany and, 256
  Calvin Tomkins and, 264
  Jan van der Marck and, 259
Clemente, Francesco, 133–37
  artworks:
    Self-Portrait, 135
    Self-Portrait #4, 136
    This Side Up, 136
  education, 134
  mythology and, 136
  Todashi Toda and, 135
color photography, 182
"Copper" Eskimos, 253
Cragg, Tony, 173–77
  artworks:
    Citta, 177
    Red Skin, 175
    Self-Portrait with Sack, 176

24-Hour Cycle, 177
biochemistry and, 177

Di Suvero, Mark, 9–12
  artworks:
    Are the Years What? (for Marianne Moore), 9
    Motu Viget, 11
  GSA contract dispute and, 10
Dine, James, 103, 105–15
  artworks:
    The Leaning Man, 114
    Red Etching Robe, 112
    Rimbaud Alchemy Series, 111
    2 Hearts (the Donut), 109
    The World (for Anne Waldman), 113
  education, 107
  London and, 108
  Martha Jackson Gallery and, 108
  New York School of Poetry and, 113
  origin of empty bathrobe theme, 112
  Petersburg Press and, 110
  theatrical events:
    Car Crash, 106
    Jim Dine's Vaudeville, 106

Estes, Richard, 50–60
  artworks:
    Ansonia, 58–59
    Automat, 53
    Baby Doll Lounge, 55–56, 60
    Downtown, 54
    Escalator, 57
    Helene's Florist, 57
    Paris Street Scene, 58–59

Geldzahler, Henry, 5, 44
General Services Administration (GSA), 9–12
Gibson, Ralph, 203–8
  artworks:
    Days at Sea, 207
    Deja-Vu, 207
    The Somnambulist, 205, 207
  Robert Frank and, 204
  San Francisco Art Institute and, 208
Grosman, Tanya, 115

Haacke, Hans, 255
Happenings, 107, 216
Haring, Keith, 310–13
  artworks:
    Untitled, 313
    Untitled Ink Drawing, 312
  communication devices and, 311
  drawing in the subway, 311
  education, 311
  graffiti, 310

transit police and, 312
  Vivienne Westwood and, 313
Hayter, Stanley William, 103
Heizer, Michael, 276–81
  artworks:
    Circular Surface Planar Displacement, 277
    Complex of the City, 281
    Displaced, Replaced Mass, 278
    Double Negative, 279
  Virginia Dwan and, 280
  Dwan Gallery and, 278
  Las Vegas and, 276
  Robert Scull and, 277
Hollis, Douglas, 33
  artworks:
    Sound Garden, 34
Holzer, Jenny, 308–10
  artworks:
    Living Series, 309–10
    Truism, 308
    Truism: Your Oldest Fears are the Worst Ones, 310
  broadcast television and, 310
  Peter Nadin and, 309
  Public Art Fund and, 310
  Times Square Lightboard and, 310

Johns, Jasper, 115–29
  artworks:
    According to What, 124
    Ale Cans, 118
    Bent Blue (Second State), 125
    The Critic Sees, 117
    Figures in Color Series, 122
    Foirades/Fizzles, 127
    High School Days, 123
    Leg and Chair, 124
    Target, 119
    Target with Four Faces, 118
    Torse, 128
    0–9, 120
  Samuel Beckett and, 126
  John Cage and, 117
  Leo Castelli and, 116
  Crommelynck brothers and, 127
  education, 117
  Gemini printshop and, 120
  Leo Steinberg and, 116
  Ken Tyler and, 120, 122

Kiefer, Anselm, 88–93
  artworks:
    Interior, 1981, 93
    Margarethe, 90
    The Meistersinger, 91
    Sand of the March, 92
  Paul Celan and, 89
  fascist architecture and, 93

Marinetti, Filippo, 215
Meyerowitz, Joel, 181–86

  artworks:
    The Arch, 185
    Bay/Sky, Provincetown, 184
    Provincetown, 184
    Red Interior, Provincetown, 184
    St. Louis and the Arch (exhibit), 185
  Cartier-Bresson and, 183
  education, 182
  Robert Frank and, 182
  St. Louis Art Museum and, 185
Michals, Duane, 193–203
  artworks:
    Chance Meeting, 197–99
    The Human Condition, 195–96
    Real Dreams (book), 193
    Things Are Queer, 200–202
    This Photograph Is My Proof, 203
  use of language, 193
Minimalism, 37
Morris, Robert, 8, 140–50
  artworks:
    Box with the Sound of Its Own Making, 142
    Grand Rapids Project, 8
    Labyrinth, 147
    Untitled 1970, 146
    Untitled 1978, 149
    Untitled (1978 for R.K.), 149
    Untitled 1965, 144
  earthwork artists and, 146
  education, 144–45
  Green Gallery and, 143
  Minimalism and, 141
  mirrors and, 144
  performance events, 145
  Yvonne Rainer and, 145
Museum of Modern Art, 179–80, 181
  exhibits:
    Family of Man, 181
    Mirrors and Windows, 179–80

Nadelman, Elie, 138
National Endowment for the Arts (NEA), 5, 12–14
National Oceanic and Atmospheric Administration
  (NOAA), 33
Nauman, Bruce, 292
"New Image Painting", 84

Oldenburg, Claes, 19–23
  Art and Technology Exhibit and, 19–20
  artworks:
    Clothespin, 23
    Giant Icebag, 21
    Lipstick, 21
  Colossal Keepsake Corporation and, 22
  Maurice Tuchman and, 19
"One-Percent Program", 10
Oppenheim, Dennis, 300–307
  artworks:
    Attempt to Raise Hell, 305–6

Oppenheim, Dennis (*cont.*)
  Decoy, 301
  Echo, 305
  Lecture Piece No. 1, 307
  Maze, 301
  Polarities, 303
  Reading Position for Second Degree Burn, 303
  Theme for a Major Hit, 306
  Wishing Well, 304
  Jack Burnham and, 300
  John Gibson Gallery and, 303
  language and, 302
  shamanism and, 300
  Sonnabend Gallery and, 303

Paik, Nam June, 231–41
  Schuya Abe and, 234
  artworks:
    Electronic Opera No. 2, 236
    Étude for Pianoforte, 233
    Homage à John Cage, 232
    The Medium Is the Medium, 236
    Moon Is the Oldest TV Set, 239
    One for Violin, 232
    Opera Sextronique, 235
    The Selling of New York, 237–38
    TV Buddha, 239
  Fred Barzyk and, 237
  John Cage and, 232
  Dada and, 233
  Galerie Parnass and, 231
  Charlotte Moorman and, 235
  musical education, 234
  René Block Gallery and, 238
  Rockefeller Foundation and, 241
  Television Laboratory and, 237
  WGBH and, 236–37
Pepper, Beverly, 24–26
  artworks:
    Perazim, 25
    Thel, 27
  at Dartmouth College, 24
  at Spoleto, 26
Performance art, 215
Philip Morris Corporation, 15–17
  exhibits of:
    Air, 16
    New Alchemy, 16
    When Attitudes Become Form, 16
  Ulrich Franzen and, 17
  George Wiseman and, 15
Photorealists, 51, 58
Pollock, Jackson, 36
Pop Art, 37
Print Council of America, 102
Printmaking:
  etching techniques, 103–4
  history, 102
  lithography techniques, 104
  silkscreen techniques, 104
  woodcut techniques, 104–5

Rasmussen, Knude, 253–54
Rimbaud, Arthur, 110
Rose, Barbara, 7
Rothenberg, Susan, 84–88
  artworks:
    Flanders, 85
    Mist from the Chest, 88
    Reflections, 87
    Rose, 86

Salle, David, 97–101
  artworks:
    Autopsy, 98
    Making the Bed, 101
    My Head, 100
    Poverty Is No Disgrace, 99
  education, 99
  Jasper Johns and, 100
  Modernism and, 101
  Whitney Museum and, 97
Samaras, Lucas, 160–68
  artworks:
    Book, 166
    Book 4, 165
    Book 6, 165
    Box 8, 163
    Phototransformation 10/28/73, 167
    Wire Hanger Chair (Open Shoe), 168
  Columbia University and, 162
  Greece and, 161
  Pace Gallery and, 166
  Rutgers University and, 162
  George Segal and, 162
Scully, Sean, 93–97
  abstract painting and, 94
  artworks:
    Desire, 96
    Enough, 95
    Flyer, 96
    Hidden Drawing No. 3, 95
  Croyden College of Art and, 94
  education, 94
  Harvard University and, 94
Sculpture off the Pedestal, 7
Seattle's Public Arts Program:
  Scott Burton, *Viewpoint,* 34
  Douglas Hollis, *Sound Garden,* 33
  Nathan Jackson, *Hatchcover,* 32
  Jack Mackie, *Dancer's Series: Steps,* 33
  National Oceanic and Atmospheric
    Administration, 33–35
Segal, George, 26–29
  artworks:
    Next Departure, 28
    The Restaurant, 28
Serra, Richard, 29–31
  artworks:
    Tilted Arc, 29
shamanism, 253
"Sharp Focus Realism" (exhibit), 50–51
Sherman, Cindy, 209–12

artworks:
  Ad for Diane B, 212
  Untitled Film Stills Series, 210
  Untitled Film Still #48, 211
  Untitled 1981, 211
  Untitled 1983, 211–12
Diane Benson and, 213
education, 210
Hallwalls and, 210
Robert Longo and, 210
Simonds, Charles, 281–92
  Artpark and, 287
  artworks:
    Cliff Dwellers, 284
    Dwelling, Houston Street, New York City, 285
    Dwellings, 290
    Floating Cities, 290
    Growth House, 288
    Hanging Gardens, 289
    Landscape/Body/Dwelling (film), 282
    La Placita, 287
    Niagara Gorge, 287–88
    Three Peoples (book), 289
  education, 282
  "Little People" and, 283
  Lower East Side and, 284
  sculptural techniques, 285
  Stanley Tankel Housing Project and, 288–89
Smith, David, 12, 138
  artworks:
    Cubi XXIII, 139
    Medals for Dishonor, 139
    Wagon II, 13
  Bolton's Landing and, 139
Smithson, Robert, 270–76
  artworks:
    Amarillo Ramp, 275
    Broken Circle, 274
    Spiral Hill, 274
    Spiral Jetty, 271–72
    Tailing Pond Project, 274
  geology and, 270
  "Greensward" plan and, 271
  Nancy Holt and, 275
  Stanley Marsh and, 275
  Frederic Law Olmsted and, 271
  Primary Structures Exhibit and, 270
Steir, Pat, 129–33
  artworks:
    The Brueghel Series (A Vanitas of Style), 131
    Form, Illusion, Myth, 132
    Self as Picasso as a Young Man #1, 133
  Landfall Press and, 129
  Agnes Martin and, 130
  monoprint process and, 133
Szarkowski, John, 179

Twombly, Cy, 77–84
  artworks:
    Aristaeus Mourning the Loss of His Bees, 83
    Bay of Naples, 79

    Epithalamion III, 1976, 82
    Fifty Days at Illiam, 84
    Untitled, Captiva Island, Florida, 1968, 79
    Untitled, 1976, 82–83
    Untitled, 1967, 79
  cryptology and, 81
  Greco-Roman myth and, 78
  Heiner Freidrich Gallery and, 81
  Homer and, 83
  Don Judd and, 80
  Rome and, 78
  Surrealism and, 77

Universal Limited Art Editions, 115

Video art, 216

Warhol, Andy, 37–50
  artworks:
    American Death Series, 39
    Before and After, 40, 41
    Bellevue, 45
    Del Monte Peach Halves, 40, 41
    Do It Yourself, 40, 42
    Lavender Disaster, 46
    Marilyn, 39
    Marilyn Monroe Diptych, 43, 44
    Mona Lisa, Thirty Are Better than One, 43
    Plane Crash, 38
    Portrait of Leo, 48–49
    Suicide, 45
  films:
    Empire, 47–48
    Haircut, 48
  gall-bladder operation and, 37
  Henry Geldzahler and, 44
  Murial Latow and, 42
  Roy Lichtenstein and, 40
  performance art, 47
  Valeria Solanis and, 38
  Stable Gallery and, 44
Winsor, Jackie, 150–60
  artworks:
    Bound Square, 156
    Burnt Piece, 159
    Double Bound Circle, 154
    Fifty-Fifty, 157
    Four Corners, 156
    Nail Piece, 152
    #2 Copper, 158
    Sheet Rock Piece, 158
    30 to 1 Bound Trees, 155
  Douglas College and, 153
  Massachusetts College of Art and, 153
  Minimal art and, 150
  Newfoundland and, 153
  Nova Scotia and, 154
  performance art, 151–52